Futuristic Trends for Sustainable Development and Sustainable Ecosystems

Fernando Ortiz-Rodriguez
Tamaulipas Autonomous University, Mexico

Sanju Tiwari
Tamaulipas Autonomous University, Mexico

Sailesh Iyer
Rai University, India

José Melchor Medina-Quintero
Tamaulipas Autonomous University, Mexico

A volume in the Advances in Systems Analysis, Software Engineering, and High Performance Computing (ASASEHPC) Book Series

IGI Global
PUBLISHER of TIMELY KNOWLEDGE

Published in the United States of America by
IGI Global
Engineering Science Reference (an imprint of IGI Global)
701 E. Chocolate Avenue
Hershey PA, USA 17033
Tel: 717-533-8845
Fax: 717-533-8661
E-mail: cust@igi-global.com
Web site: http://www.igi-global.com

Library of Congress Cataloging-in-Publication Data

Names: Ortiz-Rodriguez, Fernando, 1974- editor. | Tiwari, Sanju, 1979- editor. | Iyer, Sailesh (Sailesh Suryanarayan), 1973- editor. | Medina Quintero, José Melchor, editor.
Title: Futuristic trends for sustainable development and sustainable ecosystems / Fernando Ortiz-Rodriguez, Sanju Tiwari, Sailesh S. Iyer, and Jose-Melchor Medina-Quintero, editors.
Description: Hershey, PA : Engineering Science Reference, an imprint of IGI Global, [2022] | Includes bibliographical references and index. | Summary: "This book communicates the recent advances in ICT (Information Communications and Technology) by covering practices in several artificial intelligence (AI) fields, such as knowledge representation and reasoning, natural language processing, machine learning, and the semantic web"-- Provided by publisher.
Identifiers: LCCN 2021059212 (print) | LCCN 2021059213 (ebook) | ISBN 9781668442258 (h/c) | ISBN 9781668442265 (s/c) | ISBN 9781668442272 (ebook)
Subjects: LCSH: Environmental protection--Data processing. | Sustainable development--Data processing. | Artificial intelligence--Industrial applications.
Classification: LCC TD170.3 .F88 2022 (print) | LCC TD170.3 (ebook) | DDC 363.700285--dc23/eng/20220114
LC record available at https://lccn.loc.gov/2021059212
LC ebook record available at https://lccn.loc.gov/2021059213

This book is published in the IGI Global book series Advances in Systems Analysis, Software Engineering, and High Performance Computing (ASASEHPC) (ISSN: 2327-3453; eISSN: 2327-3461)

British Cataloguing in Publication Data
A Cataloguing in Publication record for this book is available from the British Library.

All work contributed to this book is new, previously-unpublished material.
The views expressed in this book are those of the authors, but not necessarily of the publisher.

For electronic access to this publication, please contact: eresources@igi-global.com.

Advances in Systems Analysis, Software Engineering, and High Performance Computing (ASASEHPC) Book Series

ISSN:2327-3453
EISSN:2327-3461

Editor-in-Chief: Vijayan Sugumaran, Oakland University, USA

MISSION

The theory and practice of computing applications and distributed systems has emerged as one of the key areas of research driving innovations in business, engineering, and science. The fields of software engineering, systems analysis, and high performance computing offer a wide range of applications and solutions in solving computational problems for any modern organization.

The **Advances in Systems Analysis, Software Engineering, and High Performance Computing (ASASEHPC) Book Series** brings together research in the areas of distributed computing, systems and software engineering, high performance computing, and service science. This collection of publications is useful for academics, researchers, and practitioners seeking the latest practices and knowledge in this field.

COVERAGE

- Distributed Cloud Computing
- Computer System Analysis
- Virtual Data Systems
- Human-Computer Interaction
- Parallel Architectures
- Computer Graphics
- Enterprise Information Systems
- Engineering Environments
- Computer Networking
- Software Engineering

IGI Global is currently accepting manuscripts for publication within this series. To submit a proposal for a volume in this series, please contact our Acquisition Editors at Acquisitions@igi-global.com or visit: http://www.igi-global.com/publish/.

The Advances in Systems Analysis, Software Engineering, and High Performance Computing (ASASEHPC) Book Series (ISSN 2327-3453) is published by IGI Global, 701 E. Chocolate Avenue, Hershey, PA 17033-1240, USA, www.igi-global.com. This series is composed of titles available for purchase individually; each title is edited to be contextually exclusive from any other title within the series. For pricing and ordering information please visit http://www.igi-global.com/book-series/advances-systems-analysis-software-engineering/73689. Postmaster: Send all address changes to above address.

Titles in this Series

For a list of additional titles in this series, please visit:
http://www.igi-global.com/book-series/advances-systems-analysis-software-engineering/73689

Emerging Technologies for Innovation Management in the Software Industry
Varun Gupta (Universidad de Alcalá, Madrid, Spain) and Chetna Gupta (Jaypee Institute of Information Technology, Noida, India)
Engineering Science Reference • © 2022 • 282pp • H/C (ISBN: 9781799890591) • US $270.00

Technology Road Mapping for Quantum Computing and Engineering
Brojo Kishore Mishra (GIET University, India)
Engineering Science Reference • © 2022 • 243pp • H/C (ISBN: 9781799891833) • US $250.00

Designing User Interfaces With a Data Science Approach
Abhijit Narayanrao Banubakode (MET Institute of Computer Science, India) Ganesh Dattatray Bhutkar (Vishwakarma Institute of Technology, India) Yohannes Kurniawan (Bina Nusantara University, Indonesia) and Chhaya Santosh Gosavi (MKSSS's Cummins College of Engineering, India)
Engineering Science Reference • © 2022 • 325pp • H/C (ISBN: 9781799891215) • US $270.00

Implementation of Machine Learning Algorithms Using Control-Flow and Dataflow Paradigms
Veljko Milutinović (Indiana University, Bloomington, USA) Nenad Mitić (University of Belgrade, Serbia) Aleksandar Kartelj (University of Belgrade, Serbia) and Miloš Kotlar (University of Belgrade, Serbia)
Engineering Science Reference • © 2022 • 296pp • H/C (ISBN: 9781799883500) • US $270.00

For an entire list of titles in this series, please visit:
http://www.igi-global.com/book-series/advances-systems-analysis-software-engineering/73689

IGI Global
PUBLISHER of TIMELY KNOWLEDGE

701 East Chocolate Avenue, Hershey, PA 17033, USA
Tel: 717-533-8845 x100 • Fax: 717-533-8661
E-Mail: cust@igi-global.com • www.igi-global.com

Table of Contents

Detailed Table of Contents

This paper aims to analyze the implications of organizational greening and green-lean management. It is assumed that organizational green management is a critical factor on environmental and socio-socio socio ecosystem development in any green economy. The method used is the critical analysis based on the review of the literature and linked to practice in organizational settings. The analysis concludes that the organizational greening and green-lean management use organizational resources more efficiently, enhance innovation capabilities, facilitates institutional and organizational changes, processes, and products, enter new markets, and improve the socio environmental organizational competitive advantage.

The chapter is a bibliometric analysis of research work published on the Sustainable Development Goals. The authors extracted relevant data from Scopus on 22 August 2020, which were subsequently analyzed using open-source software – Bibliometrix (an R package). They performed a textual query on Scopus for the title "Sustainable Development Goal" by selecting the 'Article Title' in the 'Documents' option for the period of 1996-2020 and retrieved 1872 documents from 865 sources. This study aims to identify the global trends in the SDG's scientific production over time with analysis covering various bibliometric indicators (i.e., sources, authors, and documents). Results reveal growth in research in SDGs in recent years, trending

scientific contributions in this field have demonstrated good results and with some modification ensures effective solution in this state of the art. The proposed GCBAAS methodology trained on remote GPU server minimized false labels in much less time. A simple pipeline has been created for training such enormous data on GPU servers for multiclass labeling. The performance analysis has been taken in a real class, which results in mean accuracy of 98%. Additionally, after collecting some more data with different augmentation techniques, it results in increased accuracy.

Small datasets are common in research areas where measurement of data is expensive or difficult (e.g., healthcare). For small datasets, training of a data model and classification using the same remains a challenge. To address the issue of limited data training and classification, the authors propose to use a novel ensemble of self-organizing maps (SOMs). SOMs are shallow neural networks, and as a result, the number of parameters is not very high. This avoids over-fitting and over-parameterization of the network. For the SOM ensemble, a novel early stopping training technique is proposed that requires a small number of samples for training. In the ensemble, one SOM is used for every class. For benchmarking the proposed technique, its performance is compared with that of other state-of-the-art techniques like deep learning, CATBOOST decision tree approach, etc. For reproducibility of results, three popular and extensively benchmarked digit image datasets have been used (i.e., MNIST, USPS+, and MADB). The proposed technique generally outperforms all other techniques with which it is compared.

Android applications require necessary permissions to perform the right function, but at the same time, there are other permissions that may affect the user privacy that need to be retrieved and revoked. Each step forward in digitization is also leading to an increase in the number of malware attacks by the malicious applications with the use of permissions. To solve the problem of APK app user data privacy and security, the authors created a system that searches for and downloads Android apps from the Android Market automatically. In addition, they produced a thorough mapping of Android application programming interface (API) calls to the needed permission(s) for each call, if any were required. Based on a static examination of each application's APK bytecode, they next ran an analysis of 141,372 Android applications to see if they had the right set of permissions. According to the research, most mobile app developers do not use the proper permission set and either over-specify or under-specify their security requirements.

COVID-19 has various symptoms, and they are cold, cough, mild or high fever, and breathing problems for severe cases. In addition, it is in talk that diabetes, chronic obstructive pulmonary disease (COPD), cardiovascular, asthma, and many diseases invite the virus. Machine learning and data analysis are considered best for predicting approaches for finding various aspects of the effect of the virus. In this chapter, the authors deal with symptoms that have been recorded in the dataset and try to find which pre-defined symptoms are considered effective or responsible in positive case of infection by coronavirus. The dataset has been taken from Kaggle. As the dataset is categorical in nature, the authors use correlation and logistic regression analysis to find the symptoms that prevail in the patient and have caused the infection in them. This is also about dimensionality reduction and feature selection where they are reducing the available features based on regression.

Machine learning is one of the most exciting technologies of the 21st century. This is true both because of the complex issues it possesses the capability to solve, as well as its wide range of application areas. One of the most apparent applications of this technology is in the form of recommendation systems. The chapter outlines a clustering-based recommendation system and contrasts the viability of such systems in comparison with traditional recommender methodologies. The utility of such systems is further highlighted by the demonstration of an application. Utilizing massive online data and the foursquare application programming interface, the colleges across the US have been clustered into groups based on the similarity indexes. This segmentation is then used to present a rudimentary recommendation system that can prove to be useful for prospective students looking to make decisions about higher education institutions to find universities similar to the ones they already have in mind.

In India, the population is large, and it is very crucial to maintain emergency vehicle services properly. Everyday many accidents occur due to lack responsibility of drivers and lack awareness of motor act laws. Here the authors have developed a very unique solution and try to find unsolved issues in more technical way. The system is about tracing follower vehicles of emergency vehicles using RFID (radio frequency identification). Camera set on an emergency vehicle will capture an image of a trespassing vehicle and read the RFID tag. It will trace the details of the vehicle using an RFID reader. There is one front camera which will check the traffic signals, and if red signal already on red state, then it will not consider the follower vehicle, but if there is green signal and vehicle is following the emergency vehicle, then RFID will scan it, trace it. E-memo is generated and sent to the nearest RTO office for penalty.

Blockchain is a distributed ledger apparatus that utilizes timestamped blocks to manage and verify various transactions between two parties in a network with the help of cryptographic algorithms. As compared to popular opinion, blockchain not only serves as a backbone to cryptocurrencies but it can also serve as a pioneering solution to various disciplinary fields such as healthcare, logistics, and education. The approach taken by this chapter to tackle the various problems faced by the healthcare sector utilizes blockchain technology. It avails smart contracts as well as the decentralized nature of the blockchain and smart tokens to ensure smooth interactions between the patients and the physicians. Protection and maintaining the integrity of healthcare data without overlooking any security threats are the primary objectives of this chapter. This chapter discusses a model that uses a cloud-based decentralized model for the storage of patient data.

This chapter analyzes the implications that the organizational environmental management and green transformational leadership have between them and with the green organizational innovation aimed to achieve sustainable development. It departs from the assumption that green organizations have implemented environmental management and green innovation enhanced by green transformational leadership to achieve organizational sustainable competitive advantage. The method employed is the analytic-reflective based on the review of the theoretical and empirical literature. It is concluded that the environmental management issues guided by a green transformational leadership are related to organizational sustainable development encouraging organizations to act in favor of greening organizations, green consumers, and achieving a sustainable competitive advantage with the Sustainable Development Goals (SDGs).

Web service technology and cloud computing offer promising solutions for developing distributed business applications published and invoked over the internet. Web service discovery finds the most relevant web services satisfying the functional needs of users. Semantic description of web services offers more accurate matching results between user query and web services descriptions. User's quality of service (QoS)-based criteria select the best web service from a set of functionally similar web services. The risk of privacy leakage of users is possible as service providers collect user personal data without any agreement. It isn't easy to assess the trustworthiness of values of QoS attributes provided by different service providers. Blockchain is an emerging technique that can overcome privacy leakage and identifying web services with fake values of QoS attributes. This chapter gives an overview of semantic web services, discovery and QoS-based selection of semantic web services, and the use of blockchain technology in the selection of web services.

Software defect prediction using data mining techniques is one of the best practices for finding defective modules. The existing classification techniques can be used for efficient knowledge discovery on normal datasets. Most of the real-world data sources are biased towards any one of the classes. This type of data source is known as class imbalance or skewed data sources. The defect prediction rate for the class imbalance datasets reduces with the increases in the class imbalance nature. To handle such type of datasets, an approach with specific designing technique is required for improved performance. In this chapter, the authors propose an algorithm known as improved integrated sampling strategy (IISS) for improved performance using noisy removal strategy for software defect prediction. The experimental analysis conducted on skewed software defect prediction datasets provides the results that IISS algorithm have performed well when compared with C4.5, C4.5+Balance, RF, and RF+Balance algorithms with various class imbalance evaluation measures.

The electronic government is a new application field for the Semantic Web, and the ontologies play a key role in the development of the Semantic Web. Consequently, if we focus on the law and the public policy that base the governments and build a documentation ontology that encompasses the electronic governments' needs of services at the front-office and mainly at the back office, we will provide a starting point to enable a broad class of semantic electronic government application. In this chapter, the authors show the e-Government Documentation Ontology as a part of the EGO model in order to describe how to access and manage official and non-official documentation across the governments. They also show a couple of applications using the documentation ontology on a semantic information retrieval system and on a semantic peer-to-peer.

Preface

Growth and Sustainability have always been the essential ingredients for balanced Lifestyle. There are tremendous challenges to maintain the quality of lifestyle and preserving the valuable resources provided by Nature. The co-existence of the key components like Technology, Innovation and Spirituality are integral to growth of mankind.

This edited book consists of Advancements in Information and Communication Technology, Engineering, Innovation, and selected quality papers from International Conference on Futuristic Trends for Sustainable Ecosystem (FTSE-2021). *Futuristic Trends for Sustainable Development and Sustainable Ecosystems* discusses recent advances and innovative research in the area of information and communication technology for sustainable development and covers practices in several artificial intelligence fields such as knowledge representation and reasoning, natural language processing, machine learning, and the semantic web.

Multidisciplinary International Conference Futuristic Trends for Sustainable Ecosystem (FTSE-2021) organized by Rai University, Ahmedabad and Tamaulipas Autonomous University (UAT), Mexico in August 26-28th, 2021 was conceptualized to address the concerns, technological factors and human angle during these challenging times. FTSE-2021 was organized under the able guidance of Chief Patron Dr. Anil Tomar (Provost), Patron Prof. Lalit Adhikari (Registrar) and Track Chairs Dr. Sailesh Iyer, Dr. Fernando Ortiz-Rodriguez (UAT, Mexico) and Dr. Sanju Tiwari (UAT, Mexico). Few selected quality manuscripts are included in the Edited Book in addition to the globally authored manuscripts

This Edited Book consists of 16 quality chapters from renowned academicians and researchers all around the globe.

Chapter 1 aims to analyze the implications of organizational greening and green-lean management. It is assumed that organizational green management is a critical factor on environmental and socio-socio socio ecosystem development in any green economy.

Chapter 2 presents a bibliometric analysis of research work published on Sustainable Development Goals. The authors extracted relevant data from Scopus on 22 August 2020, which were subsequently analyzed using open-source software – Bibliometric (an R package).

Chapter 3 explores the future of the smart healthcare networks as a combination of 5G & IoT devices which are anticipated to extend the cellular scope & handle security problems. It also provides an overview of Taxonomy, research trends, difficulties & future research themes.

Chapter 4 proposes GPU CNN Based Automated Attendance System (GCBAAS). Image content analysis and pattern recognition using deep learning accompanying with High Computing Technology are rapidly expanding areas of application. Chapter 5 proposes a novel SOM ensemble classification technique for limited data. For benchmarking of the proposed technique, its performance is compared with that of other state-of-the-art techniques like Deep Learning, CATBOOST Decision Tree approach and found to outperform all the other techniques.

Chapter 6 produces a thorough mapping of Android application programming interface (API) calls to the needed permission(s) for each call, if any were required. According to research, most mobile app developers do not use the proper permission set and either over-specify or under-specify their security requirements.

Chapter 7 uses correlation and logistic regression analysis to find out the symptoms which were prevailing in the patient and have caused the infection in them. This is also about dimensionality reduction and feature selection where we are reducing the available features based on regression. Chapter 8 provides segmentation to present a rudimentary recommendation system that can prove to be useful for prospective students looking to make decisions about higher education institutions to find universities similar to the ones they already have in mind.

Chapter 9 provides system for tracing trailing vehicles which are emergency vehicles using RFID (Radio Frequency Identification). Camera set on an emergency vehicle will capture an image of a trespassing vehicle and read the RFID tag. It will trace the details of the vehicle using an RFID reader. Protection and maintaining the integrity of healthcare data is the primary objective of Chapter 10 without overlooking any security threats. This chapter discusses s a model that uses a cloud-based decentralized model for the storage of patient data.

Chapter 11 analyzes the implications that the organizational environmental management and green transformational leadership have between them and with the green organizational innovation aimed to achieve sustainable development. Chapter 12 provides an overview of semantic web services, discovery and QoS based selection of semantic web services, and use of blockchain technology in the selection of web services.

Chapter 13 propose an algorithm known as Improved Integrated Sampling Strategy (IISS) for improved performance using noisy removal strategy for software defect prediction. The experimental analysis conducted on skewed software defect prediction datasets provides the results that IISS algorithm have performed well when compared with C4.5, C4.5+Balance, RF and RF+Balance algorithms with various class imbalance evaluation measures.

Chapter 14 propose a LOOP (Local Optimal Oriented Pattern) method is applied to discriminate ten laser printer models based on the character images. LOOP is efficient descriptor to differentiate the images. Chapter 15 proposes an algorithm which is trying to predict the future pattern for any stock based on the real-time analysis and data provided from Yahoo Finance data. The software provides pictorial and graphical representations. The objective is to provide short-term and long-term prediction competence to prepare for future potential investments.

Chapter 16 show the e-Government Documentation Ontology as a part of the EGO model, in order to describe how to access and manage official and non-official documentation across the governments. We also show a couple of applications using the documentation ontology on a semantic information retrieval system and on a semantic peer-to-peer.

This book provides an exhaustive coverage of topics such as blockchain, deep learning, and renewable energy, which makes this reference work ideal for computer scientists, industry professionals, researchers, academicians, scholars, instructors, and students.

Fernando Ortiz-Rodriguez
Tamaulipas Autonomous University, Mexico

Sanju Tiwari
Tamaulipas Autonomous University, Mexico

Sailesh Iyer
Rai University, India

Jose Melchor Medina-Quintero
Tamaulipas Autonomous University, Mexico

Chapter 1
Organizational Greening and Green-Lean Management

José G. Vargas-Hernández
https://orcid.org/0000-0003-0938-4197
Posgraduate and Research Department, Tecnológico Mario Molina Unidad Zapopan, Mexico

ABSTRACT

This paper aims to analyze the implications of organizational greening and green-lean management. It is assumed that organizational green management is a critical factor on environmental and socio-socio socio ecosystem development in any green economy. The method used is the critical analysis based on the review of the literature and linked to practice in organizational settings. The analysis concludes that the organizational greening and green-lean management use organizational resources more efficiently, enhance innovation capabilities, facilitates institutional and organizational changes, processes, and products, enter new markets, and improve the socio environmental organizational competitive advantage.

INTRODUCTION

Global business organizations are shifting from conventional financial structures to modern capacity-based economy aimed to take advantage of green economic practices. Organizations face challenges for greening innovative and sustainable activities, products, and services in any type of economy. Organizations need to change from traditional to more green practices in all the areas. Organizations going green help to manage risks more effectively, use resources more efficiently,

DOI: 10.4018/978-1-6684-4225-8.ch001

enhance innovation, enter new markets, and improve the competitive advantage. Green management and innovation protect the organizational environment

Researchers and practitioners have a growing interest in organizational greening and environmental issues and their impacts towards the marketing processes, practices, activities, products, and services. Organizational green sustainability is a challenge to cope with greener management to influence change in innovation capabilities, organizational structures, human resources, and other factors aimed to improve design and manufacture of better products and services to achieve competitive advantage. Socio-economic and environmental inclusion are the basis to shape the new green deals of socioenvironmental development at all levels of public and private institutional and organizational arrangements and structures leading them towards a more equal and sustainable society and communities (de Jong, 2021).

Some other researchers have included mediating variables between green human resources management and environmental strategy, such as organizational culture, green organizational climates, industry practices, and country-level factors and have found indirect relationship (Shatouri et al. 2013, Sharma 2009, Wagner 2015, Chou 2014). Research at individual level of driving factors of organizational greening innovation practices is more related to environmental awareness, support, commitment, and environmental behavioral intentions (Sharma, 2000; Xu et al., 2017; Zhang et al., 2015).

Research on organizational greening innovation promotes the ideology of a proper alignment of human resource management principles with the objectives of organizational green management. Green management principles, policies, strategies, and practices of organizational greening in the forms of green outcomes, products and services and production processes (Loknath and Azeem, 2017). Green management benefits by the strengths of managerial actions regarding the processes of business model greening. Research on the driving factors of organizational greening innovation practices is abundant but research on management styles leadership is scarce. The relationship between environmental leadership and organizational green innovation practices is being analyzed by empirical research.

This analysis begins with the conceptualization of organizational greening proposing an organizational model greening with a holistic objective to leverage the organizational green vision and to obliterate the old practices for the new ones in a green economy. The analysis continues stablishing the scope of the concept of organizational green management to sustain the strategic implementation for the organizational sustainable development. Following these arguments, the analysis explores some of the green issues and concerns to propose a program for greening the organizations. The implications between organizational green and lean practices are analyzed before it is concluded.

CONCEPT OF ORGANIZATIONAL GREENING

The concept of organizational greening business development introduces green innovations to promote the green economy and is shaped by environmental innovations for sustainable development, including nonagricultural economic activities, greening activities, institutional and organizational changes, processes, and products. This green economy is an economic and organizational mechanism for achieving stability by solving problems in rural development (Honcharuk, Tomashuk, 2018; Honcharuk, Tomashuk, 2017). Green organizational system assesses the environmental impacts the level of quality management programs and green project management. Green project management processes apply and adapt into structured and proactive approach for managing environment-related concerns contributing to the organizational environment strategies and policies.

Achieving organizational sustainability implies changes in the current practices of green quality management connected to green purchasing and green productivity. Implementation of green organizational systems and tools are motivated by the green revolution of practices aimed to enhance green productivity, environmental protection and performance, sustainable development, and lifestyle. Organizational environmental development green activities are increasingly being included in the job descriptions to find the fit in green recruitment agenda, defined as the hiring process of the individuals that possess the behaviors, attitudes, knowledge, skills, and capabilities identified with organizational environmental management systems.

Organizational green issues are relevant to predict the organizational environmental innovation and the facilitation of organizational greening innovation influencing pro-environmental behaviors (Norton, Parker, Zacher, Ashkanasy, 2015; Ángel del Brío, Junquera, Ordiz, 2008; Robertson, Barling, 2013). Organizational greening innovation is the ability to cope with green-oriented organizational conditions.

Drawing on traditional know-how, many countries have developed the concept of environmental sustainability with greening approaches of producing goods and services environmentally green (Katundu Imasiku, 2021). Environmental sustainability and organizational model change are the fundamental conceptual pillars to support the organizational model greening at interface to link the concepts of environmental sustainability and organizational models modified from empirical analysis to explain the change in the organizational model greening.

The organizational model greening reshapes critical components challenging the conventional organizational model in terms of processes and production of goods and services. Organizational model greening has a holistic objective to leverage the green vision and to obliterate the old practices for the new ones. The reluctance to develop organizational greener processes forces a solution to locate partners committed to green goals and environmental sustainability, reinvent the organizational model greening, realization of long-term strategy, reconfigure the organizational socio ecosystem, create a recycling system and reuse of technology, seek green innovations, improvement of processes, redesign of manufacturing plant, green product design, material recovery, etc.

Reinventing the organizational model greening and reconfiguring the organizational socio ecosystem for environmentally sustainable practices in a multilayered and multifaceted processes including iterative and overlapping managerial activities helps practitioners to support the organizational model greening. An analysis of organizational greening leads to a longitudinal view of sustainable change of the organizational model in its organizational socio ecosystem (Holt, 2011). The organizational change process based on the phase model (Lewin, 1951) is insufficient to analyze and explain multifaceted and multilayered processes of organizational greening model.

The greening organizational model change and the configured socio ecosystem results in waste elimination activities, shedding light on contingencies of organizational model greening. An emergent organizational model greening in new organization socio ecosystem is built upon a philosophy to bring a new environmental perspective. The organizational greening knowledge infuse organizational models with environmental sustainability confirming the links among the managerial agencies aimed to transform the organizational model greening and the organizational socio ecosystem implications of greening. Greening organizational model is a process based on knowledge and experience challenges opportunities and frustrations.

ORGANIZATIONAL GREEN MANAGEMENT

Organizational green management is a concept that is growing and leading to more opportunities for the creation of more competitive advantages (Russo & Fouts, 1997; Shrivastava, 1995). The green management concept is a strategy for the organizational sustainable development (Banerjee, 2002). Organizational green management is getting more active involvement to adapt environmental management, green innovation and green and environmental strategies in response to the pressures from stakeholders to address the socio ecological systems Xiaofeng Su, Anxin Xu, Wenhe Lin, Youcheng Chen, Songtao Liu, Wenxing Xu (2020)

Green management integrates the environment in organizational decision-making processes referring to the practices leading to environmentally friendly products and minimizing the impact on the environment through green production, research, development, and marketing (Peng and Lin, 2008). The green concept is the building block of the concepts of green management (McDonagh & Prothero, 1997) green marketing (Peattie, 1992), green retailing (Kee-hung, Cheng, & Tang, 2010), and green accounting (Bebbington, 2001; Owen, 1992).

The design and development of green management model expands the social and environmental responsibilities of organizations aimed to promote green purchasing and improve environmental performance. Environmental green management is an organizational action applying inventiveness for achieving sustainability and competitiveness, ensuring social responsibility, and reducing waste supported by learning initiatives aimed to endorse organizational strategic environmental goals (Pane Haden, Oyler, & Humphreys, 2009). Organizations are getting more inclined to environmental green management and organizational social responsibility (Barin Cruz & Avila Pedrozo, 2009)

Green management focuses on the influential organizational green initiatives and practices identified with the environmental sustainability and catalyze initiatives into action. Certified environmental green management systems are measures to assess activities that improve corporate environmental performance in life cycle activities such as introducing environmental management structures goals and programs. The organizational sectors have active participation to adopt environmental and green management strategies (Boiral, 2002; González-Benito & González-Benito, 2006)

Theories of psychological ownership and organizational identity adopt multi-level relationship of organizational greening management. Green management is a core strategy and conscious need in organizational greening in different regions and sectors (Yu & Ramanathan, 2015). Green management is related to the environmental policy to the establishment of green technology activities to enhance the organizational environmental performance.

Green management focuses on organizational pro-environmental behaviors modeled in organizational greening activities in the workplace achieved due to the influence of leaderships on pro-environmental behaviors. Development of innovative methods and initiatives of environment management (EM) have an impact on organizational sustainability and promotes an organizational competitive advantage (Hart, 1997; Lin, Jones, & Hsieh, 2001).

Organizations implement and integrate green initiatives with the support of green human resources practices and the promotion of technical and management skills aimed to implement an organizational green management system (Daily, Bishop, & Govindarajulu, 2009; Unnikrishnan & Hegde, 2007). Long-term trust built among the employees and the organizational management provide opportunities to express ideas and develop policies to frame the creation of green management practices.

Green management elements of the compensation program promote and improve green behaviors among managers and employees in the organizations to bring green ideas and initiatives pertaining to their activities to attain the objectives, as the basis of receiving incentives. The design and implementation of green compensation practices may lead to the achievement of organizational environmental goals in alignment of individual goals, perceptions, motivations, and capabilities of green management practices such as resource usage, reducing waste and pollution and improving environmental management systems (Florida & Davison, 2001; May & Flannery, 1995; Kitazawa & Sarkis, 2000).

Organizational greening and environmental protection are the elements of the framework of organizational green identity which brings meaning to organizational environmental management activities and individual environmental greening behaviors and green knowledge sharing. Organizational identity provides the understanding of tasks and work related to green management and environmental issues Chang, Chen, Luan, Chen, 2019). Organizations focus on green shared vision of environmental issues to implement green management aimed to strengthen the green organizational identity of members, to improve environmental behavior and green product psychological ownership (He, Z.; Xu, S.; Shen, W.; Wang, M.; Li, C. 2019).

Organizational greening and environmental protection create the organizational greening identity of members bringing meaning to the organizational environmental management activities and the individual environmental behaviors in green knowledge behaviors. Individuals working in organizations contribute at different levels to achieve environmental sustainability (Ones & Dilchert, 2012a). Members and stakeholders of organizational greening may be more satisfied and prouder of its environmental management and protection (Chen, 2011).

Knowledge of environmental sustainability can contribute to a conceptualization of an organizational model greening and to determine the essential activities to recognize the potential change in the desired vision of environmental sustainability, reinventing to leverage the green vision and reconfiguring the organizational socio ecosystem for sustainable practices. Corporate green management is related to green shared vision. Green management envisions novel ways of operating the emerging organizational socio ecosystem with the collaboration of new partners.

Inter-organizational support for organizational model greening leading to improve management practices in the use of natural resources. Organizational greening identification in relationship with members when the green product and employees adopt the green organizational identity is analyzed through green knowledge sharing and green product development performance.

Organizational model greening management concerns about the awakening of the need for environmental sustainability and identified opportunities for production, development, operations, and supply chain activities, creating new meanings of sustainability, making sense of the greening vision. The organizational management strategy realizes multiple benefits from a green approach incorporating ecological consciousness and their vision and mission statements among the employees, suppliers, consumers, investors, and the public.

Managerial agency driving organizational model greening links with environmental sustainability. The managerial agency is an approach to analyze the actors, drivers, and processes of the organizational model greening in the organizational socio ecosystems. Managerial agency in organizational model greening is a reinvented multi-layered process to leverage the green vision of sustainability and reconfiguring the organizational socio ecosystem for sustainable organizational practices. The conduction of green business practices is an organizational green management mandate where environmental sustainability is the green vision at the core of the organization, with a roadmap for managerial activities and related actions for personnel.

The link between strategic change and organizational model greening creates a conceptual framework of environmental sustainability of organizations engaged in green values, adopting a proactive approach to climate change, and driving by a competitive advantage (Esty and Winston, 2006; Holliday, 2001; Nidumolu et al., 2009). Focusing on the organizational greening initiatives must be influential to identify the activities that create the greatest impact in capturing the value of nurturing the sustainable business model. Industrial and organizational psychology contributes to organizational greening initiatives and environmental sustainability.

Organizational model greening and green business models support environmental sustainability concerning change as the expression of values and management principles, cognition of the operations of its organizational business socio ecosystem (Halme et al.,2007; Stubbs and Cocklin, 2008; Mason and Spring, 2011). Organizational green shared value for green management is related to green creativity, green radical, incremental, and reactive innovation, pro-environmental behavior, and green product performance (Chen, Chang, Yeh, Cheng, 2015; Chen, Chang, Lin, 2014; Chen, Chang, Lin, Lai, Wang, 2016; Afsar, Maqsoom, Shahjehan, Afridi, Nawaz, Fazliani, 2020; Chen, Lin, Lin, Hung, Chang, Huang, 2020; Chang, Chen, Luan, Chen, 2019).

The managerial agency in an organizational model greening is consistent with the core values and practices of organizational identity (Gioia et al., 2013). Organizational identity in business model greening requires the establishment of business model contributes to environmental sustainability through transformative activities in the business socio ecosystem and changing operations.

Differentiating the identity is crucial for sensemaking and identifying the potential development, gain acceptance for green programs from the perspective of strategy to catalyze green initiatives into organizational greening action. Green organizational identity enhances the green product psychological ownership resulting in a combination of organizational governance to improve green management performance and maintain environmentally sustainable development (Connelly, Ford, Turel, Gallupe, Zweig, 2014).

The organizational model greening perspective (Neuendorf, 2002) can be identified by some activities that challenge the organization to pursue a vision and the efforts to manage environmental concerns and issues, influence green behaviors and change the mindset and schemes of reasoning to promote green solutions and innovations. The organizational model greening recognizes the opportunities and challenges of the greening processes engaging organizational sensemaking related to vision of environmental sustainability. The activities to stablish the organizational model greening includes to recognize its potential with the new ways of thinking, establishing the vision of environmental sustainability, reinventing the organization model greening to leverage the green vision and reconfigure it for sustainable organizational practices.

Organizational model greening has the potential to develop core capabilities and create competitive advantages through environmental sustainability and develop the essential external capabilities, eco activities, green policies, and programs. Organizational model greening has the potential to promote climate change concerns and issues and influence actions of other green organizations and competitors. Recognizing the potential of organizational model greening it is necessary to develop a transformative organizational vision and agenda for greening change. Organizational greening is critical factor in survival and become more sensitive to sustainable agenda (Ng, Butt, Khong, Ong, 2013).

The organizational greening renewal should have legitimate supported by the leadership agency for the individuals and organizational teams who are committed to renew the greening practices and activities that create value to trigger the sustainable goals specified in the green vision. Individual and collective green organizational behaviors may have different environmental impact and validated in other areas of organizational green behavior, whether within a group, in a dyadic relationship (Barsade & Gibson, 2007).

Green teamwork facilitates individuals working together in the solution of environmental problems in the organization (Daily et al. 2007), which are generally considered complex and interdisciplinary. Creating the potential for greening in energy and materials, eliminating energy waste and being financially transparent may get acceptance of stakeholder´s to join the green initiative.

The greening is a commitment to advance organizational management standards for a healthier and sustainable development. The implementation of green and environmental organizational management practices at formal level initiatives, are dependent on individual, informal and discretionary eco-friendly behaviors in human resources management (Kim et al. 2019; and Pham et al. 2019). The implementation of corporate green management system it is necessary to promote technical and management skills among the employees (Daily et al., 2007; Unnikrishnan & Hegde, 2007).

Organizations committed to greening embedded operations in their mission and strategies supported by organizationally committed employees in congruence with environmentally orientation (Harvey, Williams, & Probert, 2013). The greening business agenda aimed to create and develop green companies continuously improves resource productivity, renewable energy technologies, and environmental performance to deliver environmental goods and services including recycling services and waste management.

The organizational greening trend influences green performance management on environmental policies and concerns and concentrating on environmental responsibilities. Greening business is the core of socio-economic competitiveness and socio environmentally sustainable growth based on the cost of resource inputs leading to improve the efficiency and creating and developing a competitive advantage, besides contributing to poverty alleviation, and promoting safety, energy security, health, etc. Other actions in greening can be addressed by the revenue component of the organization model defined by structures of costs and profits. Building organizational green competitive advantage is the result of organizational greening.

The organizational model greening framework is based on grounded theory to highlight the processes to transform (Strauss and Corbin, 1998) into a green pioneer from the cultural orientation (Dubose, 2000; Johansen, 1998; Toktay et al., 2006). Changes of organizational identity are required in the organizational model greening to be linked with the founding and adapting values. Greening organizations may be acting in the incorporation of environmental issues into the managerial practices at a certain questionable level identified as reactive, preventive, and proactive in the evolutionary stages of green management (Prasad & Elmes, 2005).

Emerging organizational green environmental management has positioned the greening function to improve the environmental performance of organizations (Ambec & Lanoie, 2008). Environmental management takes place at all levels of individual, team, organizational and industry in different country perspectives considering the external factors that may contribute to environmental behaviors at the workplace. Environmental management systems of organizations are shaped by green individual and team contributions (Jabbour 2011).

Organizations using more green teams have more proactive greening functions (Jabbour, 2013a). Green teams are more used in organizations that have more advanced environmental management systems (Jabbour et al. 2013, p. 65). Organizational greening practices and activities have significant implications related to green leadership, environmental behaviors, organizational responses to environmental sustainability Staats, Harland, & Wilke, 2004; Francoeur, Paillé, Yuriev, and Boiral, 34).

Organizations must optimize environmental losses by accelerating depreciation of assets of environmental relevance, proposals for the greening activities, system of environmental management, material, and non-material incentives, energy efficiency, optimization of waste management, environmental responsibility, etc. (Matukova1, Khrapkina, Bahashova, Matukova-Yaryha, and Levitskaia, 2021). Employee attitudes and behaviors are related with environmental management. Individual environmental norms explain the within-firm variance and green organizational climate explain the relationships between-firm variance Chou (2014, p. 436).

Multinational companies within the foodservice industry tend to develop organizational environmental management adopting a consistent approach with green human resources despite that international green management practices may have some differences (Chen & Chang, 2013; Hofer, Cantor, & Dai, 2012; Jabbour, Santos, & Nagano, 2010, 2012; Jabbour, 2013b; Paille´, Chen, Boiral, & Jin, 2014; Tatoglu, Bayraktar, Sahadev, Demirbag, & Glaister, 2014; Teixeira, Jabbour, & Jabbour, 2012),

GREENING THE ORGANIZATIONS

Organizations should implement a program on green purchasing, productivity, marketing, and management. Green marketing has influence relationships among the organizations creating the channel from the raw material to consumption. The management of green marketing activity incorporates new perspectives about organizational change becoming more aware of sustainability and ecological efforts on recycling of products, modifying inputs to production, etc. Green management has potential impacts on production and consumption of goods and services in their natural environment and improving the environmental footprint (Galeazzo et al., 2014; Verrier et al., 2014).

Green organizational behavior development influences the environmental behaviors in the workplace that lead to the organizational greening practices and activities. Contributions to green organizational behavior from environmental leaders tend to exhibit transformational style (Egri & Herman, 2000). Green quality management is the connecting link bridging between procurement and production greening.

Greening the business is a trend in green human resource management assigning responsibility of incorporating green human resource's philosophy in the organizational mission and policies. Green human resource management practice is an active strategy for going green at the organizations. Green human resource management initiatives and practices pursued by green organizations aimed to create green workforce in accordance with their organizational green culture

(Shoeb and Nisar 2015) Organizational greening issues are reflected in the green human resources management strategy leading to performance management system (Marcus & Fremeth, 2009). Organizational model greening requires an echo program connecting environmental sustainability with managerial performance measures and incentive systems.

Green human resources management behaviors and practices confirm the positioning of the organizational greening function aimed to improve the environmental performance sustainability (Ambec & Lanoie, 2008). Human resources management, green human resources management and environmental sustainability are identified with the concept of organizational citizenship behavior, which is an approach based on individual, voluntary and informal initiatives that are essential for corporate greening (Boiral et al. 2015, p. 532).

Organizational green behavior is performed on a voluntary basis and are related to individual actions associated with tasks at daily work (Bissing-Olson et al., 2013). Green organizational climates, effective organizational commitment and green work climate perceptions among co-workers are mediating variables between green human resources and organizational citizen behavior (Chou 2014; Temminck et al. 2015, Norton et al. 2014, p. 52; Chou, Chia-Jung 2014).

The debate on green human resources management is based on the organizational greening function and its alignment, positioning and integration with green employment engagement, organizational performance, and corporate strategy aimed to improve organizational environmental sustainability. Organizational reputation as a green employer is an attractor of new talent and high-quality staff with profile of being proficient in environmental responsibilities and activities and creators of organizational greening image (Phillips, 2007; Stringer, 2009; Wehrmeyer, 1996, p. 30).

Organizations are being more required to establish green management, green organizational image, and competitiveness (Bansal, 2005). Organizational green management affects the image of the firm, organizational reputation, and customer loyalty (Dutta, Umashankar, Choi, & Parsa, 2008). Organizational green human resources management has the challenge to attract the best talent for organizational greening development (Renwick et al., 2013, p. 2).

Environmental positioning of organizations has supportive and substantial effect of organizational greening as competitive advantage arguing that the inclusion of ecological concerns in the organizational culture develops green capabilities difficult to imitate (Harris and Crane 2002). Greening organizational culture is influenced by national cultures, economic, geographical, and other factors (Young et al. 2015). A challenge for greening the supply chain requires more flexibility of organizational culture of suppliers.

Organizational greening behavior is dependent on the involvement and leadership (Boiral et al. 2014) that implement policies and practices (Graci & Dodds, 2008). Leadership perceives external events and environmental problems as opportunities or treats to determine the commitment and actions of organizational greening behaviors (Sharma, 2000).

Greening leadership development initiatives influences the greening of organizations. Leaders influence organizational greening practices and activities of employee's pro-environmental behaviors. Organizational greening practices depend on leadership voluntary behaviors towards organizational green innovation issues that may have an impact on the voluntary workplace green behavior to improve the environmental sustainability not necessarily controlled by environmental management of formal institutions (D'Mello, Ones, Klein, Wiernik, Dilchert, 2011; Boiral, 2009; Kim, Kim, Han, Jackson, Ployhart, 2017).

The pro-environmental leadership behaviors contribute to the greening of organizations such as conservation, waste reduction and recycling, might be encouraged at the workplace to prevent environmental degradation and affect positively the climate change (Robertson, and Barling (2013).

ORGANIZATIONAL GREEN AND LEAN PRACTICES

Lean green management is related to the issues of green supply chain management, sustainability, organizational sustainable performance (Jurado et al., 2014). Khalili et al. (2016) has proposed a model to explain the interaction among lean approached in problem-solving, process, philosophy, and people, green in terms of environmental management system standards (ISO 14001) and organizational sustainable performance measures in the manufacturing-oriented organizations.

Some of the dominant practices of organizational greening activities and lean-green practices are the ones related to the reduction of water consumption, recycling and reuse of materials, resources and energy efficiency and renewal, waste reduction, prevention of pollution and exposure to unhealthy substances, reduction of regulatory risks and law requirements, and so forth… The interactions between lean-green management activities and organizational sustainable performance measure the implementation of these concepts. The overlap between green and lean focus in people and organization, waste reduction tools, lead time reduction and supply chain relationship (Dües et al. 2013). The green tools for protecting the environment are among others, green purchasing, green productivity, green marketing, green management, green supply chain

The eco-conscious strategies of organizational green management are based on the principle of caution and prevention for the formation of the ecological requirements to achieve the economic goals. Organizational eco-conscious strategies of green and lean management lead to emphasize strategic corporate social responsibilities regarding the environmental protection aimed to improve operational development and performance. SMEs improve sustainability performance using green and lean methods aimed to reduce costs, waste, better organization, lead time reduction and long-term success.

Organizational greening policies must be well-managed in accordance with the labor market policies to promote the greening of the economy and pursue creation of green jobs and ensuring skills development at all levels. Evidence from country-level policies and sectoral strategies is geared towards developing environmental sustainability, creating greener organizations, promoting green jobs, and enabling social inclusion. Active environmental engagement strategies in organizational green management can vary (Aragon-Correa et al., 2013; Boiral, 2009; Ramus & Steger, 2000; Renwick et al., 2013).

Organizational lean–green management is in interaction with strategy integration for organizational sustainable performance measures. The strategy lean-green integration and social management integration and agile and resilient practices integration. The practices and strategies of lean and green management synergies are related with organizational performance (Hallam and Contreras (2016); Bergmiller and McCright (2009); Cabral et al. (2012); D€ues et al. (2013); Galeazzo et al. (2014).

Environmental organizational citizenship behavior is a concept introduced by Boiral (2009) which plays a critical role in the organizational greening process, leading to efficiencies in time and financial returns from the environmental development (Arago´n-Correa and Rubio-Lo´pez 2007). Citizenship behavior of green organizations enhances organizational green management and attracting green consumers. (Boiral, Paillé, 2012; Boiral, Talbot, Paillé, 2015; Paillé, Chen, Boiral, Jin, 2014; Ramus, Killmer, 2007; Alt, Spitzeck, 2016; Ramus, 2001). Environmental and organizational social responsibility is a relevant trend in organizational greening development (Rahman & Post, 2012). Green reporting and social media are crucial activities to improve organizational greening performance (Too and Bajracharya, 2013 and Reilly, Hynan, 2014).

Environmental policy must disclose and monitor transparently the environmental performance for taking environmental responsibility related to greening and climate neutrality. Greening land use change has the core provision of payments for environmental services (Pagiola et al. 2020) in linear relationship between incentives and ensuing. Water policy objectives should be integrated in sectorial policies to address the quality needs, compliance for direct and greening payments based on environmental objectives (ECA (European Court of Auditors), 2017, 2014). Small and medium companies and enterprises should move towards government incentives in green management and informal pressure (Ashton, Russell, Futch, 2017).

The systematization of economic and organizational risks arising from the ecologization of agricultural production processes and the summarization of risk management stages agricultural greening to justify the used tools, is conducted by Burliai, Burliai, Revutska, Smolii, & Klymenko, 2021). Farmers are reluctant to use subsidies in green areas as the management contract to preserve the green image on environmental nature and a growth strategy in agriculture. Chemical fertilizers are restricted in areas of low input permanent pasture to catch crops. Voluntary, contractual agri-environmental agreements for rural development instead of using greening payments are used to meet water goals in nature and the environment (Buckley, Wall; Moran, O'Neil, and Murphy, 2016).

CONCLUSION

Environmentalism in organizations in times of rapid information development embrace knowledge of organizational greening and green-lean management, and knowledge sharing behavior leading to green management through a multilevel framework based on the theories of organizational identity and psychological ownership. Organizational green knowledge sharing broadens the knowledge of green management related to environmental behavior and organizational performance.

An understanding of socio ecosystem process of organizational model greening process is critical to manage organizational transformation. Organizational greening initiatives have solutions in ecological and economic terms that benefits the socioecological systems and allows growth and trigger environmental offerings. The organizational greening practices and activities have effects on sustainability that benefits the socio ecosystem.

Green management has a critical role in creating and establishing an organizational image. Individual characteristics contribute to the construction of an organizational green image and volunteering for activities contribute to the environmental concerns. Organizational policies and measures support organizational green management.

Organizational model greening must consider the links between the organizational eco-systems, managerial agency and organizational identity enable to build the capabilities needed for change towards sustainable development.

Future research on organizational green management may include financial indicators. Organizations carry out the green strategies to allocate budget resources and develop designs to promote green management

REFERENCES

Alt, E., & Spitzeck, H. (2016). Improving environmental performance through unit-level organizational citizenship behaviors for the environment: A capability perspective. *Journal of Environmental Management, 2016*(182), 48–58. doi:10.1016/j.jenvman.2016.07.034 PMID:27454096

Ambec, S., & Lanoie, P. (2008). Does it pay to be green? A systematic overview. *The Academy of Management Perspectives, 22*(4), 45–62. doi:10.5465/amp.2008.35590353

Ángel del Brío, J., Junquera, B., & Ordiz, M. (2008). Human resources in advanced environmental approaches–a case analysis. *International Journal of Production Research, 2008*(46), 6029–6053. doi:10.1080/00207540701352094

Arago'n-Correa, J. A., & Rubio-Lo'pez, E. A. (2007). Proactive corporate environmental strategies: Myths and misunderstandings. *Long Range Planning, 40*, 357–381. doi:10.1016/j.lrp.2007.02.008

Aragon-Correa, J. A., Martin-Tapia, I., & Hurtado-Torres, N. E. (2013). Proactive environmental strategies and employee inclusion: The positive effects on information sharing and promoting collaboration and the influence of uncertainty. *Organization & Environment, 40*(2), 1–23. doi:10.1177/1086026613489034

Banerjee, S. (2002). Corporate environmentalism: The construct and its measurement. *Journal of Business Research, 55*(3), 177–191. doi:10.1016/S0148-2963(00)00135-1

Bansal, P. (2005). Evolving sustainably: A longitudinal study of corporate sustainable development. *Strategic Management Journal, 26*(3), 197–218. doi:10.1002mj.441

Barin Cruz, L., & Avila Pedrozo, E. (2009). Corporate social responsibility and green management: Relation between headquarters and subsidiary in multinational corporations. *Management Decision, 47*(7), 1174–1199. doi:10.1108/00251740910978368

Barling, J., Christie, A., & Hoption, C. (2010). Leadership. In S. Zedeck (Ed.), *APA handbook of industrial and organizational psychology* (Vol. 1). American Psychological Association.

Barling, J., Weber, T., & Kelloway, K. E. (1996). Effects of transformational leadership training on attitudinal and financial outcomes: A field experiment. *The Journal of Applied Psychology, 81*(6), 827–832. doi:10.1037/0021-9010.81.6.827

Barsade, S. G., & Gibson, D. E. (2007). Why does affect matter in organizations? *The Academy of Management Perspectives, 21*(1), 36–59. doi:10.5465/amp.2007.24286163

Bebbington, J. (2001). Sustainable development: A review of the international development, business, and accounting literature. *Accounting Forum, 25*(2), 128–157. doi:10.1111/1467-6303.00059

Bergmiller, G. G., & McCright, P. R. (2009). *Parallel models for Lean and green operations*. Paper presented at the Industrial Engineering Research Conference, Miami, FL. Available at: https://pdfs.semanticscholar.org/0a67/c7f0f6d60ba0506e037aaf17cfe8b3a89e7d.pdf

Bissing-Olson, M. J., Iyer, A., Fielding, K. S., & Zacher, H. (2013). Relationships between daily affect and pro-environmental behavior at work: The moderating role of pro-environmental attitude. *Journal of Organizational Behavior, 34*(2), 156–175. doi:10.1002/job.1788

Boiral, O. (2002). Tacit knowledge and environmental management. *Long Range Planning, 35*(3), 291–317. doi:10.1016/S0024-6301(02)00047-X

Boiral, O. (2009). Greening the corporation through organizational citizenship behaviors. *Journal of Business Ethics, 87*(2), 221–236. doi:10.100710551-008-9881-2

Boiral, O., Baron, C., & Gunnlaugson, O. (2014). Environmental leadership and consciousness development: A case study among Canadian SMEs. *Journal of Business Ethics, 123*(3), 363–383. doi:10.100710551-013-1845-5

Boiral, O., & Paillé, P. (2012). Organizational citizenship behaviour for the environment: Measurement and validation. *Journal of Business Ethics, 2012*(109), 431–445. doi:10.100710551-011-1138-9

Boiral, O., Talbot, D., & Paillé, P. (2015). Leading by example: A model of organizational citizenship behavior for the environment. *Business Strategy and the Environment, 2015*(24), 532–550. doi:10.1002/bse.1835

Buckley, C., Wall, D. P., Moran, B., O'Neil, S., & Murphy, P. N. (2016). Farm gate level nitrogen balance and use efficiency changes post implementation of the EU Nitrates Directive. *Nutrient Cycling in Agroecosystems, 2016*(104), 1–13. doi:10.100710705-015-9753-y

Burliai, A., Burliai, O., Revutska, A., Smolii, L., & Klymenko, L. (2021). Organizational and economic risks of ecologization of agriculture. *Agricultural and Resource Economics: International Scientific E-Journal*, *7*(1), 96–114. doi:10.51599/are.2021.07.01.06

Cabral, I., Grilo, A., & Cruz-Machado, V. (2012). A decision-making model for Lean, Agile, Resilient and Green supply chain management. *International Journal of Production Research*, *50*(17), 4830-4845.

Chen, Y. S. (2011). Green organizational identity: Sources and consequence. *Management Decision*, *2011*(49), 384–404. doi:10.1108/00251741111120761

Chen, Y. S., & Chang, C. H. (2013). The determinants of green product development performance: Green dynamic capabilities, green transformational leadership, and green creativity. *Journal of Business Ethics*, *116*(1), 107–119. doi:10.100710551-012-1452-x

Chen, Y. S., Chang, C. H., & Lin, Y. H. (2014). The determinants of green radical and incremental innovation performance: Green shared vision, green absorptive capacity, and green organizational ambidexterity. *Sustainability*, *2014*(6), 7787–7806. doi:10.3390u6117787

Chen, Y. S., Chang, C. H., Yeh, S. L., & Cheng, H. I. (2015). Green shared vision and green creativity: The mediation roles of green mindfulness and green self-efficacy. *Quality & Quantity*, *2015*(49), 1169–1184. doi:10.100711135-014-0041-8

Chen, Y. S., Chang, T. W., Lin, C. Y., Lai, P. Y., & Wang, K. H. (2016). The influence of proactive green innovation and reactive green innovation on green product development performance: The mediation role of green creativity. *Sustainability*, *2016*(8), 966. doi:10.3390u8100966

Chou, C.-J. (2014). Hotels' environmental policies and employee personal environmental beliefs: Interactions and outcomes. *Tourism Management*, *40*, 436–446. doi:10.1016/j.tourman.2013.08.001

Connelly, C. E., Ford, D. P., Turel, O., Gallupe, B., & Zweig, D. (2014). 'I'm busy (and competitive)!' Antecedents of knowledge sharing under pressure. *Knowledge Management Research and Practice*, *2014*(12), 74–85. doi:10.1057/kmrp.2012.61

Çop, S., Olorunsola, V. O., & Alola, U. V. (2021). Achieving environmental sustainability through green transformational leadership policy: Can green team resilience help? *Bus Strat Env.*, *30*, 671–682. https://onlinelibrary.wiley.com/doi/epdf/10.1002/bse.2646 doi:10.1002/bse.2646

D'Mello, S., Ones, D., Klein, R., Wiernik, B., & Dilchert, S. (2011). Green company rankings and reporting of pre-environmental efforts in organizations. *Proceedings of the Poster session presented at the 26th Annual Conference of the Society for Industrial and Organizational Psychology*.

Daily, B. F., Bishop, J. W., & Govindarajulu, N. (2009). Conceptual model for organizational citizenship behavior directed toward the environment. *Business & Society*, *48*(2), 243–256. doi:10.1177/0007650308315439

Daily, B. F., Bishop, J. W., & Steiner, S. (2007). The mediating role of EMS teamwork as it pertains to HR factors and perceived environmental performance. *Journal of Applied Business Research*, *23*, 95–109.

de Jong, M. (2021). Inclusive capitalism. The emergence of a new purpose paradigm in economics and business administration and its implications for public policy. *Global Public Policy and Governance*, *1*, 159–174. https://link.springer.com/content/pdf/10.1007/s43508-021-00020-z.pdf doi:10.1007/s43508-021-00020-z

Dubose, J. (2000). Sustainability and performance at Interface Inc. *Interfaces*, *30*(3), 190–201.

Dües, C. M., Tan, K. H., & Lim, M. (2013). Green as the new Lean: How to use Lean practices as a catalyst to be greening your supply chain. *Journal of Cleaner Production*, *40*, 93–100. doi:10.1016/j.jclepro.2011.12.023

Dutta, K., Umashankar, V., Choi, G., & Parsa, H. G. (2008). A comparative study of consumers' green practice orientation in India and the United States: A study from the restaurant industry. *Journal of Foodservice Business Research*, *11*(3), 269–285. doi:10.1080/15378020802316570

Dvir, T., Eden, D., Avolio, B. J., & Shamir, B. (2002). Impact of transformational leadership on follower development and performance: A field experiment. *Academy of Management Journal*, *45*, 735–744.

ECA (European Court of Auditors). (2014). *Special Report. Integration of EU Water Policy Objectives with the CAP: A Partial Success*. European Court of Auditors. Available online: https://www.eca.europa.eu/Lists/ECADocuments/SR14_04/SR14_04_EN.pdf

ECA (European Court of Auditors). (2017). *Greening: A More Complex Income Support Scheme, Not Yet Environmentally*. Effective; European Court of Auditors. Available online: https://www.eca.europa.eu/Lists/ECADocuments/SR17_21/SR_GREENING_EN.pdf

Eco Measures Report. (2010). Interface Inc.

Esty, D. C., & Winston, A. S. (2006). *Green to Gold: How Smart Companies Use Environmental Strategy to Innovate Create Value and Build Competitive Advantage*. Yale University Press.

Flannery, B. L., & May, D. R. (1994). Prominent factors influencing environmental activities: Application of the environmental leadership model (ELM). *The Leadership Quarterly*, *1994*(5), 201–221. doi:10.1016/1048-9843(94)90012-4

Florida, R., & Davison, D. (2001). Gaining from Green Management: Environmental management systems inside and outside the factory. *California Management Review*, *43*(3), 64–84. doi:10.2307/41166089

Francoeur, V., Paillé, P., Yuriev, A., & Boiral, O. (2021). The Measurement of Green Workplace Behaviors: A Systematic Review. *Organization & Environment*, *34*(1), 18–42. doi:10.1177/1086026619837125

Galeazzo, A., Furlan, A., & Vinelli, A. (2014). Lean and green in action: Interdependencies and performance of pollution prevention projects. *Journal of Cleaner Production*, *85*, 191–200. doi:10.1016/j.jclepro.2013.10.015

Gioia, D.A., Patvardhan, S.D., Hamilton, A.L., & Corley, K.G. (2013). Organizational identity formation and change. *Acad. Manag. Ann.*, *7*(1), 123-193.

González-Benito, J., & González-Benito, O. (2006). A review of determinant factors of environmental proactivity. *Business Strategy and the Environment*, *15*(2), 87–102. doi:10.1002/bse.450

Graci, S., & Dodds, R. (2008). Why goes green? The business case for environmental commitment in the Canadian hotel industry. *Anatolia*, *19*(2), 251–270. doi:10.1080/13032917.2008.9687072

Hallam, C., & Contreras, C. (2016). Integrating lean and green management. *Management Decision*, *54*(9), 2157–2187. doi:10.1108/MD-04-2016-0259

Halme, M., Anttonen, M., Kuisma, M., Kontoniemi, N., & Heino, E. (2007). Business models for material efficiency services: conceptualization and application. *Ecol. Econ.*, *63*(1), 126-137.

Harris, L. C., & Crane, A. (2002). The greening of organizational culture: Management views on the depth, degree, and diffusion of change. *Journal of Organizational Change Management*, *15*(3), 214–234. doi:10.1108/09534810210429273

Hart, S. (1997). Beyond greening: Strategies for a sustainable world. *Harvard Business Review*, *75*, 66–76.

Hart, S. L. (2005). *Capitalism at the Crossroads: The Unlimited Business Opportunities in Solving the World's Most Difficult Problems*. Wharton School Publishing.

Hofer, C., Cantor, D. E., & Dai, J. (2012). The competitive determinants of a firm's environmental management activities: Evidence from US manufacturing industries. *Journal of Operations Management*, *30*(1-2), 69–84. doi:10.1016/j.jom.2011.06.002

Holliday, C., (2001). Sustainable growth the DuPont way. *Harv. Bus. Rev.*, *79*(8), 129-132.

Holt, D., (2011). Where are they now? Tracking the longitudinal evolution of environmental businesses from the 1990s. *Bus. Strat. Env.*, *20*(4), 238-250.

Honcharuk, I.V., & Tomashuk, I.V. (2017). Vplyv ekoloho-ekonomichnoho faktoru na osoblyvosti orhanizatsiino-ekonomichnoho mekhanizmu vykorystannia resursnoho potentsialu silskykh terytorii. *Ekonomika, finansy, menedzhment: aktualni pytannia nauky i praktyky*, *4*(20), 55-62.

Honcharuk, I.V., & Tomashuk, I.V. (2018). Derzhavne rehuliuvannia rozvytku resursnoho potentsialu silskykh terytorii: zahalni aspekty. *Ekonomika, finansy, menedzhment: aktualni pytannia nauky i praktyky*, *4*(32), 19-30.

Imasiku, K. (2021). *Organizational Insights, Challenges and Impact of Sustainable Development in Developing and Developed Nations*. IntechOpen. https://www.intechopen.com/chapters/73366 doi:10.5772/intechopen.93748

Jabbour, C. J. C. (2011). How green are HRM practices, organizational culture, learning and teamwork? A Brazilian study. *Industrial and Commercial Training*, *43*(2), 98–105. doi:10.1108/00197851111108926

Jabbour, C. J. C. (2013a). Environmental training in organisations: From a literature review to a framework for future research. *Resources, Conservation and Recycling*, *74*, 144–155. doi:10.1016/j.resconrec.2012.12.017

Jabbour, C. J. C. (2013b). Environmental training and environmental maturity of Brazilian companies with ISO14001: Empirical evidence. *Journal of Cleaner Production*, *43*(2), 1–8.

Jabbour, C. J. C., Almada Santos, F. C., Azevedo Fonseca, S., & Seido Nagano, M. (2013). Green teams: Understanding their roles in the environmental management of companies located in Brazil. *Journal of Cleaner Production*, *46*, 58–66. doi:10.1016/j.jclepro.2012.09.018

Jabbour, C. J. C., Santos, F. C. A., & Nagano, M. S. (2010). Contribution of HRM throughout the stages of environmental management: Methodological triangulation applied to companies in Brazil. *International Journal of Human Resource Management, 21*(7), 1049–1089. doi:10.1080/09585191003783512

Jackson, S. E., Renwick, D. W. S., Jabbour, C. J. C., & Muller-Camen, M. (2011). State-of-the-art and future directions for green human resource management: Introduction to the special issue. *German Journal of Research in Human Resource Management, 25*(2), 99–116. doi:10.1177/239700221102500203

Jackson, S. E., & Seo, J. (2010). The greening of strategic HRM scholarship. *Organizational Management Journal, 7*(4), 278–290. doi:10.1057/omj.2010.37

Johansen, D. (1998). Interface Inc: Taking the lead toward sustainability. *Corporate Environmental Strategy, 5*(3), 54. doi:10.1016/S1066-7938(00)80100-1

Jurado, M., Jose, P., & Fuentes, J. M. (2014). Lean management, supply chain management and sustainability: A literature review. *Journal of Cleaner Production, 85*, 134–150. doi:10.1016/j.jclepro.2013.09.042

Kee-hung, L., Cheng, T., & Tang, A. (2010). Green retailing: Factors for successes. *California Management Review, 52*(2), 6–31. doi:10.1525/cmr.2010.52.2.6

Kim, A., Kim, Y., Han, K., Jackson, S. E., & Ployhart, R. E. (2017). Multilevel influences on voluntary workplace green behavior: Individual diferences, leader behavior, and coworker advocacy. *Journal of Management, 2017*(43), 1335–1358. doi:10.1177/0149206314547386

Kim, Y. J., Woo, G. K., Hyung-Min, Ch., & Kullada, P. (2019). The effect of green human resource management on hotel employees' ecofriendly behavior and environmental performance. *International Journal of Hospitality Management, 76*, 83–93. doi:10.1016/j.ijhm.2018.04.007

Kitazawa, S., & Sarkis, J. (2000). The relationship between ISO 14001 and continuous source reduction programs. *International Journal of Operations & Production Management, 20*(2), 225–248. doi:10.1108/01443570010304279

Lampikoski, T. (2012). Green, innovative, and profitable: a case study of managerial capabilities at Interface Inc. *Tech. Inn. Manag. Rev., 2*(11), 4-12.

Lin, B., Jones, C., & Hsieh, C. (2001). Environmental practices and assessment: A process perspective. *Industrial Management & Data Systems, 101*(2), 71–80. doi:10.1108/02635570110384348

Loknath, Y. & Azeem, B. (2017). *Green Management-Concept and Strategies. National Conference on Marketing and Sustainable Development.*

Marcus, A. A., & Fremeth, A. R. (2009). Green management matters regardless. *The Academy of Management Perspectives*, *8*, 17–26.

Mason, K., & Spring, M. (2011). The sites and practices of business models. *Industrial Marketing Management*, *40*(6), 1032–1041. doi:10.1016/j.indmarman.2011.06.032

Matukova, H., Khrapkina, V., Bahashova, N., Matukova-Yaryha, D., & Levitskaia, A. (2021). Eco-Controlling: Prospects for the Environmental Policy in Enterprise Management. *SHS Web of Conferences, 100*. doi:10.1051/shsconf/202110005011

McDonagh, P., & Prothero, A. (1997). *Green management: A reader*. Dryden Press.

Mullen, J. E., & Kelloway, E. K. (2009). Safety leadership: A longitudinal study of the effects of transformational leadership on safety outcomes. *Journal of Occupational and Organizational Psychology*, *82*(2), 253–272. doi:10.1348/096317908X325313

Neuendorf, K. A. (2002). *The Content Analysis Guidebook.* Sage.

Ng, P. F., Butt, M. M., Khong, K. W., & Ong, F. S. (2014). Antecedents of Green Brand Equity. *Journal of Business Ethics, Springer*, *121*(2), 203–215. doi:10.100710551-013-1689-z

Nidumolu, R., Prahalad, C.K., & Rangaswami, M.R., (2009). Why sustainability is now thekey driver of innovation. *Harv. Bus. Rev.*, *87*(9), 2-9.

Norton, T. A., Parker, S. L., Zacher, H., & Ashkanasy, N. M. (2015). Employee green behavior: A theoretical framework, multilevel review, and future research agenda. *Organization & Environment*, *2015*(28), 103–125. doi:10.1177/1086026615575773

Norton, T. A., Zacher, H., & Ashkanasy, N. M. (2014). Organisational sustainability policies and employee green behavior: The mediating role of work climate perceptions. *Journal of Environmental Psychology*, *38*, 49–54. doi:10.1016/j.jenvp.2013.12.008

Ones, D. S., & Dilchert, S. (2012a). Employee green behaviors. In D. S. S. E. Jackson (Ed.), *Managing human resource for environmental sustainability* (pp. 85–116). Jossey-Bass.

Owen, D. (1992). *Green reporting: Accountancy and the challenge of the nineties*. Chapman Hall.

Pagiola, S., Honey-Roses, J., & Freire-Gonzalez, J. (2020). Assessing the permanence of land-use change induced by payments for environmental services: Evidence from Nicaragua. *Tropical Conservation Science, 13*, 194008292092267–194008292092215. doi:10.1177/1940082920922676

Paille´, P., Chen, Y., Boiral, O., & Jin, J. (2014). The impact of human resource management on environmental performance: An employee-level study. *Journal of Business Ethics, 121*, 451–466.

Paillé, P., Chen, Y., Boiral, O., & Jin, J. (2014). The impact of human resource management on environmental performance: An employee-level study. *Journal of Business Ethics, 2014*(121), 451–466. doi:10.100710551-013-1732-0

Pane Haden, S. S., Oyler, J. D., & Humphreys, J. H. (2009). Historical, practical, and theoretical perspectives on green management: An exploratory analysis. *Management Decision, 47*(7), 1041–1055. doi:10.1108/00251740910978287

Peattie, K. (1992). *Green Marketing*. Pitman.

Peng, Y., & Lin, S. (2008). Local responsiveness pressure, subsidiary resources, green management adoption, and subsidiaries performance: Evidence from Taiwanese manufacturers. *Journal of Business Ethics, 79*(1-2), 199–212. doi:10.100710551-007-9382-8

Pham, N. T., Tučková, Z., & Chiappetta Jabbour, C. J. (2019). Greening the hospitality industry: How do green human resource management practices influence organizational citizenship behavior in hotels? A mixed-methods study. *Tourism Management, 72*, 386–399. doi:10.1016/j.tourman.2018.12.008

Phillips, L. (2007). Go green to gain the edge over rivals. *People Management, 13*, 9.

Prasad, P., & Elmes, M. (2005). In the name of the practical: Unearthing the hegemony of pragmatics in the discourse of environmental management. *Journal of Management Studies, 42*, 845–867. doi:.00521.x doi:10.1111/j.1467-6486.2005

Rahman, N., & Post, C. (2012). Measurement issues in environmental corporate social responsibility (ECSR): Toward a transparent, reliable, and construct valid instrument. *Journal of Business Ethics, 105*(3), 307–319. doi:10.100710551-011-0967-x

Ramus, C. A. (2001). Organizational support for employees: Encouraging creative ideas for environmental sustainability. *California Management Review, 2001*(43), 85–105. doi:10.2307/41166090

Ramus, C. A., & Killmer, A. B. (2007). Corporate greening through prosocial extra role behaviors—A conceptual framework for employee motivation. *Business Strategy and the Environment, 2007*(16), 554–570. doi:10.1002/bse.504

Ramus, C. A., & Steger, U. (2000). The roles of supervisory support behaviors and environmental policy in employee 'eco initiatives' at leading-edge European companies. *Academy of Management Journal, 43*(4), 605–626. doi:10.2307/1556357

Reilly, A. H., & Hynan, K. A. (2014). Corporate communication, sustainability, and social media: It's not easy (really) being green. *Business Horizons, 57*(6), 747–758. doi:10.1016/j.bushor.2014.07.008

Renwick, D. W. S., Redman, T., & Maguire, S. (2013). Green Human Resource Management: A review and research agenda. *International Journal of Management Reviews, 15*(1), 1–14. doi:10.1111/j.1468-2370.2011.00328.x

Robertson, J. L., & Barling, J. (2013). Greening organizations through leaders' influence on employees' pro-environmental behaviors. *Journal of Organizational Behavior, 34*(2), 176–194. doi:10.1002/job.1820

Russo, M. V., & Fouts, P. A. (1997). A resource-based perspective on corporate environmental performance and profitability. *Academy of Management Journal, 40*(3), 534–559. doi:10.2307/257052

Sharma, S. (2000). Managerial interpretations and organizational context as predictors of corporate choice of environmental strategy. *Academy of Management Journal, 43*(4), 681–697.

Sharma, S. (2009). The mediating effect of information availability between organizational design variables and environmental practices in the Canadian hotel industry. *Business Strategy and the Environment, 18*(4), 266–276. doi:10.1002/bse.603

Shatouri, R. M., Omar, R., Igusa, K., & Flávio de São, P. F. (2013). Embracing green technology innovation through strategic human resource management: A case of an automotive company. *American Journal of Economics and Business Administration, 5*(2), 65–73. doi:10.3844/ajebasp.2013.65.73

Shrivastava, P. (1995). The role of corporations in achieving ecological sustainability. *Academy of Management Review, 20*(4), 936–960. doi:10.2307/258961

Staats, H., Harland, P., & Wilke, H. A. M. (2004). Effecting durable change: A team approach to improve environmental behavior in the household. *Environment and Behavior, 36*(3), 341–367. doi:10.1177/0013916503260163

Strauss, A. L., & Corbin, J. M. (1998). *Basics of Qualitative Research. Techniques and Procedures for Developing Grounded Theory*. Sage.

Stringer, L. (2009). *The Green workplace. Sustainable strategies that benefit employees, the environment, and the bottom line*. Macmillan.

Su, X., Xu, A., Lin, W., Chen, Y., Liu, S., & Xu, W. (2020, April-June). Environmental Leadership, Green Innovation Practices, Environmental Knowledge Learning, and Firm Performance. *SAGE Open*, *10*(2), 1–14. doi:10.1177/2158244020922909

Tatoglu, E., Bayraktar, E., Sahadev, S., Demirbag, M., & Glaister, K. W. (2014). Determinants of voluntary environmental management practices by MNE subsidiaries. *Journal of World Business*, *49*(4), 536–548. doi:10.1016/j.jwb.2013.12.007

Teixeira, A. A., Jabbour, C. J. C., & Jabbour, A. B. L. (2012). Relationship between green management and environmental training in companies located in Brazil: A theoretical framework and case studies. *International Journal of Production Economics*, *140*(1), 318–329. doi:10.1016/j.ijpe.2012.01.009

Temminck, E., Mearns, K., & Fruhen, L. (2015). Motivating employees towards sustainable behaviour. *Business Strategy and the Environment*, *24*(6), 402–412. doi:10.1002/bse.1827

Toktay, L. B., Selhat, L., & Anderson, R. (2006). Doing well by doing good: interface's vision of being the first industrial company in the world to attain sustainability. In Enterprise Transformation: Understanding and Enabling Fundamental Change. Wiley.

Too, L., & Bajracharya, B. (2015). Sustainable Campus: Engaging the community in sustainability Int. *Journal on Sustainability in Higher Education*, *16*(1), 57–71. doi:10.1108/IJSHE-07-2013-0080

Unnikrishnan, S., & Hegde, D. S. (2007). Environmental training and cleaner production in Indian industry—A micro-level study. *Resources, Conservation and Recycling*, *50*(4), 427–441. doi:10.1016/j.resconrec.2006.07.003

Verrier, B., Rose, B., Caillaud, E., & Remita, H. (2014). Combining organizational performance with sustainable development issues: The lean and green project benchmarking Repository. *Journal of Cleaner Production*, *85*, 83–93. doi:10.1016/j.jclepro.2013.12.023

Wagner, M. (2011). Environmental management activities and sustainable HRM in German Manufacturing firms. Incidence, determinants, and outcomes. *German Journal of Research in Human Resource Management*, *25*(2), 157–177. doi:10.1177/239700221102500206

Wehrmeyer, W. (1996). *Greening people: Human resources and environmental management*. Greenleaf.

Xu, J., Guan, J., & Lin, Y. (2017). Institutional pressures, top managers' environmental awareness and environmental innovation practices: An institutional theory and upper echelons theory perspective. *Management Review*, *29*(9), 72–83.

Young, W., Davis, M., McNeill, I. M., Malhotra, B., Russell, S., Unsworth, K., & Clegg, C. (2015). Changing behavior: Successful environmental programmes in the workplace. *Business Strategy and the Environment*, *24*(8), 689–703. doi:10.1002/bse.1836

Zhang, B., Wang, Z., & Lai, K. H. (2015). Mediating effect of managers' environmental concern: Bridge between external pressures and firms' practices of energy conservation in China. *Journal of Environmental Psychology*, *43*, 203–215. doi:10.1016/j.jenvp.2015.07.002

Chapter 2
Global Trends in the Research Output on Sustainable Development Goals:
A Bibliometric Analysis Using Bibliometrix R-Tool

Diksha Diksha
State Library, Chandigarh, India

Rupak Chakravarty
https://orcid.org/0000-0001-5046-1663
Panjab University, India

ABSTRACT

The chapter is a bibliometric analysis of research work published on the Sustainable Development Goals. The authors extracted relevant data from Scopus on 22 August 2020, which were subsequently analyzed using open-source software – Bibliometrix (an R package). They performed a textual query on Scopus for the title "Sustainable Development Goal" by selecting the 'Article Title' in the 'Documents' option for the period of 1996-2020 and retrieved 1872 documents from 865 sources. This study aims to identify the global trends in the SDG's scientific production over time with analysis covering various bibliometric indicators (i.e., sources, authors, and documents). Results reveal growth in research in SDGs in recent years, trending topics, most relevant sources, prolific authors in the field, validation of Lotka's law. The findings can assist future research in this field by providing a worldwide analytical overview of the SDGs.

DOI: 10.4018/978-1-6684-4225-8.ch002

INTRODUCTION

Sustainable Development Goals

The term Sustainable Development Goals was publicized in 1987 when the World Commission on Environment and Development presented a document for the United Nations titled "Our Common Future," also known as the Brundland Report. The term was popularized from this report and was translated into Spanish as desarrollo sostenible (D.S.), sustainable development, or durable development (United Nations, 1987).

According to the United Nations Sustainable Development Goals, "The Sustainable Development Goals are the blueprint for achieving a better and more sustainable future for all. They address the global challenges we face, including those related to poverty, inequality, climate change, environmental degradation, peace, and justice". The 17 goals and 169 targets of the SDGs were developed to succeed in the eight Millennium Development Goals (MDGs), which ended in 2015 (Sachs, 2012). The transition from eight Millennium development goals (MDGs) to 17 SDGs marked a shift from a vertical to a horizontal approach in addressing global issues (Stafford-Smith et al., 2017). Achieving cross-sectional progress in the ambitious 2030 plan requires the involvement of policymakers at all levels and in all sectors (United Nations, n.d.).

The 17 goals are interconnected, and all must be achieved by 2030 so that no purpose is left behind. The 17 SDGs are to end poverty in all its forms; Zero Hunger; Health; Education; Gender Equality and women empowerment; Water and Sanitization; Energy; Economic Growth; Infrastructure, Industrialization; Inequalities; Cities; Sustainable consumption and productions; Climate change; Oceans; Biodiversity, Forest, Desertification; Peace, Justice and strong institutions; Partnerships. (un.org)

Sustainable Development is a concept researched extensively over time. Our work presents a bibliometric analysis of the scientific production generated around Sustainable Development Goals (SDGs) over a time period of 1996-2020. We performed research using Bibliometrix R- Tool.

Bibliometrix: R Package

R-Studio is an open-source development environment of R and runs on Windows, Linux, and other operating systems. Many open source packages are available on Rstudio. Bibliometrix R is a specially designed package for bibliometric and scientometric studies. Massimo Aria and Corrado Cuccurullo developed the software that includes all the main bibliometric methods of analysis (Warin, n.d.). The package provides various routines for importing bibliographic data from SCOPUS, Clarivate

Analytics' Web of Science (WoS), PubMed, and Cochrane databases performing bibliometric analysis, and building data matrices for co-citation, coupling, scientific collaboration analysis, and co-word analysis (Warin, n.d.).

LITERATURE REVIEW

In a cross-sectional descriptive bibliometric study, Bibliometric analysis on scientific publication on Sustainable development goals with an accentuation on Good health and well being (2015-2019), Sustainability journal positioned first in the number of SDGs related publications. WHO was the most dynamic foundation publishing research in SDGs. The investigation additionally uncovered that the vast majority of the retrieved documents belonged to the SDG 17 (Partnership), trailed by climate action and responsible consumption and production. Affordable and clean energy had the least publications. Further investigation revealed that the European region had the highest research contribution while the eastern Mediterranean contributed the least (Sweileh, 2020)

Quality education as perceived under SDG4 (2030 Agenda) aims to "Ensure inclusively and quality education for all and promote lifelong learning". González García et al., (2020) investigated SDG4 pertaining to quality education by applying bibliometric techniques on data derived from the WoS and Scopus (240 publications), noted the highest research output in the year 2019 for the SDG4 parameter, "quality education" [n= 84]. According to the discoveries of the investigation, quality education inside the SDGs started picking up importance with the most noteworthy logical creation in 2019. The study concluded that the bibliometric approach is essential and necessary to know the reality of scientific production on the subject and to be able to make proposals and lines of research for its development.

Meschede (2020) in his study, "The Sustainable Development Goals in Scientific Literature: A Bibliometric Overview at the Meta-Level," considered 4593 research articles at the meta-level, explicitly referring to the SDGs. According to the investigation, the maximum research done in SDGs is from Life Sciences and Biomedicine and Social Sciences. The most predominant SDG among the analyzed research articles is SDG 3 ("acceptable wellbeing and prosperity"). A relatively high share of open access articles contributes to the idea of knowledge sharing for the SDGs. Nearly 37% of all articles count as international publications, i.e., as being co-authored by authors from affiliations of multiple countries.

In a study to explore the usage of bibliometric analysis software Bibliometrix R package Dervis (2020) used Graphene-one of the fastest-growing research fields in nanotechnology for the analysis. As a part of the search strategy, the authors performed a textual query with the term "Graphene" on the Web of Science. They retrieved 1155 scholarly papers from 2000-2017 with at least one author from Turkey. As per the bibliometric results, graphene within nanotechnology is growing steadily. Graphene is not only used in engineering but also used in medical technology. Furthermore, this is ongoing research exploring Open-source software and its roles in the field of information studies.

Jalal (2019) studied research collaboration between India and Bangladesh. The study author retrieved jointly 1156 papers from the web of science from 1991 to 2017. The collaboration network on co-authorship and co-occurrences was build using the Bibliometrix R package. As per the studies, significant collaborations were in medical sciences, agriculture, and biological sciences. Studies also verified Lotka's law for co-authorship data. With the help of the Bibliometrix R Package and based on the full-count method, the study revealed that the top authors have the highest publications with joint-institutional affiliations between India and Bangladesh. G.B. Nair (92), ZA Bhutta (53), and T. Ramamurthy (51) were the top three authors.

Fuad et al. (2020) studied the bibliometric profile of the journal "BioScience Trends." The authors retrieved data from the core collection database of the web of science. The authors further divided the dataset into three intervals, i.e., 2007-2010,2011-2014, and 2015-2017 respectively. They executed data analysis using Bibliometrix, VOSViewer 1.66, Orange 3.15, and CitNetExplorer. As per the results, within one decade of scientific production, BST continues to attract global researchers in life sciences dominated by authors from China and Japan. Annual Journal growth of BST is 12.83%. Research themes are consistent with the Aims and Scope of the Journal with a strong emphasis on molecular biology, biochemistry, and clinical research. Entering the second decade, strategies to promote and enlarge authors' participation from countries that are not in the current list are encouraged.

Munim et al. (2020), in their study, provide a bibliometric review of 279 studies on the applications of big data and artificial intelligence (A.I.) in the maritime industry, published in 214 academic outlets, authored by 842 scholars. Data were extracted from the Web of Science and analyzed it using the Bibliometrix tool in R software. Their study revealed the most influential articles, journals, authors, and institutions. Authors with bibliographic coupling identified four underlying research clusters: (1) digital transformation in the maritime industry, (2) applications of big data from AIS, (3) energy efficiency, and (4) predictive analytics. These clusters were analyzed in detail and extracted future research questions.

The scientific production of rural depopulation has grown in recent years. However, a global picture of the research carried out on this topic does not exist. Rodríguez-Soler et al. (2020) worked to identify the worldwide trends in rural depopulation scientific production. Data were extracted and merged from the databases: WoS and Scopus. The final sample of 1150 articles published between 1979 and 2018 was analyzed using bibliometrix an R package software. According to the results, there are two significant areas of co-citation networks: an inadequate network of collaboration between countries with some highlighted interaction, an author collaboration network with close groups of knowledge, and two main themes resulting from a co-word analysis. Authors concluded rural depopulation as not a solid field of research, and the most important themes identified are related to specific geographical areas. Authors observed growth in international collaboration along with new trends. Findings could assist future research in this or related fields by providing a worldwide overview of rural depopulation research over time.

Trivedi (2020) did a scientometric and visualization study of global Agriculture Big Data (ABD) research. The study aimed to traverse the present status of Agriculture Big Data research through network analysis and visualizations study of the ABD publications. They retrieved a sample of 379 publications from Web of Science within the period of "all years." VOSviewer, MSExcel, and R statistical software were used for data analysis. Various results are drawn based on annual scientific production, most cited papers, most cited authors, most cited affiliations, most prolific nations, author's h-index, co-authorship analysis of countries and organizations, co-citation analysis of sources, the keyword co-occurrence analysis, and density visualization. This study investigates the growth status in ABD related research, which can help the policymaker, researchers, and people of agriculture and allied sector to have a thorough understanding of Agriculture Big Data research for further study.

In Dentistry Research by Hosaain & Batcha (2020) did Scientometric studies to elaborate the scientific production of original and review articles published in the dental journals for the study period 1999 to 2018, considering qualitative and quantitative measures from all over the countries. 15870 records of 1635 journals were retrieved from WoS. Data analysis revealed that 2018 was considered the most productive with 1411 (8.84%) publications. Lang NP published the highest number of articles, 89. The findings of the analysis revealed that the single author was published 2342 papers and 13628 papers published by multiple authors. From the study, it can be concluded; this paper may be considered as a baseline study for the scientometric information related to articles on Dentistry published in the Web of Science

Laksham et al. (2020) studied the global level perspective of Coronavirus research output from 1989 to March 2020, and these analyses included year-wise research growth, global publication share, and patterns of research communication channels and the most productive journals. Using the search string of "Coronavirus" OR "COVID 19" limiting the period from 1989 to 2020, 7381 publications were retrieved from Web of Science. The analysis revealed that 2019 had the highest publications, i.e., 561 with 848 citations. Thus this article can be concluded by a collaborative author's productivity dominates compared to the single author's contribution. Based on literature analysis around the world, it is found that the 7381 publications came from 127 countries. United States (USA) is the most productive country with 2801 publications (37.9% and received 107738 Citations. India (80) has to improve in the field of Coronavirus research in the future.

Study Objectives

The main objectives of the present study are:

1. To examine the growth of SDG's research published during the year 1996-2020
2. To identify the most prolific sources in the field of SDGs.
3. To identify authors productivity in the field
4. To identify the most prolific author in the field
5. To validate the Lotka's Law for the dataset
6. To identify the trending topics in the field.

MATERIALS AND METHODS

To accomplish this purpose the analysis was carried in which the publications were measured and statistical method was applied that have made it conceivable to know the truth and present status of the topic following the guidelines and criteria of bibliometric studies. [18] Bibliometric approaches, in which various factors connected to scientific production regarding a matter are analyzed, permit us to know the truth of this subject matter. It is an extremely valuable exploration technique for approaching and inspecting a subject, permitting us to comprehend what are the principle distributions, authors, and organizations of reference, just as the fundamental patterns regarding subjects and the connections that underlie and are created between the fields of information (Bhatt et al., 2020; Di Stefano et al., 2010).

Thus, to perform the bibliometric analysis data was extracted from the database of Scopus.The search Strategy used for document retrieval was TITLE ("Sustainable development Goal*") by selecting the "Article title" in the "Documents" option. The search was carried out on 22 August 2020. The study is based on 1872 papers published in 865 Sources by authors worldwide during 1996-2020. Data was downloaded from Scopus on 22 August 2020. We used bibliometrix (R package) for data analysis and interpretation.

RESULTS

Main Information about Data

Table 1 gives explanatory information retrieved from the dataset giving brief data accumulated. Table 1 gives the absolute number of records for example 1872 recovered from 865 sources. Average years from distributions are determined to be 1.9, average references per document 9.645 and average references every year per document 2.563. Further information was disintegrated into kinds of documents for example Articles(1046), Articles in press(1), Book(20), Book chapter(138), data paper(149), Meeting Review(1), Information Paper(3), Editorial(142), Erratum(29), Letter(25), Note(117), Review(185) and Short survey (16). It was likewise seen that 7131 authors have worked at SDGs where significant author appearance could be watched for example 10498. Authors were ordered dependent on creators of single authored (446) and creators of the multi-author document (6685). Collaboration Index of the creators was additionally determined i.e. 4.89.

Table 1. Data analysis

MAIN INFORMATION ABOUT DATA	
Time	1996 - 2021
Sources (Journal, Books, etc.)	865
Documents	1872
Average years from publications	1.9
Average citations per document	9.645
Average citations per year per document	2.563
References	77810
Document Types	
Article	1046
Article in Press	1
Book	20
Book Chapter	138
Conference Paper	149
Conference Review	1
Data Paper	3
Editorial	142
Erratum	29
Letter	25

Continued on following page

Table 1. Continued

Note	117
Review	185
Short Survey	16
Authors	
Authors	7131
Author Appearance	10498
Authors of Single Authored Document	446
Authors of the multi-authored document	6685
Single Authored Documents	504
Documents per author	0.263
Authors per document	3.81
Co-Authors per Documents	5.61
Collaboration Index	4.89

Annual Scientific Production (Year Wise Distribution)

Fig1. indicates the Annual scientific production (articles published/year) in the literature about Sustainable Development Goals (SDGs). The first publication on SDGs appeared in the year 2003. During the initial years ranging from 2003 –2013, the research output on SDGs was dismal publications growth [n=1 or 2]. The situation somewhat improved in 2014 when 28 articles were published deliberating core issues and developments concerning SDGs. However, a sharp rise in research on SDGs was noted in 2019, with the highest publications being produced this year [n=480], embarking the momentum in SDG-related research.

Figure 1. Annual scientific production in SDG's year-wise

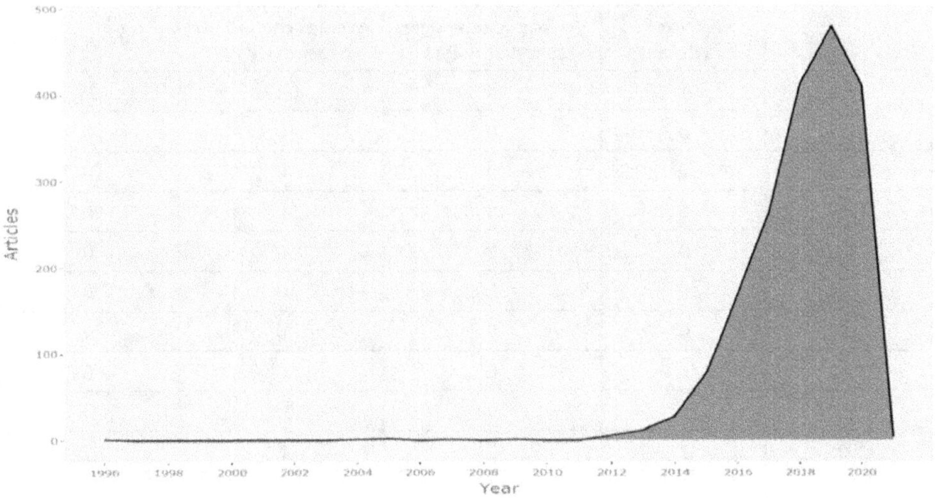

Average Citations

Table 2 give the Average citations per article (ACPA) and Average citation per year (ACPY) received by the articles (N) each year. Average citations for the articles are calculated on the basis of citable year calculated from the year the article was published to the current year. An aggregate of 1046 articles for each year distributed from 1996-2020 were retrieved. It was seen from Figure 2 and Table 2 that in 2005 the most noteworthy ACPA, just as ACPY, was observed, i.e., 323 and 21.5, individually. With the progressive reduction in resulting years, another ascent was recorded in the years 2012-13, where ACPA and ACPY for the year 2012 was 101.3 and 12.6 respectively, while for the year 2013 ACPA and ACPY was 88.8 and 12.7 respectively.

Table 2. Average citations received by the articles.

Year	N (No of articles)	Average citation per article (ACPA)	Average citation per year (ACPY)	CitableYears
1996	1	4	0.17	24
1997	0	0	0	0
1998	0	0	0	0
1999	0	0	0	0
2000	0	0	0	0
2001	0	0	0	0
2002	0	0	0	0
2003	0	0	0	0
2004	1	9	0.6	16
2005	2	323	21.5	15
2006	0	0	0	0
2007	1	0	0	13
2008	0	0	0	0
2009	1	8	0.7	11
2010	0	0	0	0
2011	0	0	0	0
2012	6	101.3	12.6	8
2013	12	88.8	12.7	7
2014	28	20.7	3.5	6
2015	79	21.8	4.4	5
2016	170	25.4	6.4	4
2017	263	11.4	3.8	3
2018	413	9.8	4.9	2
2019	480	3.5	3.5	1
2020	411	0.9		0

Figure 2. Average citations per year

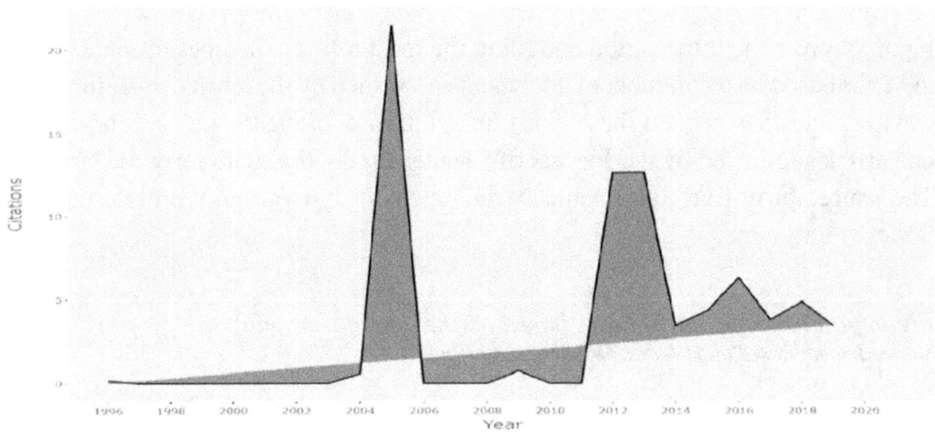

Three Field Plot

The three fields plot in Figure 3 gives a relationship among Authors, Sources, and Countries. It works dependent on the Sankey charts. As plainly delineated in the chart, The Lancet is the most well-known source among the creators Bhutta za, J.R., Murray Cji, Norheim, Allen C having a place with various countries. The diagram likewise demonstrates the commitment of the various nations with the U.S.A with the most elevated commitment. Other top nations contributing are India, Iran, Australia, Joined Realm, Canada, South Africa, Germany, China, and Switzerland.

Figure 3. Three fields plot between authors, source, and country.

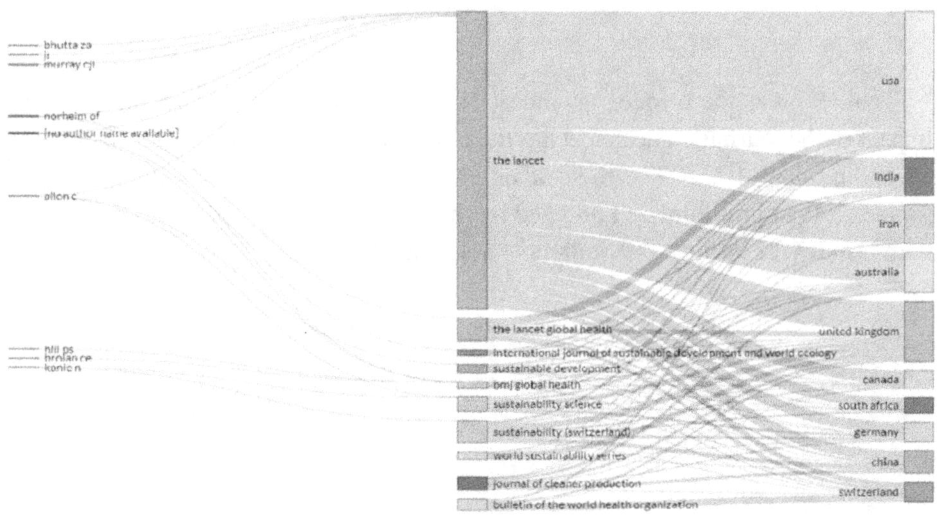

Most Relevant Sources

Figure 4 gives the information regarding the most relevant sources for publication in SDGs based on the number of documents published by the journal. Sustainability (Switzerland) has printed the highest quantum of documents, i.e., a hundred and one articles followed by the Journal of Cleaner Production with forty-one articles, The Lancet thirty-five, Sustainable Development thirty-two, and World Sustainable Series twenty-nine.

Figure 4. Most relevant sources based on the document count
Source Impact Based on H-Index, G-Index, and M-Index

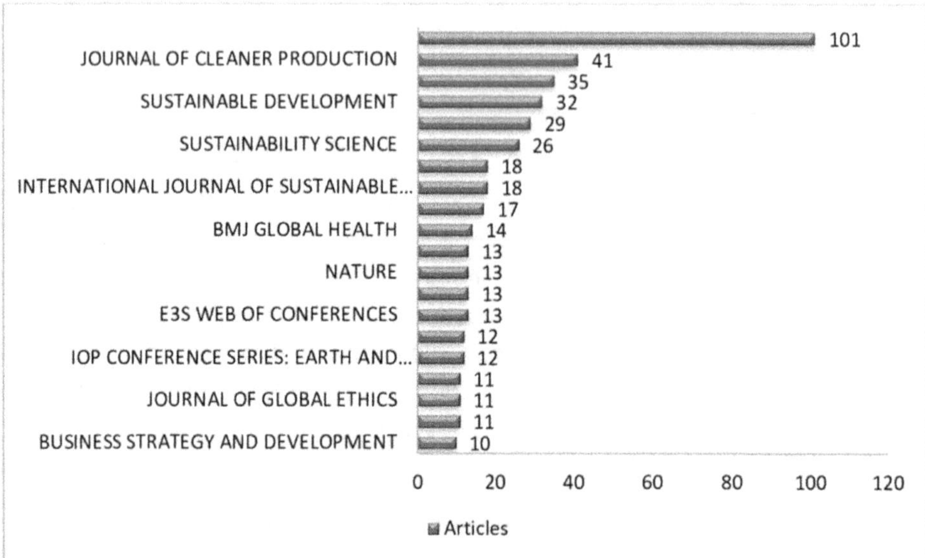

Table 3 gives the h-index, g- index, and m- index of the journals showing the impact of the journals. The lancet has the most noteworthy h-index, i.e., 16, g-index 35, and m-index 1.777777778. Sustainability (Switzerland) is next with h-index 15, g-index 24, and m-index 1. Followed by them is the Journal of cleaner production with h-index 12, g-index 19, and most noteworthy m-index 2.4.

Table 3. H-index, g-index, and m-index of the sources

Source	h_index	g_index	m_index	TC	NP	PY_start
SUSTAINABILITY (SWITZERLAND)	15	24	1.875	834	101	2013
JOURNAL OF CLEANER PRODUCTION	12	19	2.4	438	41	2016
THE LANCET	16	35	1.777777778	2438	35	2012
SUSTAINABLE DEVELOPMENT	9	20	1.5	433	32	2015
WORLD SUSTAINABILITY SERIES	5	6	1.25	59	29	2017
SUSTAINABILITY SCIENCE	9	19	1.8	382	26	2016
INTERNATIONAL JOURNAL OF SUSTAINABLE DEVELOPMENT AND WORLD ECOLOGY	6	14	1.2	222	18	2016
THE LANCET GLOBAL HEALTH	10	18	1.428571429	720	18	2014
BULLETIN OF THE WORLD HEALTH ORGANIZATION	9	16	1.5	266	17	2015
BMJ GLOBAL HEALTH	5	11	1	124	14	2016
E3S WEB OF CONFERENCES	1	1	0.333333333	2	13	2018
MARINE POLICY	8	13	1.142857143	227	13	2014
NATURE	6	13	0.666666667	1453	13	2012
PROCEEDINGS OF THE INTERNATIONAL ASTRONAUTICAL CONGRESS, IAC	1	1	0.2	3	13	2016
IOP CONFERENCE SERIES: EARTH AND ENVIRONMENTAL SCIENCE	1	1	0.333333333	2	12	2018
NATURE SUSTAINABILITY	6	11	2	142	12	2018
GLOBALIZATION AND HEALTH	4	11	0.666666667	175	11	2015
JOURNAL OF GLOBAL ETHICS	4	7	0.666666667	59	11	2015
WORLD DEVELOPMENT	5	9	0.833333333	82	11	2015
BUSINESS STRATEGY AND DEVELOPMENT	2	3	0.666666667	16	10	2018

Source Dynamics

Source dynamics in Figure 5 indicate the evolution of sources based on their publications in SDGs.

1. Sustainability (Switzerland) commenced publications in SDG's within the year 2013 and grew each year to as many as thirty five articles in 2020. Journal of cleaner production started off within the year 2016 and presented a stimulating and continuous production. Sustainable development, World sustainable series, Sustainability Science, and International Journal of sustainable development and world ecology have additionally shown a gradual yearly increase within the production.

2. The lancet started with noticeable 1st initial level production in the year 2012 however considerably reduced their productions in progressing years.

Figure 5. Evolution of sources based on their publications per year

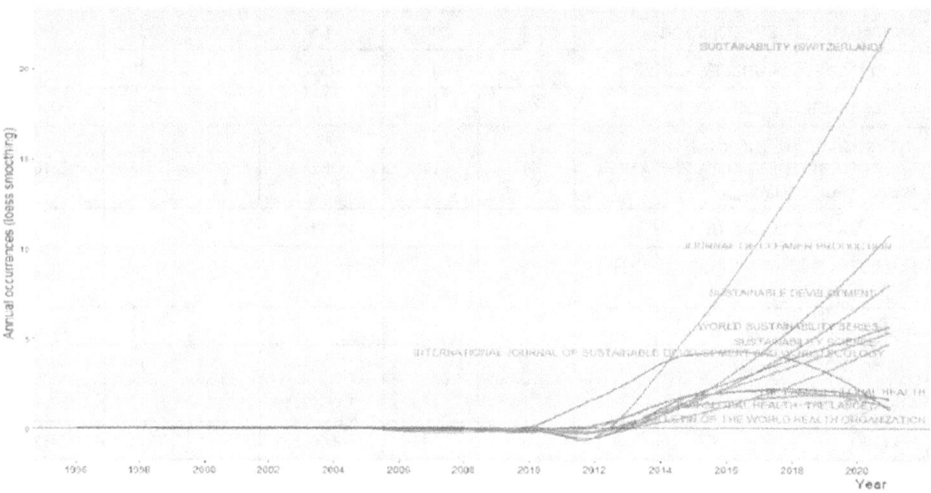

Most Relevant Authors

The table gives the foremost pertinent authors dependent on the number of publications. It is seen from the Figure 6 that the author Bhutta ZA got the got the first rank with 11 articles. The subsequent position is shared by J.R., and Kanie N with ten articles, trailed by them are Allen C, Brolan CE and Hill P.S. having nine articles, Murray CJL and Norheim OF with eight articles, Bouma J, Gupta R., Lozano R., Nilsson M., Rosa WE, and Salomon JA with seven articles and Acheampong M, Dandona L, Dubey M, Fullman M, and Gething PW with six articles each.

Figure 6. Most prolific authors

Author's Productivity through Lotka's Law

"The frequency distribution of scientific productivity," an article published by Alfred L. Lotka, an American scientist in 1926, stated the 'law of scientific productivity,' which is an inverse power function describing the rate of productivity in terms of published scientific articles over time.

The formula of Lotka's states that the number of authors producing x publications is about 1 $/x^a$ of those making one publication or:

$$Y = C/ x^a$$

Where y is the relative frequency (or relative or proportional number) of authors with x publications and C, and the exponent a are constants depending on the specific domain. Thus for $x = 1$, $C = y$. Lotka suggested that the exponent a nearly always equals *two* and the function can then be called an **inverse square law** and is commonly referred to as 'Lotka's law.' This means that the number of authors making 2 publications is *1 / 2*2 = 1 / 4 = 0.25* of those making 1 publication; those making 3 publications: *1 / 3*3 = 1 / 9 = 0.11* of those making 1 publication, etc.

The formula is a continuous function; in practical applications, it is a discrete power function, with x always a positive integer. Egghe (2005) discusses the relationship between both types of function and shows that the theoretical value of C is 0.6079 if a = two and that the observed value y for x = 1 is always lower than the theoretical value of C. Egghe (2005b) also explains that the function is scale-free and can be applied to any 'item' or 'product,' regardless its evolution in time. Using R – programming following data have been generated. ("LOTKA's Law of Productivity - Venus Iconography", 2020)

Table 4. No of articles published by number of corresponding authors

S. No	Documents written (x)	No. of Authors(y)	Proportion of Authors	X = Log x	Y = Log y	XY	XX
1	1	5588	0.784	0	3.7472564	0	0
2	2	690	0.097	0.30102996	2.83884909	0.854578628	0.0906190368
3	3	261	0.037	0.477121255	2.41664051	1.15303055	0.227644692
4	4	303	0.042	0.602059991	2.48144263	1.49397733	0.362476233
5	5	252	0.035	0.698970004	2.40140054	1.67850695	0.488559066
6	6	22	0.003	0.7781	1.34242268	1.04453909	0.60543961
7	7	6	0.001	0.84509804	0.77815125	0.657614096	0.714190697
8	8	2	0	0.903089987	0.30102996	0.271857143	0.815571525
9	9	3	0	0.954242509	0.477121255	0.455289383	0.910578766
10	10	2	0	1	0.30102996	0.30102996	1
11	11	1	0	1.04139269	0	0	1.08449873
12	27	1	0	1.43136376	0	0	2.04880221
		7131		9.032468	17.08534	7.910423	8.348381

Calculations

For the given data set,

$$a = \frac{N\sum XY - \sum X \sum Y}{N\sum X^2 - (\sum X)^2}$$

$$\mathrm{a} = \frac{94.925076 - 158.935829}{106.012572 - 81.5854782}$$

$$= \frac{64.010753}{24.4270938}$$

$= 2.62048173 \sim 2.6$

Hence, Lotka's law is validated as theCalculated value of a =2.6 is approaching theoretical value of a = 2.

Figure 7. Frequency of authors publishing corresponding articles

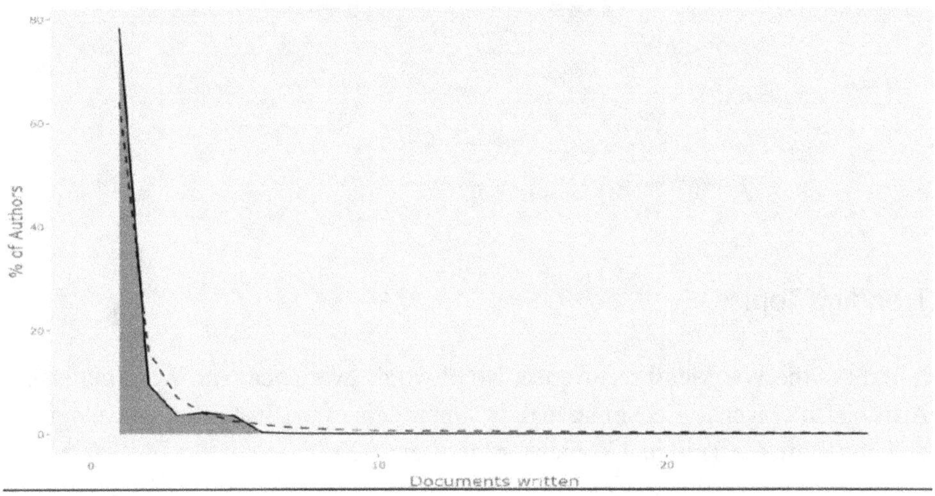

Keyword Co Occurrence

The figure represents the recurrence of keywords and associations between them. From the figure, we get the knowledge of 1046 articles. Points canvassed in them like environmental protection, climate change, poverty, economic developments as they are the most every now and again referenced keywords.

Figure 8. Keyword co-occurrence

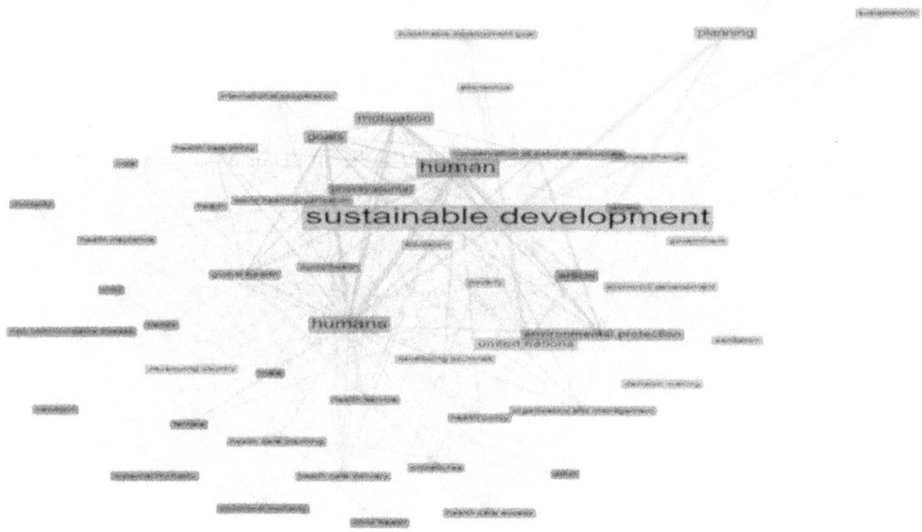

Trending Topics

In light of the Word least recurrence, No of words every year and Word mark size, Artificial Intelligence, Economic analysis, Supply chain and Sustainable Development objectives are the Trending Topics for the year 2020.

Figure 9. Most relevant sources based on the document count

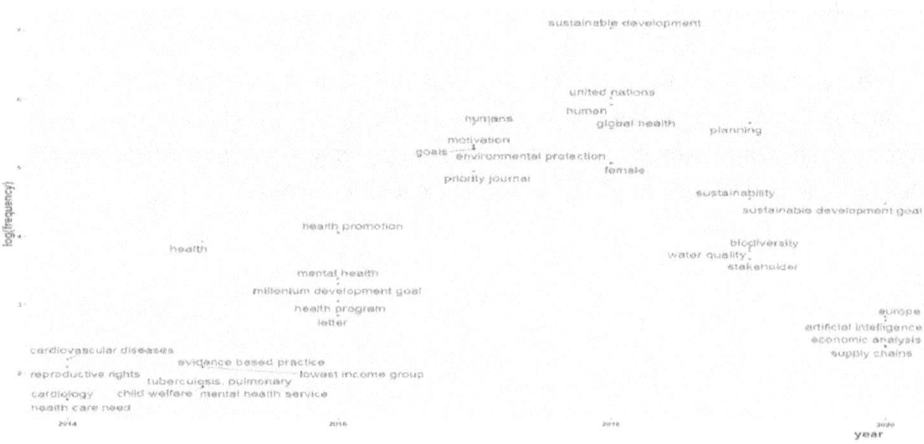

DISCUSSION

The Sustainable Development Goals are the outline for accomplishing a superior and more feasible future for all. Bibliometric investigation of SDG's would help scientists in the field to investigate the effect of SDG's by and large, effect of different authors in the field, journals in the field. This would help them to recognize especially significant papers. As concentrated above we can plainly say that Bhutta, ZA is the most productive author in the field while Sustainability (Switzerland) is the most significant journal in the field in this manner helping scientists.

CONCLUSION

Financial development, National riches, Human Development, Economic security, Freedom of discourse, Freedom of decision, Food security, Political Freedom are issues that could be settled by accomplishing SDGs consequently featuring the significance of achieving these objectives. Aim of this study was to distinguish Scientific production over the time, overall patterns, prolific authors, journals, development in the research activities in SDG's. For this reason bibliometric analysis utilizing Bibliometrix R-tool on the data retrieved from Scopus for the period 1996-2020 were performed. The inquiry Strategy utilized for report recovery was TITLE ("Sustainable advancement Goal*") by choosing the "Article title" in the "Documents" option. The pursuit is done on 22 August 2020. Examination of the information uncovered a sum of 1872 archives from 865 sources. Remarkable development in annual scientific production in SDGs was seen in the year 2019 with the most noteworthy distributions being delivered for the current year [n=480], embarking the momentum in SDG-related researches. The most imperative Annual citation per article, similarly as Annual citation per year, was viewed, i.e., 323 and 21.5, exclusively in the year 2005. Sustainability (Switzerland) has printed the utmost variety of articles, i.e., a hundred and one articles. The lancet has the most vital h-index, i.e., 16, g-index 35, and m-index 1.8 . Bhutta, ZA is the foremost pertinent authors dependent on the number of publications. Validation of Lotka's Law was made using Bibliometrix R. Further data analysis of SDGs uncovered Artificial Intelligence, Economic investigation, Supply chain and Sustainable Development goals are the Trending Topics for the year 2020. Hence, the findings can assist future research in this field by providing a worldwide analytical overview of the SDGs.

REFERENCES

Ardanuy, J. (2012). *Breve Introducción a la Bibliometría*. University of Barcelona.

Bhatt, Y., Ghuman, K., & Dhir, A. (2020). Sustainable manufacturing. Bibliometrics and content analysis. *Journal of Cleaner Production, 260*, 120988. doi:10.1016/j.jclepro.2020.120988

Derviş, H. (2020). Bibliometric Analysis using Bibliometrix an R Package. *Journal of Scientometric Research, 8*(3), 156–160. doi:10.5530/jscires.8.3.32

Di Stefano, G., Peteraf, M., & Veronay, G. (2010). Dynamic capabilities deconstructed: A bibliographic investigation into the origins, development, and future directions of the research domain. *Industrial and Corporate Change, 19*(4), 1187–1204. doi:10.1093/icc/dtq027

Fuad, A., Lee, Y-C. G., & Hsu, C-Y. (2020). Bibliometric Analysis of Bioscience Trends Journal (2007-2017): Knowledge dynamics and visualization. *Library Philosophy and Practice (ejournal)*. 4076.

González García, E., Colomo Magaña, E., & Cívico Ariza, A. (2020). Quality Education as a Sustainable Development Goal in the Context of 2030 Agenda: Bibliometric Approach. *Sustainability, 12*(15), 5884. doi:10.3390u12155884

Hosaain, S., & Batcha, D. (2020). Authorship Pattern on Dentistry Research During 1999-2018: A Scientometric Profile. *Library Philosophy and Practice (e-journal)*. 4036. https://digitalcommons.unl.edu/libphilprac/4036

Jalal, S. (2019). Co-authorship and co-occurrences analysis using BibliometrixR package: A case study of India and Bangladesh. *Annals Of Library And Information Studies, 66*, 57–64.

Laksham, S., Surulinathi, M., Balasubramani, R., & Srinivasaragavan, S. (2020). Mapping the Research output on Coronavirus: A Scientometric Study. *Gedrag & Organisatie Review, 33*(2), 163–186.

LOTKA's Law of Productivity - Venus Iconography. (2020). Retrieved 1 September 2020, from https://sites.google.com/site/venusiconography/home/research-papers/lotka-s-law-of-productivity

Meschede, C. (2020). The Sustainable Development Goals in Scientific Literature: A Bibliometric Overview at the Meta-Level. *Sustainability, 12*(11), 4461. doi:10.3390u12114461

Munim, Z. H., Dushenko, M., Jimenez, V. J., Shakil, M. H., & Imset, M. (2020). Big data and artificial intelligence in the maritime industry: A bibliometric review and future research directions. *Maritime Policy & Management*, *47*(5), 577–597. doi:10.1080/03088839.2020.1788731

Nasir, A., Shaukat, K., Hameed, I. A., Luo, S., Alam, T. M., & Iqbal, F. (2020). A Bibliometric Analysis of Corona Pandemic in Social Sciences: A Review of Influential Aspects and Conceptual Structure. *IEEE Access: Practical Innovations, Open Solutions*, *8*, 133377–133402. doi:10.1109/ACCESS.2020.3008733 PMID:34812340

Rodríguez-Soler, R., Uribe-Toril, J., & De Pablo Valenciano, J. (2020). Worldwide trends in the scientific production on rural depopulation, a bibliometric analysis using bibliometrix R-tool. *Land Use Policy*, *97*, 104787. doi:10.1016/j.landusepol.2020.104787

Sachs, J. D. (2012). From millennium development goals to sustainable development goals. *Lancet*, *379*(9832), 2206–2211. doi:10.1016/S0140-6736(12)60685-0 PMID:22682467

Stafford-Smith, M., Griggs, D., Gaffney, O., Ullah, F., Reyers, B., Kanie, N., Stigson, B., Shrivastava, P., Leach, M., & O'Connell, D. (2017). Integration: The key to implementing the sustainable development goals. *Sustainability Science*, *12*(6), 911–919. doi:10.100711625-016-0383-3 PMID:30147763

Sweileh, W. M. (2020). Bibliometric analysis of scientific publications on "sustainable development goals" with emphasis on "good health and well-being" goal (2015–2019). *Globalization and Health*, *16*(1), 68. doi:10.118612992-020-00602-2 PMID:32723366

Trivedi, G. (2020). *Visualization and Scientometric Mapping of Global Agriculture Big Data Research*. DigitalCommons@University of Nebraska - Lincoln. Retrieved 9 September 2020, from https://digitalcommons.unl.edu/libphilprac/2478/

United Nations. (1987). Informe Brundtland, "Nuestro Futuro en Común." Available online: www.un.org/es/comun/docs/?sym-bol=A/42/427

United Nations. (n.d.). *Sustainable development goals*. https://www.un.org/sustainabledevelopment/sustainable-development-goals/

Warin. (n.d.). https://warin.ca/shiny/bibliometrix/#section-bibliometrix

Chapter 3
A Review of Technologies Trending Towards 5G Networks for Smart Healthcare Using IoT

Harsh Kaushik
Maharaja Agrasen Institute of Technology, Guru Gobind Singh Indraprastha University, India

Nitish Pathak
Bhagwan Parshuram Institute of Technology, Guru Gobind Singh Indraprastha University, India

Neelam Sharma
Maharaja Agrasen Institute of Technology, Guru Gobind Singh Indraprastha University, India

ABSTRACT

Healthcare is rapidly evolving from a standard hospital and specialist-centered strategy to a dispersed patient-centered one. Communication technology among other technologies have made it possible to provide tailored and distant healthcare services. Currently, the existent 4G network and other communications technologies are generally used in smart healthcare activities. Present communication technologies are inadequate to meet the increasing demands that smart healthcare applications impose on communication networks. The rising 5G is anticipated to bolster applications, which can fulfil most of the prerequisites such as high bandwidth, ultra-high reliability, high energy efficiency, and ultra-low latency. The futurity of the smart healthcare networks is anticipated to be a combination of 5G and IoT devices, which are anticipated to extend the cellular scope and handle security problems. This chapter provides an overview of taxonomy, research trends, difficulties, and future research themes.

DOI: 10.4018/978-1-6684-4225-8.ch003

INTRODUCTION

Smart healthcare permits enhanced testing tools to give enhanced therapeutic approaches, as well as smart healthcare gadgets to improve healthcare quality by delivering vital indicators in meaningful time. The benefits of sustainable health care are to make patients lives easier by providing knowledge of medical problems as well as how to solve them.

In the event of a critical situation, smart health care allows patients to take appropriate action. Also, with the emergence of smart cities, a strong smart healthcare system is essential to ensure that people have access to health care services. Aside from wellness, one of the major contributions is the reduction of healthcare costs through prompt diagnosis. For example, the smart healthcare industry incorporating the Internet of Things (IoT) is expected to be worth $158.1 billion in 2022 (Deloitte, 2019). The Internet of Things (IoT) will transform health care and reduce the cost of medical gadgets. To allow comprehensive deployment of IoT, 5G networks will be critical (Liu et al., 2018).

IoT may help with resource management practices and organizational performance, behavioural change monitoring, remote monitoring, treatment compliance monitoring, assisted living, smarter medication, and telemedicine in smart-health care (Brito, 2018). These technologies will play a significant role in the medical industry over the forecast period. Smart antennas play a crucial role in 5G network communication (Ojaroudi, 2019; Parchin & Abd-Alhameed, 2018).

Beamforming (vertical and horizontal) is one such breakthrough, in which RF energy is concentrated in a compact beam to specifically where it would be required rather than spreading the very same energy across a large region (Mohammad et al., 2019; Ojaroudiparchin et al., 2016). Machine-to-machine (M2M) connectivity and the Internet of Things are predicted to be the fundamental foundations of smart healthcare in 5G networks (Agiwal et al., 2019).

The approaches that have been offered will encounter two major difficulties. The first is a large number of terminals, which has resulted in extremely dense networks. The second factor is energy consumption, which is a result of the type of IoT-based applications that rely on wireless sensor networks (for example, the reasonably basic battery life for specified conditions is 10 years) (Palattella et al., 2016). 5G networks are projected to deliver a greater data rate in addition to network gentrification and support for a large number of IoT devices. Figure 1 shows the generic architecture for a 5G-based smart healthcare network

Figure 1. A generic architecture for a 5G-based smart healthcare network

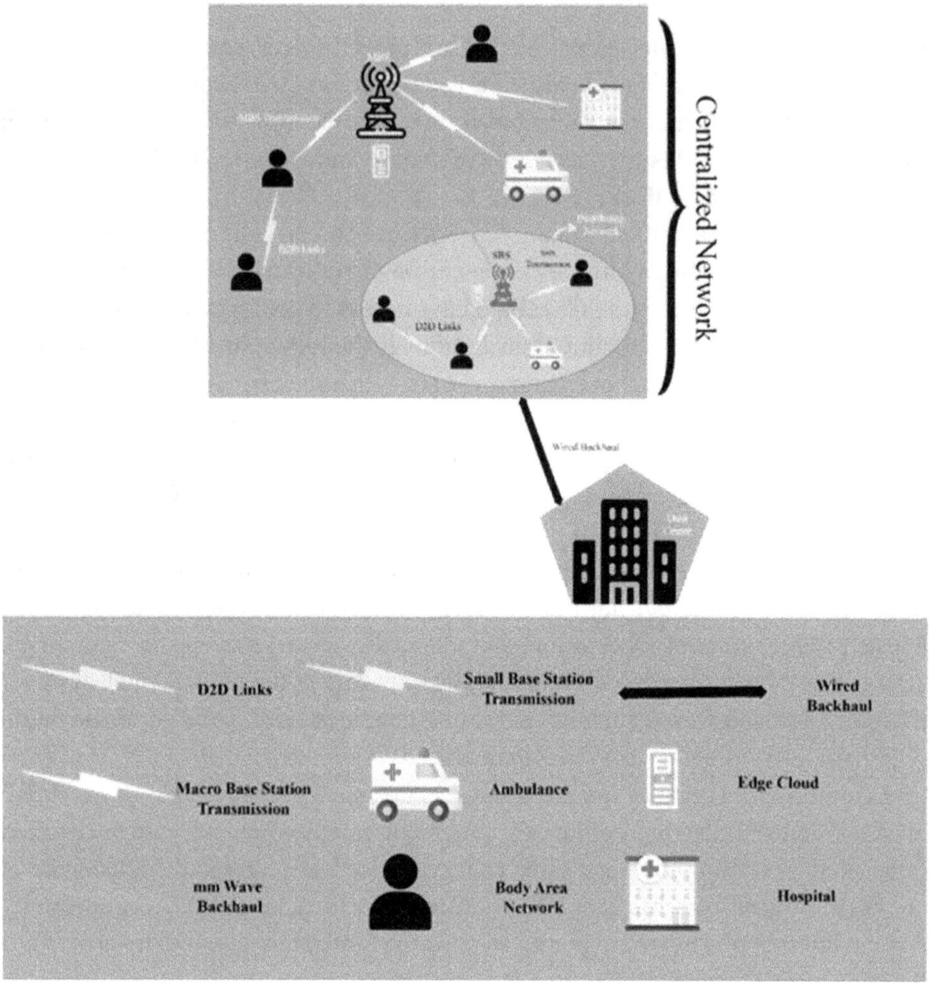

TAXONOMY

Figure 2 portrayed scientific categorization comprised of the accompanying boundaries: network types, services, applications, communication technologies, characteristics and requirements. This segment portrays the proposed scientific categorization exhaustively in the following sub-area.

Figure 2. Smart health-care taxonomy and parameters

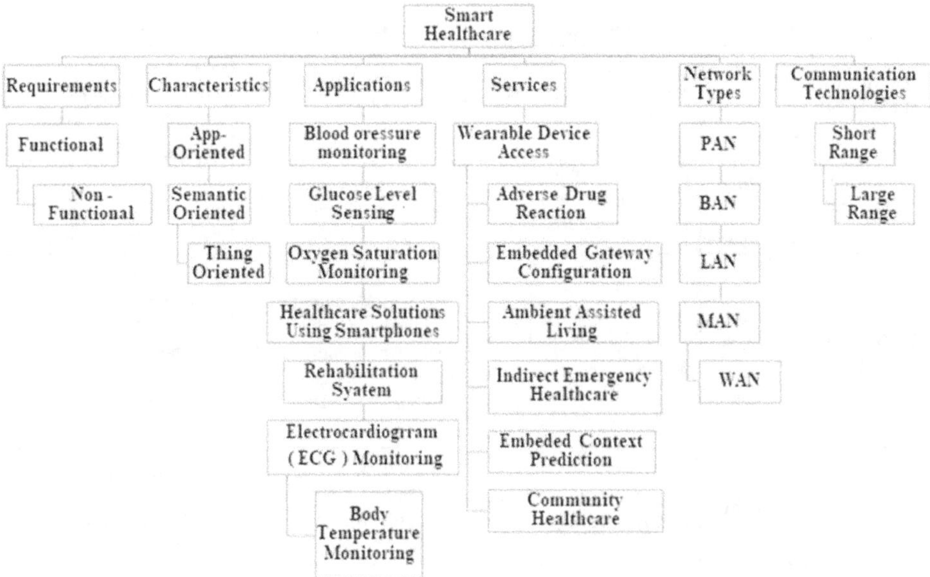

HEALTH CARE APPLICATIONS

Smart healthcare services lead to the development of technologies, which customers and patients can utilize appropriately. Internet of Things (IoT) health care applications is presented in Table 1.

1. **Scenarios for 5G Network and its Requirements**
   There are four different sorts of 5G network scenarios that may be classified EMB, MMTC, Low-Latency and High-Reliability Communications and WRAN.
   a. **Enhanced mobile broadband (EMB)**
   High-level throughput and improved traffic capacity are critical needs in this situation (Pathan et al., 2020). The scenario's major goal is to increase the network's traffic capacity and data throughput.
   b. **Massive Machine-Type Communications (MMTC)**
   The method entails a wireless sensor network, M2M connectivity, and Internet of Things (IoT) communications (Singh et al., 2015). The primary goal is energy and connectivity density. In contrast, (Al-Falahy & Alani, 2017) claims that the connection density is 106 devices/m2 and that the energy efficiency is 100 times superior to 4G networks.
   c. **Low-Latency and High-Reliability Communications**
   The possibility is connected to ultra-reliable communication and sensory Internet applications for autonomous automobiles, such as remote surgery

Table 1. Internet of Things (IoT) health care applications

Condition	Sensor Types	Operations	IoT Role
Distant surgery (Pang et al., 2014)	Augmented Reality Sensors, Surgical robot sensors	Giving feedback to user and Robot arms	Real-time data connectivity and information management
Cough detection (Hii & Chung, 2011)	A smartphone has a microphone audio system integrated.	Spectrograms were captured and analysed.	To power IoT, the software takes advantage of a smartphone's system-on-chip (SoC).
Diabetes (Istepanian et al., 2011)	Opto-physiological sensor	The sensor's output is linked to a TelosB mote, which converts the analogue signals into digital.	All wireless sensors can communicate via the 6LoWPAN and IPV6 architectural protocols. with the use of wireless nodes which are IP-based
Monitoring of Heartbeat (Baker et al., 2017)	Capacitive electrodes on electric circuits	Information is transmitted in a digital chain that is linked to a wireless transmitter.	With the support of Bluetooth and Wi-Fi, gateways connect smart gadgets.
The temperature of the body (Liu et al., 2019)	Wearables sensor of blood pressure	Temperature readings are taken from the skin	With the assistance of a gateway, smart devices are connected to WBAN.
Monitoring of Oxygen Saturation (Tabish et al., 2014)	Pulse oximeter wrist	Time-by-time pulse detection.	An integrated clinical environment that is ubiquitous

and vehicle-to-vehicle (i.e., ambulance) connectivity. For example, in (Gupta & Jha, 2015; Li et al., 2017), The needed outage probability is about equal to 10^{-7} and the end-to-end delay is 1ms.

d. **Wireless Regional Area Networks (WRAN)**
WRAN is the most recent 5G network configuration. The primary goal of this case is to develop a new application for rural, sparsely populated places. Long-distance connectivity with cells larger than 50 kilometres in diameter is the most important need in this case (Mangla et al., 2022).

TECHNOLOGY TRENDS TO ACHIEVE THE REQUIREMENTS IN THE 5G NETWORK

As a result, an overview of several smart healthcare scenarios and their needs are summarised in Table 2.

Table 2. A summary of a variety of smart healthcare scenarios

Scenario	Drives	Communication Technologies	Required Latency	Required Data Rate
M2M Wearables	Connection for data gathering	LoRa (sensor applications) Zigbee (data collection) NB-IoT (interconnected devices) Bluetooth (D2D sensors)	10-799 ms	Few Kbps to Mbps
Digital Hospital	Inside building communication	Wi-Fi	10-100 ms	Few Mbps
Emergency Medical Services	Emergency Communication and High-speed reply	LTE LTE-A LTE-A Pro	20–100 ms	From 100 Mbps to 3 Gbps
Remote Surgery	URLLC service between many locations	5G	20-30 ms	Few Gbps
Tactile Communication	URLLC (Ultra-reliable and low latency communications), eMBB (enhanced Mobile Broadband)	Wi-Fi, Bluetooth, 5G, 4G	sub-ms	Few Gbps

1. **Massive MIMO (Multiple-Input Multiple-Output) and 3D MIMO**
   Shannon's theorem for network capacity states that the highest limit of a communication link's capacity in bits per second (bps) is constant, and this rule might be used to investigate how to get higher data rates (Kartun-Giles et al., 2018).
   The bandwidth of a MIMO configuration is estimated to be as follows:

   $$C = \mathit{min}(M, N).B\log^{2}(1 + SNR)$$

   M and N: Number of antennas
   B: Bandwidth of the channel
   SNR: Signal-to-Noise Ratio
2. **Millimetre-Wave Communications**
   The data rate is increased when the channel has a large bandwidth (Goel et al., 2020). However, a higher operating frequency is essential to strengthening the

channel's bandwidth. By adopting higher frequencies in the millimetre wave spectrum, will resolve the second trend for the next generation network (i.e., 5G).

3. **Small Cells, Ultra-Dense Networks, and Heterogeneous Networks**
   Due to various improved frequency re-use, the adoption of tiny cells will boost system traffic capacity. Another trend for the next-generation network 5G is the utilization of tiny cells in ultra-dense networks.
4. **Device-to-Device (D2D) Communications**
   Another trend for the next-generation network, 5G (Li et al., 2018), is D2D communication. The approach can be applied in two distinct ways. The first method involves one terminal cooperating to improve communication between the base station and terminals. It consists of direct connection among devices (i.e. users) without data traffic passing via any infrastructure node and is usually anticipated to be a significant foundation for improving system efficiency. The second technique involves direct communication between two terminals without the use of a base station. Device-to-device communication can boost network dependability, system performance, reduce operating costs, and save power.
5. **Cognitive Radio**
   It is another essential innovation to be considered for the following era arrange 5G (Hossain & Hasan, 2015). Cognitive radio technology might be an essential solution for enabling the WRAN scenario (Liu & Zhang, 2019). A cognitive system has two modes of operation. Spectrum overlay is the first, while spectrum underlay is the second. It is categorized into three separate systems, one primary and one secondary. The major network, which includes the principal radio base station and customers, controls the licenced band. The secondary system and the major connection share the spectrum resources. For a WRAN situation, the Cognitive Network is more applicable.
6. **Artificial Intelligence (AI) and Machine Learning (ML)**
   The use of artificial intelligence and machine learning (Verma et al., 2019) to make the network smarter is still in its initial stages for smart healthcare applications in the wireless network. The 5G network's network design, topology, and propagation models, as well as the node mobility, can be sophisticated. As a result, AI and machine learning may play critical roles in helping and controlling various network resources to make the network smarter for smart healthcare applications (El-Gayar et al., 2020).
7. **Unresolved Problems and Challenges**
   Aside from the advancements highlighted above, deploying 5G for smart health care faces various hurdles and open research concerns. Table 3 shows the benefits, characteristics, and research directions. I discuss numerous research issues and potential research prospects in this section.

Table 3. Research issues

Features	Advantages	Research Challenges	Key Requirements
Achieving Security (Nguyen et al., 2015)	Provides a secure (attack-free) environment for service deployment.	Cloud-based service deployment and integration must be secure at both the device and the network levels. Threat detection at both the insider and outer levels before execution. Intelligent security solutions that assist in maintaining data integrity and avoiding delays.	Proof of vulnerabilities at a different level in the system that can be recognised. can serve as entrance points for various assaults
Achieving compatibility (Noura et al., 2019)	An important platform for leveraging multiple protocols to communicate across different IoT devices.	Devices for retailer-secured administrations	Incorporation and communication (i.e., CoAP, IP) for IoT devices require adaptable, ubiquitous, and integrated models.
Analysis of Big data (Hossain & Muhammad, 2017)	Process data obtained from reliable sources to improve network performance. (i.e., using an intelligent approach to analyse patient data helps reduce network congestion).	Limitations of effective tools for processing large amounts of data provided by network devices There are insufficient centralised and dispersed resources.	Processing requires a centralised large data centre. Appreciation from the general public on how to make the most use of existing resources safely.
Performing IoT Connectivity (Palattella et al., 2016)	Ensure the connectivity of IoT devices across many domains.	How can I ensure that multiple devices from different domains are connected in a high-mobility environment? How can resources in an ultra-high-density network be optimised? How can an ultra-high-density network achieve energy efficiency?	The use of the spectrum in conjunction with a cost-effective approach for IoT device connectivity. Intelligent algorithms ensure that types of devices in the network are connected to separate domains. Analysis techniques are used to accommodate diverse workloads while also increasing resource availability.

CONCLUSION

Smart healthcare and IoT applications will benefit greatly from next-generation (5G) networks. From some kind of cost and performance standpoint, smart healthcare and IoT applications play a critical role in the 5G network. This study included a wide range of applications.

Short and long-range communication technologies were compared from several angles.

For smart health care, consider the range, frequency, power consumption, and data rate. In addition, four alternative cases are examined based on various 5G network needs (e.g., improved mobile broadband, low-latency and high-reliability communications, huge machine-type communications, etc.). and Internet access for wireless regional area networks), as well as discussing other technological advancements for achieving these objectives in a 5G network and are sufficient.

Lastly, In the context of the 5G network, the writers have explored unsolved issues and potential research paths in smart health care. This gives possibilities for scholars who are interested in beginning research in the subject of 5G-based smart healthcare employs a variety of methods, including machine learning, scheduling, and routing networking and transfer.

REFERENCES

Agiwal, M., Saxena, N., & Roy, A. (2019). Towards connected living: 5G enabled internet of things (IoT). *IETE Technical Review*, *36*(2), 190–202. doi:10.1080/02564602.2018.1444516

Al-Falahy, N., & Alani, O. Y. (2017). Technologies for 5G networks: Challenges and opportunities. *IT Professional*, *19*(1), 12–20. doi:10.1109/MITP.2017.9

Baker, S. B., Xiang, W., & Atkinson, I. (2017). Internet of things for smart healthcare: Technologies, challenges, and opportunities. *IEEE Access: Practical Innovations, Open Solutions*, *5*, 26521–26544. doi:10.1109/ACCESS.2017.2775180

Brito, J. M. (2018). Technological Trends for 5G Networks Influence of E-Health and IoT Applications. *Int. J. Health Med Commun.*, *9*(1), 1–22. doi:10.4018/IJEHMC.2018010101

Deloitte. (2019). *Global Health Care Outlook Shaping the Future*. Deloitte.

El-Gayar, O. F., Ambati, L. S., & Nawar, N. (2020). Wearables, Artificial intelligence, and the Future of Healthcare. In *AI and Big Data's Potential for Disruptive Innovation* (pp. 104–129). IGI Global. doi:10.4018/978-1-5225-9687-5.ch005

Goel, R., Jain, A., Verma, K., Bhushan, M., Kumar, A., & Negi, A. (2020). Mushrooming Trends and Technologies to Aid Visually Impaired People. In *International Conference on Emerging Trends in Information Technology and Engineering (ic-ETITE'20)*. IEEE.

Gupta, A., & Jha, R. K. (2015). A survey of 5G network: Architecture and emerging technologies. *IEEE Access: Practical Innovations, Open Solutions*, *3*, 1206–1232. doi:10.1109/ACCESS.2015.2461602

Hii, P. C., & Chung, W. Y. (2011). A comprehensive ubiquitous healthcare solution on an AndroidTM mobile device. *Sensors (Basel)*, *11*(7), 6799–6815. doi:10.3390110706799 PMID:22163986

Hossain, E., & Hasan, M. (2015). *5G cellular: Key enabling technologies and research challenges*. arXiv:1503.00674.

Hossain, M. S., & Muhammad, G. (2017). Emotion-aware connected healthcare big data towards 5G. *IEEE Internet Things J.*, *5*(4), 2399–2406. doi:10.1109/JIOT.2017.2772959

Istepanian, R. S., Hu, S., Philip, N. Y., & Sungoor, A. (2011). The potential of Internet of m-health Things "m-IoT" for non-invasive glucose level sensing. *Proceedings of the 2011 Annual International Conference of the IEEE Engineering in Medicine and Biology Society*, 5264–5266. 10.1109/IEMBS.2011.6091302

Kartun-Giles, A., Jayaprakasam, S., & Kim, S. (2018). Euclidean matchings in ultra-dense networks. *IEEE Communications Letters*, *22*(6), 1216–1219. doi:10.1109/LCOMM.2018.2799207

Li, C. P., Jiang, J., Chen, W., Ji, T., & Smee, J. (2017). 5G ultra-reliable and low-latency systems design. *Proceedings of the 2017 European Conference on Networks and Communications (EuCNC)*, 1–5.

Li, S., Da Xu, L., & Zhao, S. (2018). 5G Internet of Things: A survey. *J. Ind. Inf. Integr.*, *10*, 1–9.

Liu, S. H., Lin, C. B., Chen, Y., Chen, W., Huang, T. S., & Hsu, C. Y. (2019). An EMG Patch for the Real-Time Monitoring of Muscle-Fatigue Conditions During Exercise. *Sensors (Basel)*, *19*(14), 3108. doi:10.339019143108 PMID:31337107

Liu, X., Jia, M., Zhang, X., & Lu, W. (2018). A novel multichannel Internet of things based on dynamic spectrum sharing in 5G communication. *IEEE Internet Things J.*, *6*(4), 5962–5970. doi:10.1109/JIOT.2018.2847731

Liu, X., & Zhang, X. (2019). NOMA-based Resource Allocation for Cluster-based Cognitive Industrial Internet of Things. *IEEE Transactions on Industrial Informatics*, *16*(8), 5379–5388. doi:10.1109/TII.2019.2947435

Mangla, Kumar, Mehta, Bhushan, & Mohanty. (2022). *Real-Life Applications of the Internet of Things: Challenges, Applications, and Advances*. CRC Press.

Mohammad, A., Virdee, B. S., See, C. H., Abd-Alhameed, R. A., Falcone, F., & Ernesto, L. (2019). High-Gain Metasurface in polyimide on-chip Antenna Based on cRLH-tL for Sub-terahertz integrated circuits. *IETE Technical Review*, *36*, 190–202.

Nguyen, K. T., Laurent, M., & Oualha, N. (2015). Survey on secure communication protocols for the Internet of Things. *Ad Hoc Networks*, *32*, 17–31. doi:10.1016/j.adhoc.2015.01.006

Noura, M., Atiquzzaman, M., & Gaedke, M. (2019). Interoperability in internet of things: Taxonomies and open challenges. *Mobile Networks and Applications*, *24*(3), 796–809. doi:10.100711036-018-1089-9

Ojaroudi, P. (2019). Recent developments of reconfigurable antennas for current and future wireless communication systems. *Electronics (Basel)*, *8*(2), 128. doi:10.3390/electronics8020128

Ojaroudiparchin, N., Shen, M., & Pedersen, G. F. (2016). Wide-scan phased array antenna fed by coax-to-microstrip lines for 5G cell phones. *Proceedings of the 2016 21st International Conference on Microwave, Radar and Wireless Communications (MIKON)*, 1–4.

Palattella, M. R., Dohler, M., Grieco, A., Rizzo, G., Torsner, J., Engel, T., & Ladid, L. (2016). Internet of things in the 5G era: Enablers, architecture, and business models. *IEEE Journal on Selected Areas in Communications*, *34*(3), 510–527. doi:10.1109/JSAC.2016.2525418

Pang, Z., Tian, J., & Chen, Q. (2014). Intelligent packaging and intelligent medicine box for medication management towards the Internet-of-Things. *Proceedings of the 16th International Conference on Advanced Communication Technology*, 352–360. 10.1109/ICACT.2014.6779193

Parchin, N. O., & Abd-Alhameed, R. A. (2018). A compact Vivaldi antenna array for 5G channel sounding applications. *Proceedings of the 12th European Conference on Antennas and Propagation (EuCAP 2018)*, 9–13. 10.1049/cp.2018.1205

Pathan, SBai, ABhushan, M. (2020). A Study on Health Care using Data Mining Techniques. *Journal of Critical Reviews*, *7*(19), 7877–7890. doi:10.31838/jcr.07.19.896

Singh, V. J., Bhushan, M., Kumar, V., & Bansal, K. L. (2015). Optimization of Segment Size Assuring Application Perceived QoS in Healthcare. *Lecture Notes in Engineering and Computer Science, 2217*(1), 274-278.

Tabish, R., Ghaleb, A. M., Hussein, R., Touati, F., Mnaouer, A. B., Khriji, L., & Rasid, M. F. A. (2014). 3G/WiFi-enabled 6LoWPAN-based U-healthcare system for ubiquitous real-time monitoring and data logging. *Proceedings of the 2014 Middle East Conference on Biomedical Engineering (MECBME)*, 277–280. 10.1109/MECBME.2014.6783258

Verma, K., Bhardwaj, S., Arya, R., Ul Islam, M. S., Bhushan, M., Kumar, A., & Samant, P. (2019). Latest tools for data mining and machine learning. *International Journal of Innovative Technology and Exploring Engineering, 8*(9S), 18-23.

APPENDIX: ABBREVIATIONS

The following abbreviations are used in this review paper.

BAN Body Area Network
CoT Cloud of Things
D2D Device-to-Device
GSM Global System for Mobile
IoT Internet of Things
LTE Long-Term Evaluation
LTE-M Long-Term Evaluation Advance
LoraWAN Long Range Wide Area Network
Lora Long Range
M2M Machine-to-Machine
mMTC Massive Machine-Type Communication
MTCs Machine-Type Communications
MBS Macro Base Station
MIMO Multiple-input multiple output
NFV Network Function Virtualization
NB-IoT Narrowband Internet of Thing
NFC Near Field Communication
SNR Signal-to-Noise Ratio
SBS Small Base Station
SDN Software Defined Network
TCP Transmission Control Protocol
URLLC Ultra-Reliable and Low Latency Communication
WLAN Wireless Local Area Network
WRAN Wireless Regional Area Networks
WBAN Wireless Body Area Network
WiMAX Worldwide Interoperability for Microwave Access
Wi-Fi Wireless Fidelity
3GPP Third Generation Partnership Project
4G Fourth Generation
5G Fifth Generation Mobile Network

Chapter 4
E-Attendance System Using Machine Learning

Yash Jitendra Modi
Vishwakarma Government Engineering College, India

Deesha Patel
Vishwakarma Government Engineering College, India

Kaushik Rana
Vishwakarma Government Engineering College, India

Mitesh Sagar
Vishwakarma Government Engineering College, India

Tirth Trivedi
Vishwakarma Government Engineering College, India

ABSTRACT

Attendance and punctuality are vital attributes of students. The mundane and repetitive manual task of taking attendance by the pedagogue results in a loss in productivity, diverting attention from key issues. To surmount this probe of performing this trivial task, the authors proposed GPU CNN-based automated attendance system (GCBAAS). Image content analysis and pattern recognition using deep learning accompanying high computing technology are rapidly expanding areas of application. Several scientific contributions in this field have demonstrated good results and with some modification ensures effective solution in this state of the art. The proposed GCBAAS methodology trained on remote GPU server minimized false labels in much less time. A simple pipeline has been created for training such enormous data on GPU servers for multiclass labeling. The performance analysis has been taken in a real class, which results in mean accuracy of 98%. Additionally, after collecting some more data with different augmentation techniques, it results in increased accuracy.

DOI: 10.4018/978-1-6684-4225-8.ch004

INTRODUCTION

In the era of cloud computing and quantum computers, managing attendance using a manual attendance system is a tiresome and dreadful task. Manually registering attendance and maintaining paper records takes a lot of time and energy. In electronic times, we can take advantage of fast-growing technology. Also, by using a various biometric method such as fingerprint, iris or voice recognition, some error has been found and would not always show the same accuracy. Cloud storage devices can substitute the papers and manual attendance systems can be replaced by face recognition systems. Cloud computing is gaining popularity as it is cost-saving, increasing productivity, speed and efficiency, performance, and security.

With the help of technology, we can overcome many limitations of the manual attendance system. In a manual attendance system, tasks are quite tedious and time-consuming. Preserving records on papers will prompt human errors. Computers perform more efficiently than humans in various tasks like Mathematical calculations, keeping track of the record, memory management, and many more. Managing a record of attendance is at very ease with very little error margin. Using API is like Cherry on top as it requires simpler code for performing high computing tasks with lower cumulative latencies.

The use of machine learning is quite often. Implementing machine learning use cases into practical applications and their different scenarios is essential. Nowadays machine learning will provide an outstanding career which will eventually lead to benefits for the company to resolve the issues. Eminent decisions to solve the problem can be taken after analyzing all the different foresight and selecting the best choice. By providing the raw data, machines can give us valuable results by forming the pattern. Using machine learning besides cognitive technologies can aim to make quicker decisions and take actions faster (H. Rathod, 2017).
The main contributions of this work can be summarized as follows:

- ·Point out different features in a CNN-based framework for feature extraction and face prediction. Since different face contains different face features, it is important to gather all the face features as much as possible to minimize the error.
- Takes more than 20 tensors of a single person as input and train the network in the way that it extracts the face features to recognize the student.

As the brain by firing or activating neurons the brain can differentiate between faces, similarly a neural network a deep learning algorithm based on the human brain can be used for facial recognition (Kar, 2012). Convolutional Neural Network (ConvNet/CNN) is a specific deep learning algorithm designed for operating with two-dimensional image data. These machine learning algorithms will help to automatically mark the attendance by recognizing the face of the student. A CNN is a neural network that takes an image as an input, assigning learnable weights and biases to various objects in the image and be capable to distinguish between two images.

RELATED WORK

The conventional method used for taking attendance is merely a waste of time and effort. So, there have been proposed many solutions to solve this problem, especially the use of the biometric feature is much secured and fast. (Mohamed, 2012) has proposed a fingerprint-based attendance system. Herewith use of fingerprint scanner attendance is marked. The reason for the more demand for fingerprint verification is that each fingerprint satisfies uniqueness, resistance, permanence, and easily taking. But it might have some disadvantages like sometimes fingerprints can be damaged when a student might get injured at that finger and this moment student might not allow marking his or her attendance. So, face recognition is the best option to minimize the high errors of injuries because of the high accuracy of face recognition by the pre-train model. In (H. Rathod, 2017) has demonstrated the use of other biometric features like facial recognition for automating attendance system using SVM classifier to acquire desired results. Here in (N. Gupta, 2020) shown another face recognition-based attendance management system by using the LBPH model. For the face detection system, they used haarcascade and for face recognition open computer vision LBPH model. In (MuthuKalyani.K, 2013) proposed the use of CCTV for taking attendance more smartly. Here the CCTV is fixed at the entrance, which will detect and recognize the face and marks the attendance which can also be view using an android enhances smartphone. E-attendance system using machine learning approach can be categorized into two groups. The first class includes the algorithm to make the enhancement of training model, while the second class of algorithm focus on the improving the predication based on the training model.

In the first part of the algorithm, the face of all the students will be trained on the GPU with the help of the CNN HOG model. In this algorithm firstly the noise is removed from the images, marks different features of the face, and generates a pickle file for the further use of predication in the next class. In the second part of the algorithm, the image of the group photo taken in the classroom is given as an input to the pickle file and get the present student's list as well as it provides the TTS interface. Using festival TTS all the present student's name is announced in last certain minutes to avoid the wrong interpretations.

There are lots of images of the single classroom like 20 images of a single student and there are more than 60 students so it might more than 1200 images and which is not suitable to train the model on the computer. So, GPU is the best option to train the model with minimum time requirements.

PROPOSED TECHNIQUE

In the proposed technique, we are automating the attendance system by using machine learning algorithms. The system is designed to work on a Raspberry Pi Camera (Mehta, 2016). The Pi camera is placed in the classroom in such a way that it can click an image of the classroom to recognize different people present in the classroom. The image clicked in a time interval of 20 minutes in a class of 60 minutes. The model which is initially trained earlier with a training image of 20 different images of the same person will check if the face detected in an image is pre-trained or not. If the face was detected then the enrollment number will be stored in the sheet. The paper will address the problems facing in the detection of multiple faces and its integration with the cloud and a remote GPU.

In the initial stages, the main issue was to extract only faces from the images that are collected for the training purpose. Many different methods have been analyzed, but we found using MIT Licensed library face_recognition [Adam Geitgey]. Using the simple line of codes, we can detect faces and store them in the folder having a unique name. The next problem is accuracy, for which we use CNN HOG deep learning algorithm to maximize the accuracy. The solution is proposed as follows.

System Setup

This consist of mainly 3 steps:

1. Database collection:
   Initially, for the collection of data of each student, we have collected 30 images and store them with the folder named as the student's name. Now, this data is prepared for further pre-processing of an image.

Figure 1. Data base collection

› Desktop › Student data

Name	Status	Date modified	Type
170173107001		15-03-2021 17:02	File folder
170173107002		15-03-2021 17:02	File folder
170173107004		15-03-2021 17:02	File folder
170173107006		15-03-2021 17:02	File folder
170173107010		15-03-2021 17:02	File folder
170173107012		15-03-2021 17:02	File folder
170173107013		15-03-2021 17:02	File folder
170173107014		15-03-2021 17:02	File folder
170173107015		15-03-2021 17:02	File folder
170173107020		15-03-2021 17:02	File folder
170173107025		15-03-2021 17:02	File folder
170173107034		15-03-2021 17:02	File folder

2. Device Used:
    a. Raspberry Pi 4 Model-B (8 GB RAM)
    b. SanDisk 64GB Class 10 MicroSD Memory Card - With Noobs
    c. DELL GPU SERVER EMC PowerEdge R740
    d. Raspberry Pi Camera Module V2 - 8 Megapixels

Figure 2. System setup i.e., integration of all devices

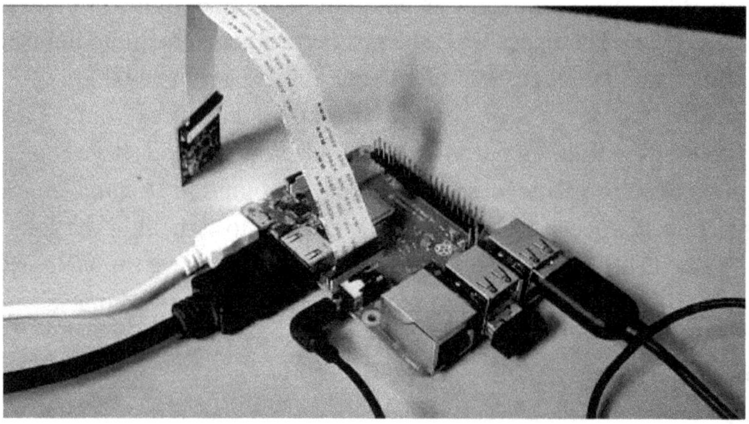

3. Model Training
   The model has to be created with the help of some machine learning technique, which later on used for testing purposes. The accuracy of the different models is analyzed, and then the most effective algorithm is selected.

Some part of coding is as follow:

```
for (i, imagePath) in enumerate(imagePaths):
        print(“[INFO] processing image {}/{}”.format(i +
1,len(imagePaths)))
        name = imagePath.split(os.path.sep)[-2]
        image = cv2.imread(imagePath)
        rgb = cv2.cvtColor(image, cv2.COLOR_BGR2RGB)
        boxes = face_recognition.face_locations(rgb,
                model=”hog”)
        encodings = face_recognition.face_encodings(rgb,
boxes)
        for encoding in encodings:
                knownEncodings.append(encoding)
                knownNames.append(name)
print(“[INFO] serializing encodings...”)
data = {”encodings”: knownEncodings, “names”: knownNames}
f = open(“encodings.pickle”, “wb”)
f.write(pickle.dumps(data))
```

Figure 3. Processing image

```
[INFO] processing image {1}/{20}
20/20 [==============================] - 3s - loss: 0.1242 - acc: 0.8421
[INFO] processing image {2}/{20}
20/20 [==============================] - 2s - loss: 0.1114 - acc: 0.8543
[INFO] processing image {3}/{20}
20/20 [==============================] - 2s - loss: 0.1153 - acc: 0.8814
[INFO] processing image {4}/{20}
20/20 [==============================] - 2s - loss: 0.1024 - acc: 0.9011
[INFO] processing image {5}/{20}
20/20 [==============================] - 1s - loss: 0.1011 -  acc: 0.9100
[INFO] processing image {6}/{20}
20/20 [==============================] - 1s - loss: 0.0011 -  acc: 0.9201
```

System Architecture

- **Image Capture:**
  The camera is fitted in the classroom in such a way that it covers the whole class. Every person's face would be visible. Within some time, interval it is going to click the image again. If the same person's image is there then it will be discarded (Sunil Aryala, 2019). Attendance will be marked only if the person is visible in both images. If it is present in any one image then it will be marked to check on it.
- **Face Detection:**
  For the detection of faces, we are using the face_recognition library which is used to automatically find the facial features of a person in an image. This will let you do recognition of faces on a folder of images using command line. Here during encoding purpose HOG is passed as an argument that is mainly used for extracting multiple faces in a single image.
- **Face Recognition:**
  The distance between landmarks is measured in the recognition phase, and this is compared to a test image containing unique known encoded image landmarks. Face recognition entails first extracting and then identifying characteristics. Regardless of lighting, expression, illumination, ageing, transformations (translate, rotate, and scale image), or position, the image in the face must be identified. After analyzing different algorithms for face recognition, we ended up using CNN HOG deep neural network algorithm. Histogram of Oriented Gradient (HOG) is used as the feature extractor (Ahamed, 2018), and CNN is used as the training algorithm for classifying the images (S. Sawhney, 2019).

Registering Attendance

- **Google sheet API initializing:**
  API has simplified our work by acting as an interface between multiple software applications, allowing the code to automatically share information and/or capabilities. Google Spreadsheets Python API Simple interface for working with Google Sheets. Here cloud storage is used do that it can be used remotely (Ofualagba Godswill, 2018).

```
for name, (top, right, bottom, left) in predictions:
for dt in range(1,5):
                    if(sheet.row_values(dt+1)[0] == number):
                     sheet.update_cell(dt+1,4,'P')
```

Here this chunk of code from the GPU updates the datasheet in google and mark the 'P' for the present students.

Figure 4. Attendance sheet

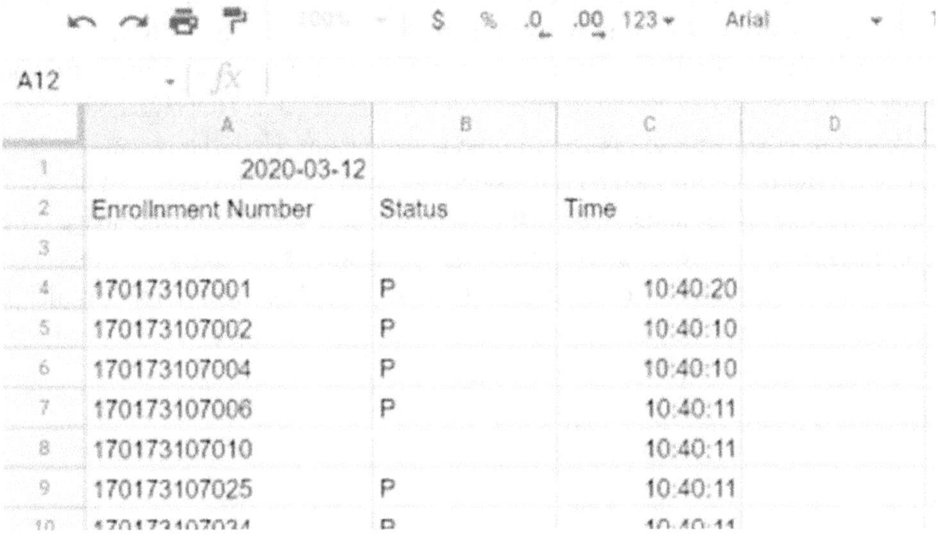

	A	B	C	D
1	2020-03-12			
2	Enrollment Number	Status	Time	
3				
4	170173107001	P	10:40:20	
5	170173107002	P	10:40:10	
6	170173107004	P	10:40:10	
7	170173107006	P	10:40:11	
8	170173107010		10:40:11	
9	170173107025	P	10:40:11	
10	[illegible]	[illegible]	[illegible]	

- **Managing Attendance:**
  The new sheet (Figure 4) with sheet name as the date is created in which the ID of the person will be written in the column, which has the column name as the time at which it is clicked.

DETAILED WORKING

Data Collection and Database Creation

Data is captured through high-definition cameras placed in Raspberry Pi in the classroom. Dataset is created before the image recognition process. Here, the dataset contains images of the students with different angles and poses. Minimum 20-25 images are captured with different angles and pose these images are then used in the training of the model. Here, student names and enrollment numbers are also used. These images are used to extract unique facial features of each face and then this information is then used in the image recognition process. This data is stored in the pickle data format. This pickle data formatted is Python-specific. Other restriction imposed by JSON (which can't represent pointer sharing) or XDR does not affect the pickle data. Python object structure can easily use for serializing and de-serializing.

Technical Process

The system was first built using the KNN model, but that is not as efficient to detect all images available in the image. This system is to be placed in the classroom so, we have to detect all humans available in that class. Here, a more efficient algorithm is the CNN-HOG method to detect the faces available in the image. CNN-HOG is more efficient in detecting faces available in the images than KNN. The core principle of CNN-HOG is to extract features of the images first then recognize the image. HOG is used as a feature extractor in this model. The HOG method is used to know whether a person is there in the image or not. Here the process flow is shown in figure 5. This method is the step-by-step method. The first input image is taken then the HOG model reduces the dimensionality available in the image. The HOG model is very good to detect the edges so, it detects the edges and prepares a histogram for the same, and based on that we can say whether a person is available in the image or not.

Figure 5. Process flow diagram

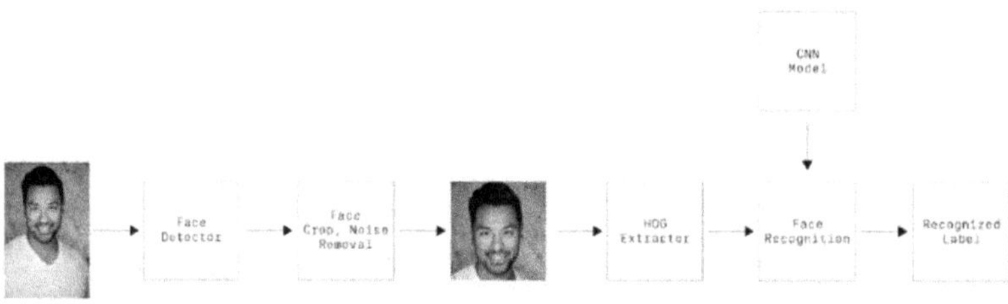

What is a Histogram of Oriented Gradients (HOG)?

A HOG is a feature descriptor that is commonly used to detect an object from given frame. This feature descriptor aids in the detection of multiple faces in the image. To transmit intensity gradients or edge orientations, a HOG relies on the property of components inside a picture, as shown in Figure 6. For each block, gradients are calculated inside the image. A block is viewed as a pixel grid, with gradients consisting of the magnitude and direction of change in the intensity of each pixel within the block. The feature extractor algorithm, HOG receives all of the samples. The descriptors are per-pixel gradient vectors whose magnitude and direction are determined using the formulas 1 and 2.

$$g = \sqrt{g_x^2 + g_y^2} \tag{1}$$

$$\theta = \arctan \frac{g_x}{g_y} \tag{2}$$

Where Gx and Gy are the horizontal and vertical components of the pixel intensity change, respectively. These were calculated across a block of pixels that was divided into bins, with each bin describing a gradient's directional angle and value. The procedure for obtaining features is depicted in figure 6. Now, this Histogram is further normalized together to minimize the effect of light conditions. This will lessen the variation in accuracy caused by changing lighting conditions. Now this HOG vector for multiple faces is further passed to trained by CNN model.

Figure 6. HOG feature extractor

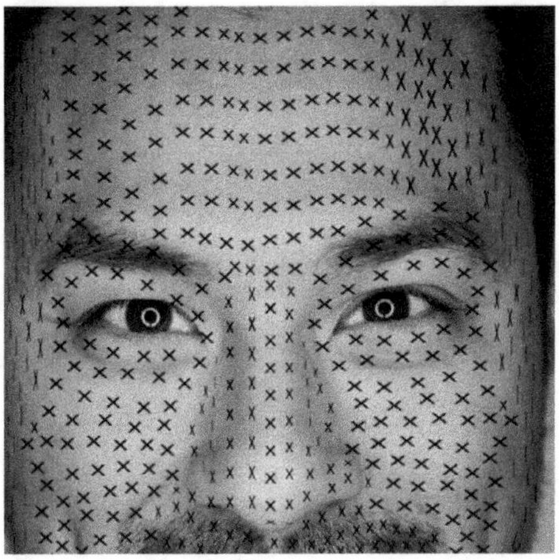

Training Algorithm

Many deep learning approaches are utilised for facial recognition. Among all the approaches, CNN has been widely implemented as the training algorithm in a wide range of machine learning applications. The construction of CNN is critical for directing overall productivity and time feeding. A CNN is composed of multiple layers, each of which can contribute to the formation of a convolutional Neural Network. The neural network's layers (Figure 7) are as follows:

- **Input layer**
  The raw input feature data is transmitted through this layer as a one-dimensional tensor.

- **Convolutional Layer:**
  Convolution layers use a predefined Filter or Kernel matrix to convolve the feature space.
- **Rectiðed Linear Unit [ReLU] Layer**
  The activation function was applied to retain the identical feature data and not being changed in the convolutional layer.
- **Pooling Layer**
  It is mostly used for down-sampling operations in spatial dimensions. For the most part, max pooling is employed for pooling.
- **Fully Connected Layer**
  The score value for each associated class was generated in this final stage, and the utmost score value denotes the associated class.

Figure 7. Convolution neural network layer

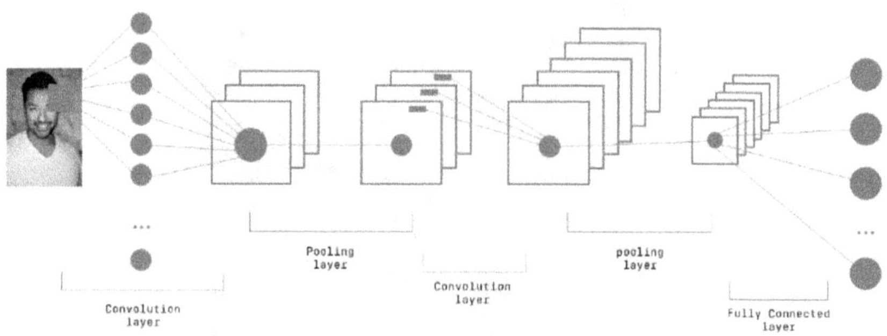

Role of GPU

CNN is a heavy neural network. It requires a lot of time to compute the whole network. Also, the data collected for training purposes is much, which requires much time. So, for training this bulky model, we used DELL GPU SERVER EMC PowerEdge R740. This high computing server will complete tasks in much less time than using a local server.

Recognition and Visualization

Recognition of face will be done based on the previously mentioned method. Here HOG has been used to determine all of the faces in a captured image. Different HOG vector is being created for multiple faces which will further passed to the training algorithm. CNN algorithm will try to detect each face present in the captured image.

This model was tested by training on the data collected earlier to check the accuracy. The accuracy was found to be 96% in the fully lit area. This accuracy can vary depending on a variety of factors, such as the image quality, the pose variations of the faces in the images, and so on. The detected people's attendance will be marked in the google sheet.

RESULT AND FINDINGS

This attendance system is a very efficient and accurate method to take attendance in university environments. This system helps to detect students present in the classroom very effectively. This system is automatic and fulfills our core goal, which is to remove manual operation in attendance marking. The collected data is trained with high accuracy and gives good accuracy on validation data (Figure 8).

This system uses google sheets as well, which also reduces our overhead of space as it uses cloud storage advantages anyone can access the data at any time

Figure 8. Training loss and accuracy vs. epoch curve

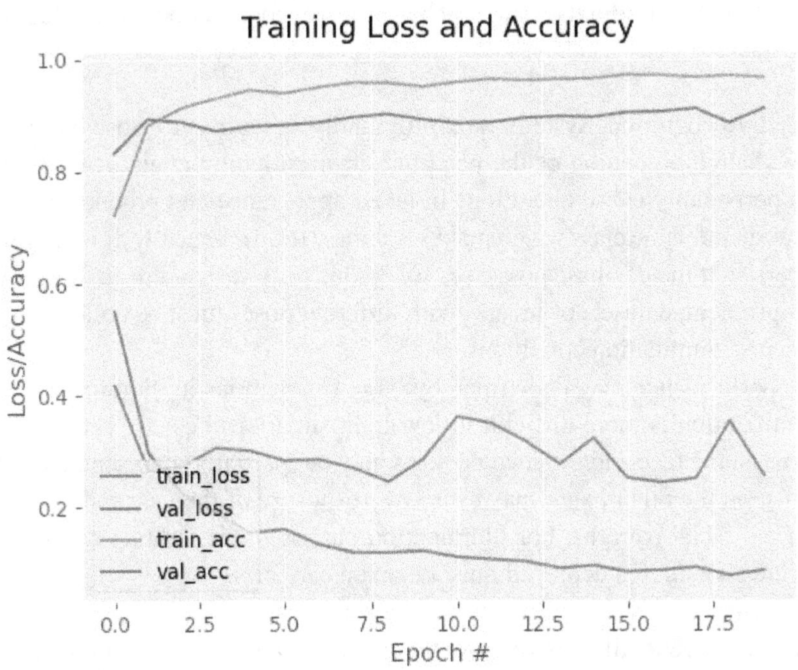

This system uses google sheets as well, which also reduces our overhead of space as it uses cloud storage advantages anyone can access the data at any time.

Here, data available in google sheets are in the form of excel sheets. With the help of the excel sheet, anyone can perform the required analysis on that sheet and get the desired result which will be very difficult to do if one is using the manual system or some other format.

This system is very scalable as it uses raspberry pi and cloud storing features as well as GPU. We can expand it with very ease. We can connect different raspberry pi with GPU and connect all of them to one server, and we fully automated the whole attendance marking system of a particular university.

RESEARCH CHALLENGES AND FUTURE WORK

Various problems starting with the resolution, pose variant, occlusion and many others have been solved up to some level, but a more efficient solution is required. These are mentioned problems that lead to challenges.

1. Secured Framework for data management
   Privacy is a severe threat. So, there must be a proper framework to store the data. Also, the flow of the data must be optimized for the continuous integration of data.
2. Pose
   Facial Recognition Systems are profoundly receptive to pose variations. The changing position of the head or varying rotating angle of the camera can perpetually cause alterations in facial appearance and produce contrasts between the same class which makes accuracy drop drastically. It will be more uncertain if the database possesses solely one pose variant image. To get over this problem, datasets of images with different poses must be collected.
3. Varying illumination conditions
   The performance may be harmed by large fluctuations in illumination. Face identification is more difficult in low-light situations because shadows may occur on the face and/or facial patterns may be (partially) undetectable. More lights, on the other hand, may cause overexposure of the face and (partially) conceal facial patterns. For illumination purposes, the lighting condition of the area where it is deployed must be taken care of.
4. Occlusion
   In an image generally, some part of the face is usually blocked by others, which is termed as occlusion. Here the whole image is not visible so, the distinguishing feature of the face cannot be identified clearly. One of the most complex

Figure 9. Research challenges

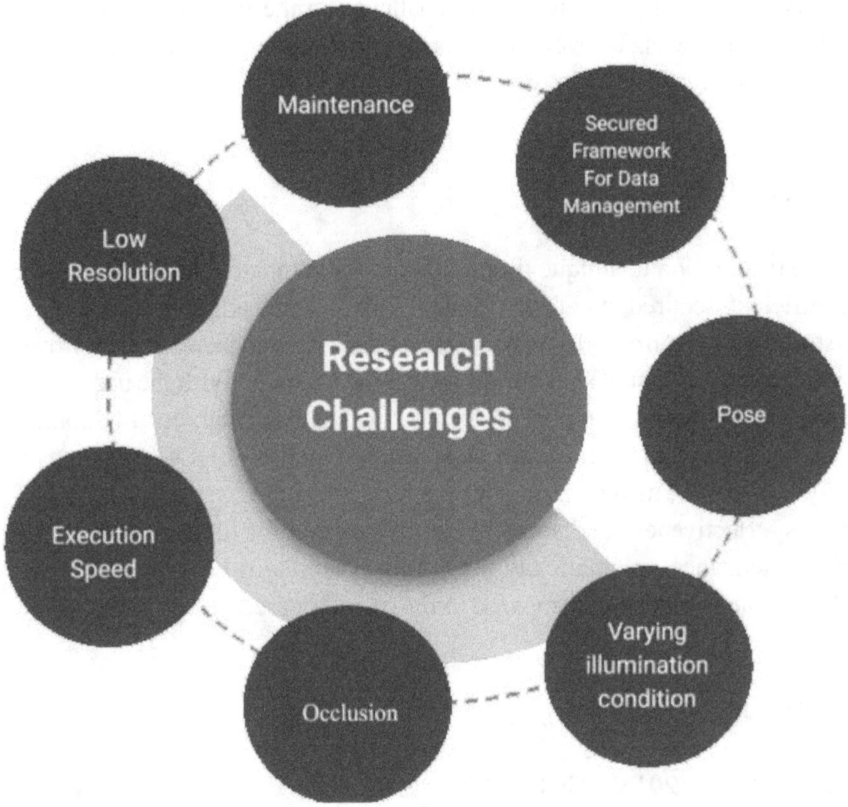

problems in face recognition systems is occlusion. A more advanced feature extractor is to be used to detect each part despite some occlusion in the image.

5. Execution Speed
   As CNN takes much time to recognize the image, some time-savvy and more efficient algorithms must be used for the proposed system.
6. Maintenance
   Hardware devices require frequent maintenance. Regular check on the functionality of each device must be checked. So, a proper execution plan must be prepared.
7. Low Resolution
   The minimum resolution must not degrade from 16*16. A picture with a resolution less than 16*16 is called a low-resolution image. This type of image is found from standalone security cameras. As this camera is at a much farther distance from the face, they will cover a very small human face area, which is of very low resolution even less than 16 * 16. This type of image is

not helpful for feature extraction and importantly many features get lost many times. This will be a significant challenge in the process of identifying the features of faces. To overcome this, the proper requirement of the device is to be analyzed and used in the system.

CONCLUSION

The face recognition technique proposed here for automated attendance systems is more efficient, secured, and less complex. This system cannot completely identify each student present in class but can identify most of the students. Many improvements are required on account of space complexity and accuracy. With the use of GPU, the time required to train the model is decrease and satisfactory result is obtained for face recognition on different poses and variation by use of HOG feature extractor. This system is built using a modular approach, so improvisation on accuracy with more cost-effectiveness and security can be targeted for future work. As a result, keeping attendance automatically using this system will be extremely beneficial and less prone to errors as compared to the manual process.

REFERENCES

Ahamed, H. A. (2018). HOG-CNN Based Real Time Face Recognition. In *International Conference on Advancement in Electrical and Electronic Engineering*. IEEE.

Gupta, N. P. S. (2020). Automated Attendance System Using OpenCV. In *2020 8th International Conference on Reliability, Infocom Technologies and Optimization (Trends and Future Directions) (ICRITO)*, (pp. 1226-1230). IEEE.

Kar, N. A. (2012). Study of implementing automated attendance system using face recognition technique. *International Journal of Computer and Communication Engineering*, 100.

Mehta, P. A. (2016). An Efficient Attendance Management Sytem based on Face Recognition using Matlab and Raspberry Pi 2. *International Journal of Engineering Technology Science and Research*, 71-78.

Mohamed, C. V. (2012). Fingerprint attendance system for classroom needs. In *2012 Annual IEEE India Conference (INDICON)* (pp. 433-438). Kochi: IEEE. 10.1109/INDCON.2012.6420657

MuthuKalyani, K V. (2013). Smart application for AMS using face recognition. *Computer Science & Engineering: An International Journal.*

Ofualagba Godswill, O. O. (2018). Automated Student Attendance Management Using Face Recognition. *International Journal of Educational Research and Information Science*, 31-37.

Rathod, H. Y. W. (2017). Automated attendance system using machine learning approach. In *2017 International Conference on Nascent Technologies in Engineering (ICNTE)* (pp. 1-5). IEEE. 10.1109/ICNTE.2017.7947889

Sawhney, S. K. K. (2019). Real-Time Smart Attendance System using Face Recognition Techniques. In *2019 9th International Conference on Cloud Computing, Data Science & Engineering (Confluence)* (pp. 522-525). IEEE.

Sunil Aryala, R. S. (2019). Automatic Attendance System Using Deep Learning. *International Conference on Sustainable Computing in Science, Technology and Management (SUSCOM).*

Chapter 5
A Novel SOM Ensemble Classification Technique for Limited Data

Abdul Mueed Hafiz
https://orcid.org/0000-0002-2266-3708
Institute of Technology, University of Kashmir, Srinagar, India

ABSTRACT

Small datasets are common in research areas where measurement of data is expensive or difficult (e.g., healthcare). For small datasets, training of a data model and classification using the same remains a challenge. To address the issue of limited data training and classification, the authors propose to use a novel ensemble of self-organizing maps (SOMs). SOMs are shallow neural networks, and as a result, the number of parameters is not very high. This avoids over-fitting and over-parameterization of the network. For the SOM ensemble, a novel early stopping training technique is proposed that requires a small number of samples for training. In the ensemble, one SOM is used for every class. For benchmarking the proposed technique, its performance is compared with that of other state-of-the-art techniques like deep learning, CATBOOST decision tree approach, etc. For reproducibility of results, three popular and extensively benchmarked digit image datasets have been used (i.e., MNIST, USPS+, and MADB). The proposed technique generally outperforms all other techniques with which it is compared.

DOI: 10.4018/978-1-6684-4225-8.ch005

INTRODUCTION AND RELATED WORK

One of the important issues in statistical modeling is indirect measurement of quantities, or virtual sensors (Ingrassia & Morlini, 2005). It involves the prediction of variables whose measurement is expensive (e.g., viscosity, or the concentration of certain chemical species, or some mechanical features, by using other easy to measure variables (e.g., temperature or pressure) (de Veaux & Ungar, 1996). Modeling problems of this category are found in research areas like health sciences where the size of the dataset is often small. If one is tempted to use Artificial Neural Networks (ANNs) as the classifier, it may be so that the size of the dataset is small compared with the number of weights of the neural network. In this case, the neural network may be considered to be over-parameterized (Ingrassia & Morlini, 2005). Given the era of deep learning (LeCun, Bengio, & Hinton, 2015), deep neural networks like Convolutional Neural Networks (CNNs) (Krizhevsky, Sutskever, & Hinton, 2012) are used as classifiers. Normally, CNNs use huge datasets for training and are trained over thousands of epochs. Under such circumstances, using suitable neural networks for small datasets remains an issue. Some techniques are available in the literature to address this issue. They include the *few shot learning* (Chen, Liu, Kira, Wang, & Huang, 2019; Snell, Swersky, & Zemel, 2017; Wang, Yao, Kwok, & Ni, 2019; Yoo, Fan, Boddeti, & Kitani, 2018) technique again based on deep learning. In it, a pre-trained deep neural network (CNN) is fine-tuned on the small dataset and is subsequently used for classification. But this technique often suffers from problems arising out of transfer learning and long training periods. Further, training a deep neural network on a small dataset leads to problems such as over-fitting and over-parameterization of the deep neural network. This is because the number of parameters (e.g. weights) is huge in comparison to the limited data available. In this regard, we propose to use an ensemble of Self Organizing Maps (SOMs) for classification of small datasets.

Some recent techniques have been successfully applied in this area like the work in (Oh, Park, & Ye, 2020) which involve using modified deep learning models for classification of data using small datasets. However, as mentioned earlier deep learning models have very large number of parameters leading to need for large computational resources and extended training times, and their usage for small data has not been investigated extensively. In spite of these issues, there is a small body of work involving modified/light CNN models used for classification of small datasets, as done in (Oh et al., 2020). The authors therein have used a modified CNN for COVID patient classification. In this work, we compare our approach with the same technique (Oh et al., 2020) besides other state-of-the-art techniques.

Self Organizing Maps are a type of shallow neural networks (Kohonen, 1998) which can approximate an unlimited number of input patterns by a finite set of weights. SOM learning is efficient, as well as effective as it is suitable for high-dimensional vectors. Also, SOM learning shows more robustness to initialization. SOM based clustering has evolved into a well-researched field (Sun, Wang, Wei, & Zhang, 2018). SOMs have been used diversely in various research areas including dimensional affect recognition (Bugnon, Calvo, & Milone, 2018). SOMs have also been used in combination with K-Nearest Neighbor (KNN) Classifiers (Silva & Del-Moral-Hernandez, 2011) and Support Vector Machines (SVMs) (Patra & Bruzzone, 2014). In (Silva & Del-Moral-Hernandez, 2011) the authors combine the SOM and the KNN classifier, into a classifier called *SOM-KNN*, and use this classifier for digit recognition of images. The authors claim that the new classifier is much faster than traditional methods while being competitive with them. In (Patra & Bruzzone, 2014), SOMs and the SVM Classifier are used to increase training efficiency for remote sensing image classification. The technique uses the SVM and SOM properties respectively, for ascertaining whether uncertain and diverse images are to be included in the training set. The result of using this technique is fast training convergence even for poor quality initial training images. However, these hybrid classifiers are difficult to tune as they have to optimize two different classifiers.

SOMs (Nilashi et al., 2020; Shirani Faradonbeh, Shaffiee Haghshenas, Taheri, & Mikaeil, 2020; Yao, Zhu, & Zhao, 2020) are suitable for ensembles as the trained network weights mimic the input. Recently deep SOMs have been used (Aly & Almotairi, 2020; Sakkari & Zaied, 2020a, 2020b; Wickramasinghe, Amarasinghe, & Manic, 2019). However we do not go in that direction because they require more data and our study was used to investigate SOMs for limited data. In (Tan, Chen, Zhou, & Zhang, 2005), the researchers use a separate SOM for each class of a face-recognition task. We also propose to use an ensemble of class specific SOMs for the task of classification of small datasets. Our approach consists of a training technique which has the following properties. It is simple, is single step and needs minimal data. The SOMs used are three-layer neural networks. Our training technique aims to reduce the complexity of training neural networks having one-hidden layer. Also, it uses the universal approximation capability of SOMs (Huang & Babri, 1998). Our technique and the technique of (Tan et al., 2005) cannot be compared because of the following reasons. Our technique uses a single-step training approach for multiple training patterns per class. The approach of (Tan et al., 2005) uses a many-step training approach for a single training pattern per class. Also, the weight update principle is different in both. Our output is calculated with summation, by taking advantage of a blank background for sample digit images, while as their output is calculated by a distance function and a soft k-NN technique.

We also compare our approach with a robust Decision Tree based approach called CATBOOST (Prokhorenkova, Gusev, Vorobev, Dorogush, & Gulin, 2018). Although CATBOOST (Prokhorenkova et al., 2018) is effective for medium sized datasets, we investigate its efficacy for limited data. CATBOOST (Prokhorenkova et al., 2018) also suffers from issues like need for large computational resources and extended training times.

We wanted to work on popular datasets for which extensive benchmarking was available. We used small subsets from three popular digit image datasets viz. MNIST (LeCun, Bottou, Bengio, & Haffner, 1998), MADB (Abdelazeem & El-Sherif, 2008) and USPS (Roweis). We benchmarked our proposed approach with various other state-of-the-art approaches like a modified CNN used for small COVID related datasets (Oh et al., 2020), robust techniques like CATBOOST (Prokhorenkova et al., 2018) which is an advanced Decision Tree based technique, shallow Feed Forward Neural Networks trained using Scaled Conjugate Gradient (SCG) Backpropagation Algorithm (Møller, 1993) which has small number of training parameters, Support Vector Machine (SVM) Classifier (Cortes & Vapnik, 1995) and K-Nearest Neighbor (KNN) Classifier (Cunningham & Delaney, 2007). It should be noted that we did not use *few shot learning* (Chen et al., 2019; Snell et al., 2017; Wang et al., 2019) for benchmarking of our technique because the same uses pre-trained deep neural networks for transfer learning. Fine tuning and finding optimum hyper-parameters is an issue for the same. And these pre-trained neural networks are trained on RGB images which are larger than digit images. Also, digit images have only one channel as they are grayscale while as RGB images have 3 channels. Hence, again the input to deep networks used is of 3 channel type (RGB). There are also other techniques used for classification of small datasets as found in literature (Caiafa, Sun, Tanaka, Marti-Puig, & Solé-Casals, 2021; Masquelin, Cheney, Kinsey, & Bates, 2021; Prihasto & Azhar, 2021; Riekert, Riekert, & Klein, 2021). But they are used sparingly and are not found generally in literature.

PROPOSED TECHNIQUE

Training and classification in our approach are given below.

Training

Our technique uses learning by reinforcement along with penalty. No features are extracted. The binary or grayscale images are fed directly to the neural networks. Positive reinforcement, or rewarding behavior takes place in training for similar feature association, and a penalty is imposed for dissimilarity. A separate network is trained for each class. The class-specific networks are naïve at start of training, i.e. all weights are zero. Each network consists of N neurons in first layer, N neurons in second layer and 1 output neuron, where N is number of elements if flattened data vector (of a single digit image sample). The technique was conceived for binary patterns. However, grayscale image classification using our technique is also presented towards the end of the paper. The grayscale images are 8-bit i.e. each pixel in the grayscale images can have one of 256 values (decimals 0 to 255). A modified log-sigmoid activation function is used in second layer neurons. The training algorithm is discussed below.

1. The kth binary training pattern (or sample) P_k where $k = 1, 2, \dots K$, is presented simultaneously to the first layer and the second layer of neural network for class c, where $c = 1, 2, \dots C$; K is the total number of training patterns (or samples) presented per class; C is the total number of classes which is 10 for digit classification.
2. The jth second layer neuron is chosen, where $j = 1, 2, \dots, N$.
3. A variable l_{ijk} associated with the ith first layer neuron and j^{th} second layer neuron is updated for k^{th} iteration as per following rule:

"If the pattern values presented to the ith first layer neuron and the jth second layer neuron are same, then l_{ijk} is incremented by one, else it is decremented by one."

4. Steps 2 and 3 are repeated for all N first layer neurons and the jth second layer neuron.
5. 5. A modified log-sigmoid function is used to compute the weight w_{ijk} as follows:

$$w_{ijk} = \frac{2}{1 + exp\left(-2 \times l_{ijk} \times \frac{5}{K}\right)} - 1 \tag{1}$$

The above function exponentially crushes l_{ijk} to a value between -1 and +1.

6. Steps 2 to 5 are repeated for all the N second layer neurons.

7. A network convergence check is performed as follows:

$$\mathrm{d} = euclidean\ distance\left(\mathbf{w}_{j=1}^{k}, \mathbf{w}_{j=1}^{k-1}\right) \tag{2}$$

where $\boldsymbol{w}_{j=1}^{k}$ is the updated weight vector of first neuron in second layer after k^{th} pattern has been learnt.

8. If d is less than a certain threshold $\boldsymbol{A}$, network for class c is assumed to have converged and
9. training is stopped. The network for class c may stop learning before all the (K) training patterns are used. Using a very small a value to $\boldsymbol{A}$ leads to under-training, and using a large value of $\boldsymbol{A}$ leads to over-fitting. The novelty of the proposed training approach is that here SOMs converge for a small value of $\boldsymbol{A}$. Hence, the SOMs need very less training.
10. Finally, the entire weight matrix is normalized and inverted as:

$$\boldsymbol{w} = 1 - \frac{\boldsymbol{w}}{max(\boldsymbol{w})} \tag{3}$$

Training of the class specific networks is followed by classification (or testing).

Classification

A test pattern (or sample) $\boldsymbol{P}$ to be classified, is presented to all the trained neural networks. There are C neural networks ($C = 10$ for digit classification). The output neuron in c^{th} neural network (where $c = 1, 2, \ldots C$) has a score which is computed as per the following summation function:

$$score_c = \sum_{i=1}^{N}\sum_{j=1}^{N} \boldsymbol{w}_{ij}^{c} \times \boldsymbol{b}_j \tag{4}$$

where

$$\boldsymbol{b}_j \begin{Bmatrix} B, if\ \boldsymbol{P}_j = 1 \\ -1, if\ \boldsymbol{P}_j = 0 \end{Bmatrix}$$

$\boldsymbol{B}$ may take a small positive value e.g. $\boldsymbol{B} = 5$.

The neural network with maximum score is said to have classified the test pattern. To the best of our knowledge this is the first technique to use an ensemble of SOMs in this manner.

EXPERIMENTATION

An ARM Dual Core processor, 16 GB RAM System was used for benchmarking of the proposed technique along with others. The various techniques were implemented in Python. Tensorflow (Abadi et al., 2016) was used when required. The performance of the proposed approach was compared with various state-of-the-art techniques like a recent deep learning technique based on a modified CNN used for small COVID related datasets (Oh et al., 2020), robust techniques like CATBOOST (Prokhorenkova et al., 2018) which is an advanced Decision Tree based technique, shallow Feed Forward Neural Networks trained using Scaled Conjugate Gradient (SCG) Backpropagation Algorithm (Møller, 1993) which has small number of training parameters, Support Vector Machine (SVM) Classifier (Cortes & Vapnik, 1995) and K-Nearest Neighbor (KNN) Classifier (Cunningham & Delaney, 2007). The SCG trained NN (SCGNN) used was a shallow neural network with one hidden layer (to avoid over-parameterization). Besides the SCGNN, the other choices for comparison were made because they belong to statistical unsupervised learning (viz. KNN) and supervised learning (viz. SVM). Hence a diverse set of techniques were used for benchmarking. The datasets were specifically chosen on grounds of ease of reproducibility in results. The performances of various techniques using the datasets used are also shown below. Various important parameters used for the proposed technique are also mentioned.

MNIST AND MADB DATASETS

The MNIST dataset of handwritten digits (LeCun et al., 1998) consists of 60,000 training instances and 10,000 testing instances and is available online. Each set consists of an equal number of samples drawn from 2 distinct populations: employees of the Census Bureau, and students of high school. Each digit is a 28x28 grayscale image. Another database, viz. the Modified Arabic Digits Database (MADB) is also available online (Abdelazeem & El-Sherif, 2008; Abdleazeem & El-Sherif, 2008). It has been constructed on lines of the structure of the MNIST database, but for Arabic numeral digits. It also consists of 60000 instances for training and 10000 instances for testing. The digit images are binarized and fed to the classifiers without feature extraction. The performance of various techniques on the MNIST database is shown in Table 1. The performance of various techniques on the MADB database is shown in Table 2.

Table 1. Performance of various techniques on the MNIST digit image database (Testing Set has 2000 Samples, i.e. 200 samples per class)

Technique	Training Samples	Accuracy (%)
Modified CNN (Oh et al., 2020)	600	87.61
CATBOOST (Prokhorenkova et al., 2018)	600	86.98
SCGNN (Møller, 1993) (Hidden Layer Neurons =10, Error =0.001, Sigma =5.0e-5, Lambda =5.0e-7)	600	82.30
SVM (Fletcher, 2009) (C = 4.0, Gamma = 0.0003)	600	85.43
KNN (Cunningham & Delaney, 2007) (K=4, Euclidean Dist. Function)	600	83.85
Proposed Technique (A=0.350, B=2.750)	**528**	**87.69**

Table 2. Performance of various techniques on the MADB digit image database (Testing Set has 1000 samples i.e. 100 samples per class)

Technique	Training Samples	Accuracy (%)
Modified CNN (Oh et al., 2020)	600	87.82
CATBOOST (Prokhorenkova et al., 2018)	600	87.31
SCGNN (Møller, 1993) (Hidden Layer Neurons =15, Error =0.001, Sigma =5.0e-5, Lambda =5.0e-7)	600	80.50
SVM (Fletcher, 2009) (C=5.0, gamma=0.0003)	600	85.90
KNN (Cunningham & Delaney, 2007) (K=4, Euclidean Dist. Function)	600	84.10
Proposed Technique (A=0.3, B=4.0)	**592**	**87.95**

On both the datasets, the proposed technique converges successfully and shows better performance than other state-of-the-art techniques. It should be noted that our technique converges using lesser number of training samples (especially in MNIST) as compared to those of other techniques.

USPS+ Dataset

The USPS+ Digit Image Database (Roweis) which is available online, is used next for benchmarking. USPS+ has 1100 instances for each digit. Each digit image is a 16×16 pixel grayscale image and is stored as a 256 element array. Each array element takes values from 0 to 255 as is the case for 256 gray-level images. First, we binarize the grayscale images of the database and then use them for training and testing. Next, we use the grayscale images for training and testing as well. For both cases (binary and grayscale images), the techniques are same. It should be noted, that we obtained same accuracy for binary as well as grayscale images as shown in Table 3.

Table 3. Performance of various techniques on the USPS+ digit image dataset (Testing Set has 2000 Samples i.e. 200 samples per class)

Technique	Training Samples	Accuracy (%)	
		Multi-level Data	Binary Data
Modified CNN (Oh et al., 2020)	800	**81.91**	81.71
CATBOOST (Prokhorenkova et al., 2018)	800	80.92	81.12
SCGNN (Møller, 1993) (Hidden Layer Neurons=10, Error=0.01, Sigma=5.0e-5, Lambda=5.0e-7)	800	74.90	73.45
SVM (Fletcher, 2009) (For Multi-level Data, C = 20.0, gamma = 0.0025; For Binary Data, C = 2.0, gamma = 0.0025)	800	80.35	81.70
KNN (Cunningham & Delaney, 2007) ([For Multi-level Data: K=3; For Binary Data: K=4], Euclidean Dist Function)	800	79.60	79.95
Proposed Technique (A = 0.240, B = 2.0)	**793**	81.85	**81.85**

It is observed from Table 3, that our technique generally outperforms other techniques on the data used. The training of the proposed SOM ensemble stops after 793 training images are used for training.

It is also observed from Tables 1 to 3 that the performance of our technique is generally better than other techniques with which it is compared.

CONCLUSION

When size of the dataset is small e.g. in medical sciences research, choice of the classifier is somewhat restricted. This is because of limited training data available. When using neural networks, CNNs are also a choice. They are used in a framework known as *few shot learning*. However, this mode of classification suffers from problems arising from transfer learning, and large episode training. It may lead to over-fitting and over-parameterization of the neural network respectively. Some modified CNNs for limited data classification are available limitedly in literature and have been used for benchmarking in this work. Their usage in this area is not extensively investigated in the available literature. To address the various issues associated with limited data classification, we propose to use an ensemble of Self Organizing Maps (SOMs). SOMs are shallow neural networks and have universal approximation properties. In this paper, a novel SOM training technique is presented. The number of neural networks used in the ensemble is equal to number of classes. Digit images from 3 databases (viz. MNIST, MADB and USPS+) have been used for training and classification. The databases are used because most of them are popular, are available online for public, and have been benchmarked extensively. The proposed technique is benchmarked against other state-of-the-art techniques like modified CNN, CATBOOST Decision Tree technique, Scaled Conjugate Gradient (SCG) Backpropagation trained Neural Network (or SCGNN), SVM and KNN. It is found experimentally that the proposed technique generally outperforms others on all the 3 databases. The proposed approach provides a reliable means of classifying a small dataset without any prior experience.

Conflict of Interest

The authors declare no conflict of interest.

REFERENCES

Abadi, M., Agarwal, A., Barham, P., Brevdo, E., Chen, Z., Citro, C., . . . Devin, M. (2016). *Tensorflow: Large-scale machine learning on heterogeneous distributed systems.* arXiv preprint arXiv:1603.04467.

Abdelazeem, S., & El-Sherif, E. (2008). *Modified Arabic Digits Database*. School of Science and Engineering, Department of Electronics Engineering, The American University in Cairo, Egypt. https://datacenter.aucegypt.edu/shazeem/

Abdleazeem, S., & El-Sherif, E. (2008). Arabic handwritten digit recognition. *International Journal on Document Analysis and Recognition, 11*(3), 127–141. doi:10.100710032-008-0073-5

Aly, S., & Almotairi, S. (2020). Deep Convolutional Self-Organizing Map Network for Robust Handwritten Digit Recognition. *IEEE Access: Practical Innovations, Open Solutions, 8*, 107035–107045. doi:10.1109/ACCESS.2020.3000829

Bugnon, L. A., Calvo, R. A., & Milone, D. H. (2018). Dimensional Affect Recognition from HRV: An Approach Based on Supervised SOM and ELM. *IEEE Transactions on Affective Computing*, 1–1. doi:10.1109/TAFFC.2017.2763943

Caiafa, C. F., Sun, Z., Tanaka, T., Marti-Puig, P., & Solé-Casals, J. (2021). Machine Learning Methods with Noisy, Incomplete or Small Datasets. *Applied Sciences (Basel, Switzerland), 11*(9), 4132. doi:10.3390/app11094132

Chen, W.-Y., Liu, Y.-C., Kira, Z., Wang, Y.-C. F., & Huang, J.-B. (2019). *A closer look at few-shot classification.* arXiv preprint arXiv:1904.04232.

Cortes, C., & Vapnik, V. (1995). Support-vector network. *Machine Learning, 20*(3), 273–297. doi:10.1007/BF00994018

Cunningham, P., & Delaney, S. J. (2007). *k-Nearest Neighbour Classifiers*. Academic Press.

de Veaux, R. D., & Ungar, L. H. (1996). Neural Networks in Applied Statistics. *Technometrics*, 215–218.

Fletcher, T. (2009). *Support Vector Machines Explained.* www.cs.ucl.ac.uk/sta_/T. Fletcher/

Huang, G.-B., & Babri, H. A. (1998). Upper bounds on the number of hidden neurons in feedforward networks with arbitrary bounded nonlinear activation functions. *IEEE Transactions on Neural Networks, 9*(1), 224–229. doi:10.1109/72.655045 PMID:18252445

Ingrassia, S., & Morlini, I. (2005). Neural network modeling for small datasets. *Technometrics, 47*(3), 297–311. doi:10.1198/004017005000000058

Kohonen, T. (1998). The self-organizing map. *Neurocomputing, 21*(1-3), 1–6. doi:10.1016/S0925-2312(98)00030-7

Krizhevsky, A., Sutskever, I., & Hinton, G. E. (2012). *Imagenet classification with deep convolutional neural networks.* Paper presented at the Advances in Neural Information Processing Systems.

LeCun, Y., Bengio, Y., & Hinton, G. (2015). Deep learning. *Nature*, *521*(7553), 436–444. doi:10.1038/nature14539 PMID:26017442

LeCun, Y., Bottou, L., Bengio, Y., & Haffner, P. (1998). Gradient-based learning applied to document recognition. *Proceedings of the IEEE*, *86*(11), 2278–2324. doi:10.1109/5.726791

Masquelin, A. H., Cheney, N., Kinsey, C. M., & Bates, J. H. T. (2021). Wavelet decomposition facilitates training on small datasets for medical image classification by deep learning. *Histochemistry and Cell Biology*, *155*(2), 309–317. doi:10.100700418-020-01961-y PMID:33502624

Møller, M. F. (1993). A scaled conjugate gradient algorithm for fast supervised learning. *Neural Networks*, *6*(4), 525–533. doi:10.1016/S0893-6080(05)80056-5

Nilashi, M., Ahmadi, H., Manaf, A. A., Rashid, T. A., Samad, S., Shahmoradi, L., Aljojo, N., & Akbari, E. (2020). Coronary Heart Disease Diagnosis Through Self-Organizing Map and Fuzzy Support Vector Machine with Incremental Updates. *International Journal of Fuzzy Systems*, *22*(4), 1376–1388. doi:10.100740815-020-00828-7

Oh, Y., Park, S., & Ye, J. C. (2020). Deep learning covid-19 features on cxr using limited training data sets. *IEEE Transactions on Medical Imaging*, *39*(8), 2688–2700. doi:10.1109/TMI.2020.2993291 PMID:32396075

Patra, S., & Bruzzone, L. (2014). A novel SOM-SVM-based active learning technique for remote sensing image classification. *IEEE Transactions on Geoscience and Remote Sensing*, *52*(11), 6899–6910. doi:10.1109/TGRS.2014.2305516

Prihasto, B., & Azhar, N. F. (2021). Evaluation of Recurrent Neural Network Based on Indonesian Speech Synthesis for Small Datasets. *Advances in Science and Technology (Owerri, Nigeria)*, *104*, 17–25. . doi:10.4028/www.scientific.net/AST.104.17

Prokhorenkova, L., Gusev, G., Vorobev, A., Dorogush, A. V., & Gulin, A. (2018). CatBoost: Unbiased boosting with categorical features. *Advances in Neural Information Processing Systems*, *31*, 6638–6648.

Riekert, M., Riekert, M., & Klein, A. (2021). Simple Baseline Machine Learning Text Classifiers for Small Datasets. *SN Computer Science*, *2*(3), 178. doi:10.100742979-021-00480-4

Roweis, S. (n.d.). *USPS Handwritten Digit Dataset*. http://www.cs.nyu.edu/~roweis/data.html

Sakkari, M., & Zaied, M. (2020a). A Convolutional Deep Self-Organizing Map Feature extraction for machine learning. *Multimedia Tools and Applications*, *79*(27), 19451–19470. doi:10.100711042-020-08822-9

Sakkari, M., & Zaied, M. (2020b). A Convolutional Deep Self-Organizing Map Feature extraction for machine learning. *Multimedia Tools and Applications*, *79*(27-28), 1–20. doi:10.100711042-020-08822-9

Shirani Faradonbeh, R., Shaffiee Haghshenas, S., Taheri, A., & Mikaeil, R. (2020). Application of self-organizing map and fuzzy c-mean techniques for rockburst clustering in deep underground projects. *Neural Computing & Applications*, *32*(12), 8545–8559. doi:10.100700521-019-04353-z

Silva, L. A., & Del-Moral-Hernandez, E. (2011). *A SOM combined with KNN for classification task*. Paper presented at the The 2011 International Joint Conference on Neural Networks.

Snell, J., Swersky, K., & Zemel, R. (2017). *Prototypical networks for few-shot learning*. Paper presented at the Advances in Neural Information Processing Systems.

Sun, S., Wang, S., Wei, Y., & Zhang, G. (2018). A Clustering-Based Nonlinear Ensemble Approach for Exchange Rates Forecasting. *IEEE Transactions on Systems, Man, and Cybernetics. Systems*.

Tan, X., Chen, S., Zhou, Z.-H., & Zhang, F. (2005). Recognizing partially occluded, expression variant faces from single training image per person with SOM and soft k-NN ensemble. *IEEE Transactions on Neural Networks*, *16*(4), 875–886. doi:10.1109/TNN.2005.849817 PMID:16121729

Wang, Y., Yao, Q., Kwok, J., & Ni, L. M. (2019). *Generalizing from a few examples: A survey on few-shot learning* arXiv: 1904.05046.

Wickramasinghe, C. S., Amarasinghe, K., & Manic, M. (2019). Deep Self-Organizing Maps for Unsupervised Image Classification. *IEEE Transactions on Industrial Informatics*, *15*(11), 5837–5845. doi:10.1109/TII.2019.2906083

Yao, P., Zhu, Q., & Zhao, R. (2020). Gaussian Mixture Model and Self-Organizing Map Neural-Network-Based Coverage for Target Search in Curve-Shape Area. *IEEE Transactions on Cybernetics*, 1–13. doi:10.1109/TCYB.2020.3019255 PMID:32991301

Yoo, D., Fan, H., Boddeti, V. N., & Kitani, K. M. (2018). *Efficient k-shot learning with regularized deep networks*. Paper presented at the Thirty-Second AAAI Conference on Artificial Intelligence.

Chapter 6
An Android App Permission Analysis for User Privacy and Security

Vaibhav Tyagi
G. L. Bajaj Institute of Technology and Management, Greater Noida, India

Vasu Agarwal
G. L. Bajaj Institute of Technology and Management, Greater Noida, India

Rupali Singh
G. L. Bajaj Institute of Technology and Management, Greater Noida, India

Shubhankar Pandey
G. L. Bajaj Institute of Technology and Management, Greater Noida, India

Ashish Tripathi
G. L. Bajaj Institute of Technology and Management, Greater Noida, India

ABSTRACT

Android applications require necessary permissions to perform the right function, but at the same time, there are other permissions that may affect the user privacy that need to be retrieved and revoked. Each step forward in digitization is also leading to an increase in the number of malware attacks by the malicious applications with the use of permissions. To solve the problem of APK app user data privacy and security, the authors created a system that searches for and downloads Android apps from the Android Market automatically. In addition, they produced a thorough mapping of Android application programming interface (API) calls to the needed permission(s) for each call, if any were required. Based on a static examination of each application's APK bytecode, they next ran an analysis of 141,372 Android applications to see if they had the right set of permissions. According to the research, most mobile app developers do not use the proper permission set and either over-specify or under-specify their security requirements.

DOI: 10.4018/978-1-6684-4225-8.ch006

INTRODUCTION

A constant increment has been observed in the Android market share. In the present case scenario, Android dominates with over 2.6 billion users on a global scale (European Commission,). As per research data of worldwide smartphone shipment OS market share forecast by the International Data Corporation (IDC), Android maintains its dominance over iOS by attaining 86.1% share at the end of 2019 whereas iOS could only make it up to 13.9% (IDC, 2021; Iqbal, 2022).

Along with the augmentation in Android usage, the mobile malware attacks, threat to user's security and privacy, the threat of malicious apps, etc. are also booming. As a result, concern towards the user's private and sensitive information has also gained attention. The incrementation in the number of permissions of Android has been incapable of solving the security issues, as it represents an increase in the Android attack surface (Neisse et al., 2016).

When a user installs an Android app, one of the main concerns is for the app to be safe regarding the user's privacy. Malicious behavior of a particular app can also be a mutual feature of another app.

Smartphone systems have different infrastructures and security specifications as opposed to conventional operating system platforms. New corresponding safety approaches are therefore also required (Tang et al., n.d.). By default, none of the Android apps possess the entitlement to carry out any operation that could affect the operating system (OS), user's information and other apps. Though, this could be misused with the help of Android permissions.

To develop a more secure app, it is the prime responsibility of the app developer to keep only a precise set of permissions which are necessarily required. All these permissions are needed to be mentioned in the manifest file of the app. As you can see in Figure 1

If the user wishes to install an app, then he needs to peculiarly grant for all the permissions which are mentioned in the manifest file of the application. Though, it has been observed during the research that most of the apps are behaving differently as defined in their manifest file. Along with being precise about the permissions, the developer also needs to keep in mind that, if necessary, permission is missed then it could lead to malfunction of the app.

Figure 1. A screenshot of manifest file (Developers)

android_manifest.xml

```
<?xml version="1.0" encoding="utf-8"?>
<manifest
xmlns:android="http://schemas.android.com/apk/res/android"
   package="com.example.myapp"
   android:versionCode="1"
   android:versionName="1.0" >
     <uses-feature android:name="android.hardware.sensor.compass"
             android:required="true" />
    <application >
      <activity android:name="com.example.myapp.MainActivity" >
      </activity>
       <activity android:name=" MainActivity" >
    </application>
    <uses-permission
android:name="android.permission.SEND_SMS"/>

</manifest>
```

In this paper, we have provided with a solution in the form of Protection API to solve the problem of the malicious attacks and the misuse of the Android app permissions by malicious applications. Keeping in mind the issues discussed above, this paper has been further divided into 7 sections. Introduction being Section 1. Section 2 is of background research. While Section 3 displays the Literature Review. The section 4 comprises of the permissions, section 5 of Analysis methodology. And Section 6 is of limitations.
Finally, the paper is concluded in Section 7.

BACKGROUND

Android's worldwide share is growing day by day all year long. These days many people move from using websites to apps. According to Mary Meeker's 2017 research, the number of hours spent online continues to increase each year, but the gap between the web and mobile is increasing (MindSea Team,). In 2016, Americans used to spend 3+ hours a day on cell phones (which is 10 times more than 2008) and about 2.2 hours a day on a desktop or laptop (has shown no change since 2008). The annual growth in the number of downloads of mobile apps has steadily increased. There were 178 million downloads of the app in 2017. This figure is expected to increase to 258 million by 2022, a 45 percent increase in five years (Statista, 2022). Also, as the number of Android apps is growing so much attention is being given to other considerations as well such as security and privacy concerns, malfunctioning of the applications etc. Past research on the privacy issues of mobile users have largely concentrated on location monitoring and sharing (Barkhuus & Dey, 2003; Consolvo et al., 2005). Although location sharing is an essential aspect of mobile privacy, only two out of 134 Android apps correctly use location permissions (Statista, n.d.). Just like the sharing of location protection, user safety is now a major issue and the likelihood of a hacking of Android apps is also increasing. To help users avoid privacy violations, several tools have been developed by Google for Android privacy. Most of the work focused on defining unsafe activities without addressing how people would make better safety choices (Barkhuus & Dey, 2003; Consolvo et al., 2005). Howell et al. (Felt et al., n.d.), however, recommended the development of a sensor control widget that would alert the user visually when a sensor such as the camera is working. And Roesner et al. (Enck et al., 2009a) suggested user-driven entry, which transforms consent into real user behavior rather than asking users to review alerts.

Mobile apps not only provide functionality such as end- user computers but also extend their experience by incorporating new hardware features (e.g. sensors) integrated into the mobile app without the need for external devices. Many times, applications access phone information such as phone numbers, photos, etc. without the Android phone user taking permission to view the same. This work is performed to identify Android applications that do not have the same methods as described in Google Play Store description and do not authorize users to access user content. In the current context, this issue is novel and that is why it draws the attention of the researcher. To solve this problem, many recent studies have already been done and tested. Some are discussed in the literature review section in this research paper.

The applications many times ask for the permissions which are not even required by the app for its functionality (Fuchs et al., 2009), such applications many times misuse Android permissions. Some of the Android permissions, misuse and use are discussed in the table 1.

Table 1. Use and misuse of some of the Android app permissions (Alenezi & Almomani, 2017; Enck et al., 2009b; Howell & Schechter, 2010; Roesner et al., n.d.)

Android Permission	USE	MISUSE
Bluetooth	Connecting to Bluetooth devices and also pairing a new Bluetooth device	Used by Malicious application for pairing with unknown device and sending data to it or blocking the Bluetooth connectivity
Access_Network_State	Allows application to view detailed information about available network's type, the roaming or local network, the number of unsuccessful attempts to connect and the type of linked devices of network	Malicious applications maintain a user profile with respect to details about his network.
Access Wi-Fi	This enables applications to access Wi-Fi network content.	Apps may be misused by using this knowledge to hack the Wi-Fi network and transfer user details.
Read Contacts	Allows to read and write contact list	Permits the malicious application to burgle the personal info of user
Phone Call	Permission to do phone calls and control outgoing calls	Make phone calls to anonymous number without user's knowledge
Internet	This helps apps so that they're able to access the internet through the network socket's opening for data transfer.	Through submitting user's private data to unknown URLs, this permission may be misused.
Send SMS, Receive SMS	Enables SMS-related tasks	Support the malicious program read, write and send the malware author's personal information to users
Read & Write External Storage	Reading and writing of external storage is allowed	Malware apps have the ability to read user's confidential data and its malicious code can be written on external storage
Camera	Allow to access camera	Can be used by malware apps to capture photos and record whenever they want.
Read Phone State	Provide access to essential phone info, such as IMEI / IMSI user ID, phone number, voice mail box, network provider, SIM ID, etc.	It gives the malware author the ability to keep track of your phone and can use this knowledge to include your computer in malicious activities.
Access Location	Permissions to view location-related Cell Device Information	Malicious Applications can do location-based sniffing
Read and Write history bookmarks	Permission to access read and write History Bookmarks files of Internet Browsing Activities	It can be used by apps to spy on the user browsing activity

LITERATURE REVIEW

By adding an additional layer to the existing Android framework to improve consumer privacy, Khatoon et al. (Su et al., 2019) have suggested a privacy framework for Android applications. The mechanism includes a privacy-based Android notifications scheme where a user can keep track of each app. If a client accesses a critical permission, the Android program must generate a warning at any time to inform the user that a certain file has been accessed by an application. In this way, a user can keep track of all the apps that need special permission from the Android framework to access confidential data.

Liu et al. (Micro,) have developed a user-friendly method named Droidtector, based on standard item set-mining, to identify privileged Android Applications permissions. Droidtector can be used online or offline, and users can choose which mode they prefer based on their condition and circumstance. The program accessed the Google Play website in an online module to scan for the App with the user input name. Otherwise, an APK file will be downloaded by the program and the type of the software will be registered. The authorization list of the Software can then be lowered by application, which allows the monitoring device to run and view the user's detection results as well as store the results in the database. Offline modules do not need an Internet link, until detecting, users should have an APK file and know its type. The actual process of detection is identical to that of an internet program, except the test system must first look up the device category in the database. If there is no device information in the database, the test system will work with the category identified by the customer. Otherwise, the until detecting, users should have an APK file and know its type. The actual process of detection is identical to that of an internet program, except the test system must first look up the device category in the database. If there is no device information in the database, the test system will work with the category identified by the customer. Otherwise, the recognition will be done with the category stored in the database.

Ngobeni et al. (Bendre, 2018) suggested a model that would provide context protection without preventing the customer from doing their normal tasks on their mobile device. The proposed security policy model provides the detection of ransomware and the unreadability of the data to malicious apps as well.

A system named PriGuard to dynamically secure consumer privacy information based on Binder contact interception technologies and feature selection algorithms was developed by Jia et al. (Khatoon & Corcoran, 2017). Using the Binder protocol, programs usually call system services remotely, then enter the app and collect information from system services. PriGuard detects Binder messages by redirecting the Native Layer Binder interface function, thereby intercepting the Remote Procedure

Call (RPC) of the system services program, which can then dynamically track application activities that access data privacy.

Bao et al. (Wu & Liu, 2019) suggested a framework based on the idea that programs with similar functionality typically share common permissions derived from the APIs they use. A shared filtering technique was developed to suggest permissions for a new device. On F-Droid 's 936 Android apps, which is a directory of free and open-source smartphone applications, the solutions suggested were also tested.

A technique called COVERT has been developed by Bagheri et al. (Ngobeni & Mhlongo, 2019) to analyze the compositional limitations in inter-apps for Android. To allow gradual review of applications as they are downloaded, updated and deleted, COVERT 's approach is dynamic. It analyses the reverseengineered source code of each application statically and extracts applicable security parameters in a format suitable for standardized tests.

A Crowrdoid framework that is behavior-based malware identification was introduced by Burguera et al. (Jia et al., 2015). It has two functions: a crowd networking application that can be built on user devices, and a remote malware identification server for others. Through uploading the behavioral data, the crowdsourcing software communicates with the remote server and this data can be a software log file. Easy configuration details, behavioral information, and a list of applications installed are provided in the software log file. This data is processed on the remote server to produce feature vectors that can later be analyzed by a 2-mean partition cluster to determine if the program is malicious or benign. A request report is generated and stored in the remote-server database.

Faruki et al. (Bao et al., 2016) have introduced an AndroSimilar approach that uses the collection of statistically unplausible features by means of the digest hashing function of resemblance. Obfuscated malicious code, control flow obfuscation, application renaming, and garbage method inclusion are all easily understood by the proposed solution. The signature produced by the proposed solution is powerful enough to recognize unknown samples obfuscated by multiple code obfuscation techniques that are not detected by conventional Antivirus software. Manually analyzing the suspected samples detected by the method in which existing Android malware signatures verify their resemblance. In combating repackaged applications, this suggested method of signing is successful.

Rashidi et al. (Bagheri et al., 2015) proposed the RecDroid approach is a system for resource accessing authorization control by crowdsourcing. RecDroid seeks to help users make the best option when choosing whether to refuse or grant a request when permission is asked. RecDroid is a method of approval that compiles the user's consent responses to the customer's consent demands, even to determine the extent of the user interface that this solution uses and to determine whether or not to reject or allow an acceptable response to the permission request. RecDroid focuses on a small community of seedexpert users who will produce accurate suggestions for a small number of applications. In addition, this approach uses a game-theoretic Bayesian to suppress malicious responses and track malicious clients.

Russello et al. (Burguera et al., 2011) propose FireDroid which is a policybased architecture will enforce security controls by interleaving process system calls. In this way, an application monitor is created to track all processes that are spawned in the Android operating system and can approve or deny them based on humanly controlled policies. The FireDroid will detect whether an app is performing unauthorized or potentially dangerous behavior at runtime by interrupting the device calls that the software performs. Regardless of whether the malware is repackaged or a different form of current malware: FireDroid will identify and enforce the necessary protection measures until the malware performs dangerous device calls. FireDroid's goal is to render it fully seamless for the applications together with Android.

Yan et al.(Faruki et al., 2013) introduced DroidScope, a Virtual Machine Introspection (VMI) for Android applications, is a dynamic analysis platform. In contrast to other dynamic analysis systems, it is not placed inside the emulator, but generates semics at the Dalvik and OS levels by residing beyond the emulator. Through this the privilege escalation attacks can be observed even in the kernel. It also hinders the task of chaotic study of the attackers. The DroidScope is based on a QEMU simulator, which likewise renders a collection of APIs accessible to human analysts to change the research required.

The RiskMon method that incorporates runtime habits and user expectations of approved users is suggested by Jing et al.(Rashidi et al., 2015) to establish a baseline risk assessment under which applications can collect appropriate behaviors. The expectations of the apps' users are the operating system's core component. The user expectations of the programs deployed are initially obtained and the placement of authorization sets is based on their relevance to the respective software. After that, the risk management standard for the apps, based on the collected data of the customer, is developed, or generated. Finally, the system rates apps deployed, based on the possibility of experiences with the platform, using the baseline generated.

Zhou et al.(Russello et al., 2013) presented a survey on some of the emerging uncontrolled Android app markets and on Google Play to forecast software security. The DroidRanger has a two-stage analysis to find current zero-day malware and recorded malware. To detect recognized malware, they use a behavioral foot printing scheme dependent on authorization. The other argument is to apply a filtering method based on heuristics to define the underlying behaviors of a zero-day malware.

PERMISSIONS

The Android operating system (OS) employs a permission-based paradigm to limit an application's action and to notify the user of the app's possible conduct. In its AndroidManifest.xml file, an application defines the permissions it seeks. When an application is installed, the user is provided with a list of rights that the program asks. The user has the option of installing or not installing the program depending on the list of rights it claims to require. Because installation is done on an all-or-nothing basis, the user cannot individually approve or reject particular permissions that the program requires. The permissions that an application possesses stay static after it is installed.

Permissions are used to govern how essential resources on the phone are used. On an Android phone, for example, the application must have the ACCESS COARSE LOCATION or ACCESS FINE LOCATION permission to access the GPS resource. The Android OS may issue a Security Exception or simply refuse to give the desired resource if an application asks the GPS resource without having the proper authorization. The Android API and other classes on the phone are used to access these permission-protected resources. Each Android API call has a mapping to the related permission(s), if any, that it needs to function properly. The ACCESS FINE LOCATION permission in an app's AndroidManifest.xml file grants access to a variety of Android API functions that use the GPS resource. Over-provisioning of permissions enhances an application's attack surface, giving an attacker greater powers if the app is hacked.

For Android 2.2.1, researchers at Berkeley (Yan & Yin, 2012) developed a comprehensive list of intents, content providers, and Android API calls that require permissions. They did not correctly map instances of back-end server calls that many of the public API classes interact with in order to get access to permissionprotected resources, despite their comprehensive methodology. Many of these back-end server objects can only be created if the application is running as a system process. The Android server package is responsible for all of the new permission-protected calls we discovered. We examined the Android source code for permission names and tested the calls that included permission names. There are another 146 Android API calls that require at least one permission, according to our research. These calls must be made either through Java reflection or by adding these classes to the Android.jar file used by the Eclipse Integrated Development Environment. We also discovered 217 additional Android API calls in the source code that do a permission check but require a permission only available to system processes. We integrated our mapping with Berkeley's to create a more comprehensive map. We created a software that analyses an application and utilizes mapping to assess if it has more rights than it needed and/or less permissions than it needs.

ANALYSIS METHODOLOGY

We downloaded 141,372 apps from the official Android Market between March 2011 and November 2011. We created a software that searches the Android Market for free apps and downloads them automatically. Because of its capacity to interface with the market, we used the Android-market-api (Jing et al., 2014). We updated our software to download the apps we were looking for and ran many instances of it at the same time. For searching, we utilized a dictionary file, downloaded recently submitted apps, downloaded by application category, and used all permutations of a bespoke character set with fixed length strings.

An APK file is a compressed file that contains all of the materials that go into making an Android app. Apktool (Zhou et al., 2012) is a free open-source tool that decomposes an APK file into its individual resources before disassembling the Android application's classes. dex file into the smali format. A unique class exists for each application. Apktool disassembles the dex file into a directory tree of smali files. The smali files include a human-readable representation of the Dalvik bytecode, including the instructions, registers, objects, and literal values used as inputs. Methods, constructors, instance variables, data types, and variable name assignment are all included in the smali files.

Permissions are used to protect sensitive activities on the phone. To verify if the set of permissions requested by an application is correct, we utilized the mapping of Android API calls to permissions. There are a lot of permission-protected method calls in classes that aren't part of the public Android API but are on the phone. Our software looks through all of the application's smali files to get a list of method calls. To get the associated permissions, each method call discovered in the application is compared to each method call in our list of permission-protected Android API calls.

To comply with the concept of least privilege (Felt et al., 2011) the software calculates the tightest permission set the application should have based on its functioning. The produced strictest permission set is then compared to the permission set specified in the AndroidManifest.xml file for the application. Based on its functionality, the software evaluates if the application has additional rights, lacks permissions, or has exactly the permission set that it requires. Our software identifies the usage of Java reflection and dynamic class loading, in addition to permissions analysis. These characteristics can aid in the obfuscation of the program's behavior (android-market-api,).

ANALYSIS RESULTS

As shown in Table 2, the results are based on running our static analysis software on 141,372 Android apps. If an application has at least one permission that it does not require, we consider it to have excessive permissions. Lacks permission(s) is a category for apps that should have one or more permissions but do not. Using this categorization, an application can have both additional permissions and no permissions at the same time (s). Exact permission(s) are for apps that require a certain set of permissions.

Table 2. Absolute numbers and proportions for characteristics and behaviors of Android applications

	Number of applications	Percent
Extra Permissions	76,366	54.01%
Lacks Permissions	70,618	49.95%
Exact Permissions	28,199	19.95%
Reflective method invocation	67,958	48.07%
Class Loader Usage	15,580	11.02%
Total	141,372	100%

For java language apps, reflective method invocation is used. Method invoked by reflect. Method.invoke(). Class Loader is used in java language-based applications. The Android API uses the Class Loader class or one of the other five classes for dynamic class loading.

LIMITATIONS

The static analysis code mapping can help software developers figure out what permissions are necessary for each Android API call, if any are required at all. Our tool does not check for reachability in the code to see if any of the permission-protected API calls are really accessible. As a result, when inaccessible code is included in an application, it may cause false positives in terms of permissions required. The static code analysis excludes dynamic and foreign bytecode, such as Java Native Interface calls, which might result in false negatives in terms of needed permissions.

CONCLUSION

The widespread adoption of Android may be seen in our daily lives. Every day, users utilize a wide range of Android apps. During the installation of these applications on each device, various permissions are requested. These permissions are mostly used to get access to data that is necessary for the app's operational functioning. These permits, on the other hand, might be unclear. Each Android app requests a set of needed permissions, which if denied, cause the programme to fail to install or operate. Some of these needed permissions are superfluous, posing a risk to the user's data privacy and security. App developers must pay close attention to the permissions they request in order for the app to work properly. They must ensure that the user is only requested for permissions that are relevant and necessary. The privacy and security of user data should be the most important consideration when creating a set of permissions. We downloaded over 141,000 apps and analyzed a significant section of the Android market. Despite the fact that the Android Market is continuously introducing new apps, we have produced a comprehensive overview of the Android Market that highlights the predominance of specific app features and behaviors. Hundreds of new permission-protected API calls were also discovered, enhancing our ability to properly assess the right permission configuration for an application. Our findings reveal that the majority of developers do not utilize the tightest permission set based on their code functionality, but instead over-specify their permission needs, increasing the attack surface and increasing the security risks associated with an application breach.

REFERENCES

Alenezi, M., & Almomani, I. (2017). Abusing Android permissions: A security perspective. *2017 IEEE Jordan Conference on Applied Electrical Engineering and Computing Technologies (AEECT)*, 16. 10.1109/AEECT.2017.8257772

android-market-api. (n.d.). *Android Market for all Developers!* http://code.google.com/p/android-market-api/

Bagheri, H., Sadeghi, A., Garcia, J., & Malek, S. (2015, September 1). COVERT: Compositional Analysis of Android Inter-App Permission Leakage. *IEEE Transactions on Software Engineering*, *41*(9), 866–886. doi:10.1109/TSE.2015.2419611

Bao, L., Lo, D., Xia, X., & Li, S. (2016). What Permissions Should This Android App Request? *2016 International Conference on Software Analysis, Testing and Evolution (SATE)*, 36-41. 10.1109/SATE.2016.13

Barkhuus, L., & Dey, A. (2003). Location-based services for mobile telephony: a study of users' privacy concerns. *Proceedings of the International Conference on Human-Computer Interaction.*

Bendre, R. (2018). Ignoring app permissions could hurt your privacy more than you think. *BGR.in*. Retrieved from https://www.bgr.in/features/ignoring-app-permissions-could-hurt-your- privacy-more-than-you-think-550748/

Burguera, I., Zurutuza, U., & Nadjm-Tehrani, S. (2011). Crowdroid: Behavior-Based Malware Detection System for Android. *SPSM*, *11*, 15–26. doi:10.1145/2046614.2046619

Consolvo, S., Smith, I. E., Matthews, T., LaMarca, A., Tabert, J., & Powledge, P. (2005). Location disclosure to social relations: why, when, & what people want to share. *Proceedings of the ACM CHI Conference on Human Factors in Computing Systems*. 10.1145/1054972.1054985

Developers, A. (n.d.). *App Manifest Overview*. Retrieved from https://developer.android.com/guide/topics/manifest/manifest-intro

Enck, W., Ongtang, M., & McDaniel, P. (2009a). On lightweight mobile phone application certification. *Proceedings of the ACM Conference on Computer and Communication Security (CCS).*

Enck, W., Ongtang, M., & McDaniel, P. (2009b). Understanding Android Security. *IEEE Security & Privacy Magazine*, *7*(1), 50–57. doi:10.1109/MSP.2009.26

European Commission. (n.d.). *JRC Publications Repository*. Retrieved from https://ec.europa.eu/jrc/en/publication/dealing-user-privacy-mobile-apps-issues-and-mitigation

Faruki, P., Ganmoor, V., Laxmi, V., Gaur, M., & Bharmal, A. (2013). AndroSimilar: robust statistical feature signature for Android malware detection. *SIN 2013 - Proceedings of the 6th International Conference on Security of Information and Networks*. 10.1145/2523514.2523539

Felt, Ha, Egelman, Haney, Chin, & Wagner. (n.d.). *Android Permissions: User Attention, Comprehension, and Behavior*. Academic Press.

Felt, A. P., Chin, E., Hanna, S., Song, D., & Wagner, D. (2011). Android permissions demystified. *Proceedings of the 18th ACM conference on Computer and Communications Security, CCS 11*, 627-638.

Fuchs, A., Chaudhuri, A., & Foster, J. (2009). *SCanDroid: Automated Security Certification of Android Applications. Technical report*. University of Maryland.

Howell, J., & Schechter, S. (2010). What you see is what they get. *Proceedings of the IEEE Workshop on Web 2.0 Security and Privacy (W2SP).*

IDC. (2021). *Smartphone Market Share*. Retrieved from https://www.idc.com/promo/smartphone-market-share/os

Iqbal, M. (2022). *App Download Data (2022).* Business of Apps. Retrieved from https://www.businessofapps.com/data/app-statistics/#:~:text=As%20of%20the%20first%20quarter,from%20178%20billion%20in%202017

Jia, P., He, X., Liu, L., Gu, B., & Fang, Y. (2015). A framework for privacy information protection on Android. *2015 International Conference on Computing, Networking and Communications (ICNC)*, 1127-1131. 10.1109/ICCNC.2015.7069508

Jing, Y., Ahn, G.-J., Zhao, Z., & Hu, H. (2014). RiskMon: Continuous and automated risk assessment of mobile applications. *CODASPY 2014 - Proceedings of the 4th ACM Conference on Data and Application Security and Privacy*, 99-110. 10.1145/2557547.2557549

Khatoon, A., & Corcoran, P. (2017). Android permission system and user privacy — A review of concept and approaches. *2017 IEEE 7th International Conference on Consumer Electronics - Berlin (ICCEBerlin),* 153-158. 10.1109/ICCE-Berlin.2017.8210616

Micro, T. (n.d.). *Threat Encyclopedia*. Retrieved from http://about-threats.trendmicro.com/us/library/image-gallery/12-most-abused-android-app-permissions

MindSea Team. (n.d.). *25 Mobile App Statistics To Know In 2022*. MindSea. Retrieved from https://mindsea.com/app-stats/

Neisse, R., Steri, G., Geneiatakis, D., & Fovino, I. N. (2016). *A privacy enforcing framework for Android applications* (Vol. 62). Computers and Security. https://www.sciencedirect.com/science/article/pii/S0167404816300840

Ngobeni, A., & Mhlongo, S. (2019). Towards Enhancing Security in Android Operating Systems – Android Permissions & User Unawareness. *2019 2nd International Conference on Computer Applications & Information Security (ICCAIS),* 1-6. 10.1109/CAIS.2019.8769531

Rashidi, B., Fung, C., & Vu, T. (2015). *RecDroid- An Android Application Security Framework through Crowdsourcing*. Academic Press.

Roesner, Kohno, Moshchuk, Parno, Wang, & Cowan. (n.d.). *User-Driven Access Control: Rethinking Permission Granting in Modern Operating Systems*. Academic Press.

Russello, G., Jimenez, A., Naderi, H., & van der Mark, W. (2013). FireDroid: hardening security in almost-stock Android. *ACM International Conference Proceeding Series*, 319-328. 10.1145/2523649.2523678

Statista. (2022). *Number of mobile app downloads worldwide from 2016 to 2021.* Retrieved from https://www.statista.com/statistics/271644/worldwide-free-and-paid- mobile-app-store-downloads/

Statista. (n.d.). https://www.statista.com/statistics/271644/worldwide-free-and-paid-mobile-app-store-downloads/

Su, M., Chen, S., Wu, T., Chang, H., & Liu, Y. (2019). Permission Abusing by Ad Libraries of Smartphone Apps. *2019 Eleventh International Conference on Ubiquitous and Future Networks (ICUFN)*, 475-477. 10.1109/ICUFN.2019.8806157

Tang, Jin, He, & Jiang. (n.d.). *Extending Android Security Enforcement with A Security Distance Model.* Academic Press.

Wu, S., & Liu, J. (2019). Overprivileged Permission Detection for Android Applications. *ICC 2019 - 2019 IEEE International Conference on Communications (ICC),* 1-6. 10.1109/ICC.2019.8761572

Yan, L., & Yin, H. (2012). DroidScope: Seamlessly Reconstructing the OS and Dalvik Semantic Views for Dynamic Android Malware Analysis. *USENIX Security Symposium.*

Zhou, Y., Wang, Z., Zhou, W., & Jiang, X. (2012). Hey, You, Get Off of My Market: Detecting Malicious Apps in Official and Alternative Android Markets. *Proceedings of the 19th Network and Distributed System Security Symposium NDSS 2012.*

Chapter 7
An Approach to Identifying Pre-Existing Chronic Conditions for Corona Virus Positive Patients Using Regression

Anshu S. Agarwal
AURO University, India

Alex V. Patel
Yosha Techologies, India

Sailesh Iyer
https://orcid.org/0000-0002-0619-9829
Rai University, India

Akash Saxena
Rai University, India

ABSTRACT

COVID-19 has various symptoms, and they are cold, cough, mild or high fever, and breathing problems for severe cases. In addition, it is in talk that diabetes, chronic obstructive pulmonary disease (COPD), cardiovascular, asthma, and many diseases invite the virus. Machine learning and data analysis are considered best for predicting approaches for finding various aspects of the effect of the virus. In this chapter, the authors deal with symptoms that have been recorded in the dataset and try to find which pre-defined symptoms are considered effective or responsible in positive case of infection by coronavirus. The dataset has been taken from Kaggle. As the dataset is categorical in nature, the authors use correlation and logistic regression analysis to find the symptoms that prevail in the patient and have caused the infection in them. This is also about dimensionality reduction and feature selection where they are reducing the available features based on regression.

DOI: 10.4018/978-1-6684-4225-8.ch007

INTRODUCTION

COVID- 19 was first detected in Wuhan, China in the month of December 2019, (Salehi, Baglat, & Gupta, 2020) from where it spread to all across the world. WHO declared it as a pandemic in March 2020 There are many symptoms, viral origin, transmission, identification and diagnosis for potential treatment (Valencia, 2020). Until now, we have not been able to find a better and confirmed cure for the corona virus treatment. (Muhammad, Islam, Sharif, & Ayon, 2020) Although every country has taken lots of measures and educated their citizens to take of them and their family including lock down in various stages. Stay home, work from home, no outings have become new normal now along with use of mask to cover nose and mouth and sanitize/wash your hand regularly. This lockdown has changed the pattern of season, effected country as a nation in terms of economy crisis, effected individuals in terms of their expenses and income imbalance resulting into frustration, depressing life. However, its not only negative impact has positive impact also. People have spent quality time with their family, rejuvenated their hobbies etc. (Chatterjee, Sujath, & Hassanien, 2020). The symptoms data have been taken from Kaggle website to anticipate the symptoms responsible for positive results. Some of the fields taken into considerations can be pneumonia, asthma, hypertension, cardiovascular, tobacco, diabetes etc. Moreover, it is important to study the effect of such diseases on patients whether covid positive or not. This will also help us to understand the impact of various diseases on the patients with respect to covid.

Artificial Intelligence (AI) is considered to be of great help in predicting the impact of all such symptoms/diseases in the patient.in computer edge world today Machine Learning is a solution to almost all crucial problem where we need to work with real time situation. AI can help us to predict and analyse covid situation in a better way without much working on the field. Once ML/AI had been used to predict the result story telling or data analysis can help us better understand the data and the relation of every filed with each other. (Prakash, Imambi, Mohammed Ismail, Kumar, & Pavan, 2020) Machine Learning and Data Analysis helps us to understand the impact of corona virus on people-based age, gender, pre medical history etc. Based on these results we can predict the mortality rate/ risk (Pourhomayoun & Shakibu, 2020), recommend precautions, avoid coming in contact with infected person etc. it can help to predict epidemic spread situation using mathematical models. Along with Deep Learning, ML is used to exponential behaviour and predicting future reachability of COVID-19. Polynomial Regression, Root Mean Square Error are taken into consideration for evaluating losses due to the virus. (Singh, Sonbhadra, & Agarwal, 2020) (Salehi, Baglat, & Gupta, 2020)

Data Mining Techniques, Machine Learning and AI Techniques and various algorithms such as, Linear Regression, Logistic Regression, Support Vector Machine,

Decision Tree, Random Forest, Naïve Bayes, KNN etc. are becoming helpful in predicting lots of possibilities to manage spread and diagnosis of corona virus. (Muhammad, Islam, Sharif, & Ayon, 2020) Machine Learning Model performance are evaluated using various metrics like: Kappa, Accuracy, F1 – Score, Precision, Recall, ROC, AOC, Principal Component Analysis for feature extraction, Chi Square test to check overall significance etc. (Gawriljuk, et al., 2020)

Here we have taken database named "covid.csv" from Kaggle that has various variables as id, sex, patient_type, entry_date, intubed, age, pregnancy, diabetes, tobacco etc. These variable/columns have values as 1- yes, 2- No and 97-99 as no data available. We have pre-processed the data and eliminated many variables not relevant for the research and the rows found irrelevant had been deleted/omitted.

RELATED WORK

Authors have used Linear Regression to show the relation between dependent and independent variable. A Multi-layer Perceptron is implemented then to indicate any Feedforward Artificial Neural Network. The boxplot and sieve diagram is used to plot the confirmed, recovered and death cases in India. VAR model is used to find result of multivariate time series from the data (Chatterjee, Sujath, & Hassanien, 2020).

Here author Aims at developing data mining techniques for predicting covid recovery considering data set of south Korea obtained from Kaggle. KNN, SVM, Decision Tree, Logistic Regression, Naïve Bayes and RF has been considered for developing the model. The model is created to predict the number of days taken or required by patient to recover from virus infection based on age groups with low, medium and high risk. and after model comparison its found that Random Forest generates the best result for predicting recovery of patients. the fields considered for model building are Sex, age, Infection case, No_day, State (Muhammad, Islam, Usman, & Ayon, 2020)

In the time of Coronavirus infection spread during 2019-2020 there were numerous amounts of people affected by it. India was having 10,000 people effected and 300 people were dead. (Kumar, Harshavardhan, Bhukya, & Prasad) The Generative Adversarial Network is an advanced neural network approach that is used for analysing images and multimedia data. It is used for analysing the diagnosis and prediction of covid-19 patients from the available dataset. It uses advanced Deep Learning over CT scan, X-Ray etc. Using GAN, the result generated can be better as compared to classical NN. GAN uses two different neural NN where one generates artificial image and other compares actual with generated. K- Means Clustering Technique is a way through which we can generate non-overlapping clusters. Two variables and two background influential characteristics are used to get an insight of covid-19 positive patients with their fatality rate and average of population tested in the country where there was high expenditure over public health. And its found that the death rates I clusters varied and the reason was non availability of testing kits, un reported deaths etc. the result was generated using SPSS. (Chandu)

Machine Learning is used to analyse the impact of age on covid positive results. Random Forest Classifier and Regressor, SVM, KNN, Decision Tree, Multilinear Regression, Naïve Bayes and many other ML Techniques were used and result is generated to see which age group is affected with the virus most. In addition, the result shows that Random Forest Classifier and Regressor outperformed all considered techniques. In addition, it is found that age group of 20-50 years were more prone to get infected. The feature selection from the Kaggle dataset that are considered are: State, Cured, Date, Month, Confirmed National/ International and Deaths. The results are evaluated on the basis Coefficient of Determination and Accuracy. (Prakash, Imambi, Mohammed Ismail, Kumar, & Pavan, 2020) Logistic Regression is used to work with 16 factors like age, sex, symptoms etc to present a tool for identifying COVID-19 cases so that quick decision can be made by health personnel and managers (Fleitas, et al., 2020). Some authors have also used SPSS for analysing the data of COVID where they have used Confidence Interval and ODD Ratio along with p-value to justify the result generated. Post COVID symptom are analysed to find the prolonged infection after the infection where it is found that Female persons had high risk factors for long post COVID symptoms (Sultana Sarmir, et al., 2021)

Clustering is another way through which we can find the pattern of spread of infection in areas. (Azarafzaa, Azaraf, & Akgun, 2021)the data considered is from Iran where there were 7300 deaths in Iran. Data Mining and Clustering can be used for pattern recognition at Province Level. And its found that Tehran city is the majorly responsible city to spread the infection.

AI is used for predicting health risk, mortality risk for covid-19 patients. It helps to understand the severity of infection in the patient, based on which hospitals and health department can decide which patient needs immediate attention, hospitalization facility so that proper care and medication can be provided on time. Some such AI Models are SVM, NN, Logistic Regression, KNN, Decision Tree, Random Forest etc. and it's found that Mortality Rate can be predicted with 93% Accuracy. (Pourhomayoun & Shakibu, 2020) Features responsible for infection is also identified in the same study. Confusion Matrix is considered for evaluating the model. Other statistical analysis also helps us to detect the early onset or pandemics like COVID-19, Ebola Virus etc. analysing such data can help the government and other authorities to put a check on its spread before its severe. (Sun, Chen, & Viboud, 2020)

A system is built to detect corona virus. ML again plays an important role in performing the result. Clinical data is considered to generate database of symptoms, diagnosis etc. to perform feature engineering. (Khanday, Rabani, Khan, Rouf, & Mohi Ud Din, 2020)Comparative Analysis is done over the result generated by various techniques such as Logistic Regression, SVM, Random Forest, Bagging, Decision Tree, Naïve Bayes, Scholastic Gradient Boosting etc. and the result shows that Logistic Regression and Multinomial Naïve Bayes performs best with consideration of feature selection as Recall, Precision, Accuracy and F1 Score.

Random Forest plays a crucial role in visualizing the covid data and predict lots of information from it. Fine-tuned model of Random Forest using AdaBoost algorithm can be used to predict severity of disease. Model generates 94% Accuracy with 0.86 F1 Score. (Iwendi, et al., 2020)The research shows that there is a relation between age and death of the patients (Davies, et al., 2020) (Balabdaoui & Mohr, 2020) and mostly patients are between 20-70 years. Various symptoms, travel history, age, gender, country etc. are considered for the study. Evaluation Matrix that are used for finalising the outcomes are Accuracy, Precision, Recall, F1 Score. In addition, it is found that Male patients had more mortality rate.

Convolutional Neural Network is best fit for extracting high resolution features from images like CT scan, X Ray etc. hence adding to clinical researches a lot. Further Deep Learning and AI is used to develop automatic detecting method for corona virus outbreak so that needed precautions and measures be taken on right time. (Salehi, Baglat, & Gupta, 2020) (Somasekar, Kumar, Sharma, & Ramesh, 2020)

DATASET

The dataset is taken from Kaggle and can be found on the following link (https://www.kaggle.com/datasets/tanmoyx/covid19-patient-precondition-dataset?select=covid.csv). Its data fetched from data released by Mexican Government and had been translated from Spanish to English. The dataset has 23 columns and 563201 unique values/rows. The columns present are: id, sex, patient_type, date_died, intubes, pneumonia, pregnancy, copd etc.

Pre-processing: it is an important step in Machine Learning to filter the noisy data or unwanted data. The dataset contains missing or NA having value as 97,98,99 which have been removed from the dataset using "is.na" and "na.omit" function in R Studio. After removing the missing values 51778 rows are available for analysis and prediction. Further co relation and multiple correlation is applied to check the validity of data. The data set is divided into training and test data having 41422 and 10356 rows respectively. The dataset taken is a feature vector which is a symptom being considered to evaluate pre chronic symptoms.

Also, multicollinearity should not exist in logistics regression hence we have checked for multicollinearity of the model as shown below. We see that all the values below 5 so it can be confirmed that multicollinearity is not an issue in our model.

Box 1. vif(model)

obesity	pneumonia	diabetes	hypertension	intubed	inmsupr	contact_other_covid
1.030060	1.061001	1.152299	1.160155	1.040113	1.001727	1.011190

RESULTS AND DISCUSSIONS

Basic step in any Machine Learning method is data filtration and finding out the relation and dependency of data over each other. For this correlation is considered to be of great help where we try to find the dependency of multiple variables at the same time. We have also considered stepwise correlation step to identify the one-to-one dependency of predicted variable with a dependent variable and it is found that pneumonia, diabetes, hypertension, obesity, intubed, inmsurp and contact with others are found correlated. From the correlation, analysis we found that sex, copd, asthama, cardiovascular, renal_chronic, other and tobacco has negative correlation and hence we eliminate these variables. Below table shows the correlation matrix for our dataset so that we can filter out the unwanted variables.

Figure 1. Correlation coefficient for all parameters

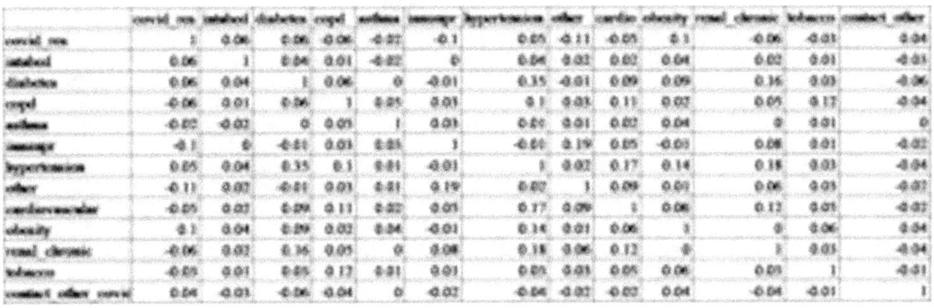

After one-to-one correlation, multiple correlation is applied to check the result again and confirm the dependent variables. The result generated is shown below in the form of Table 2. Multiple Correlation is used to show the relation of one variable with other two or more variables. Here one predicted variable is taken as covid_res, which specifies where covid test result is positive or negative. We have filtered result labelled as awaiting, as it was of no relevant.

Figure 2. Multiple correlation coefficient for all parameters

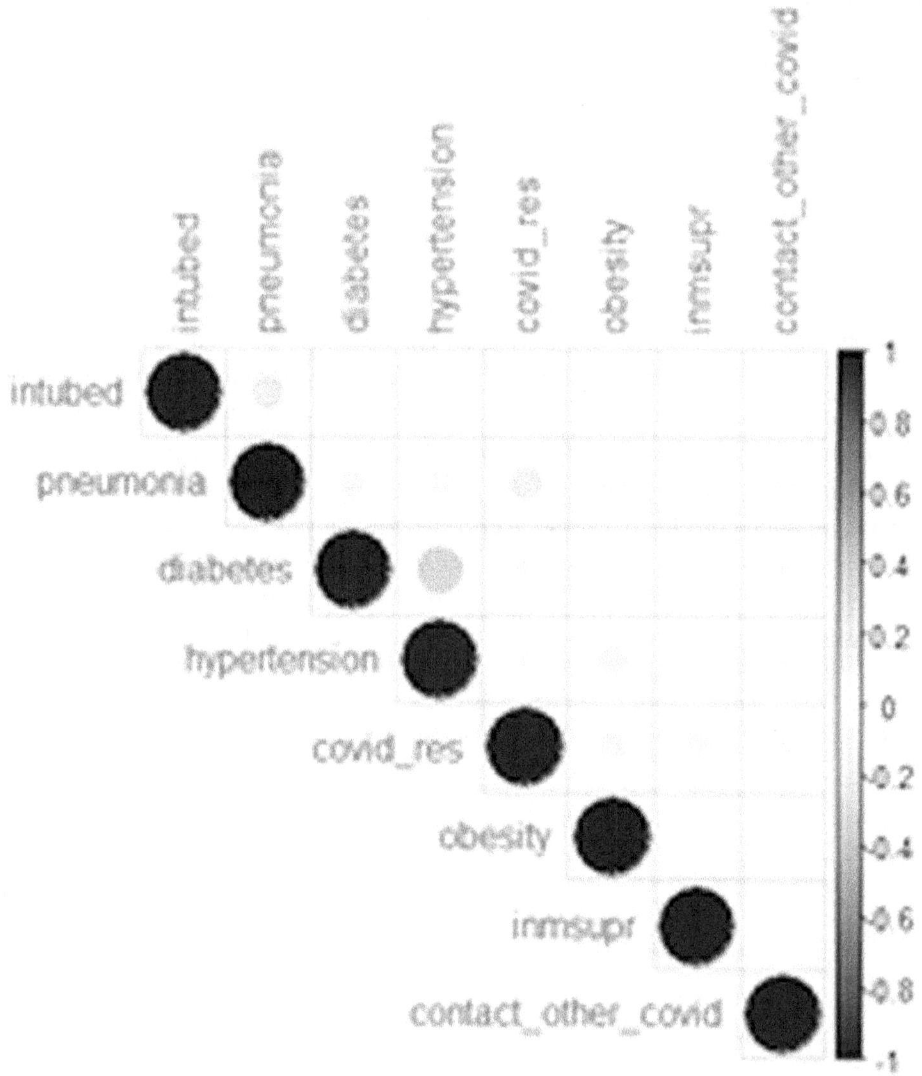

From the multiple correlation coefficient analysis, it is found that many variables are not strongly related to each other but they are all related to covid_result with varying values of correlation between covid results and its relation with other diseases that a patient had before being tested positive for corona virus infection. Multiple linear regression had been applied for the same to predict the values of independent variable. Then calculate the coefficient of correlation between the predicted and observed values of the dependent variable.

Figure 3. Correlation matrix based on significance of correlations, P-values

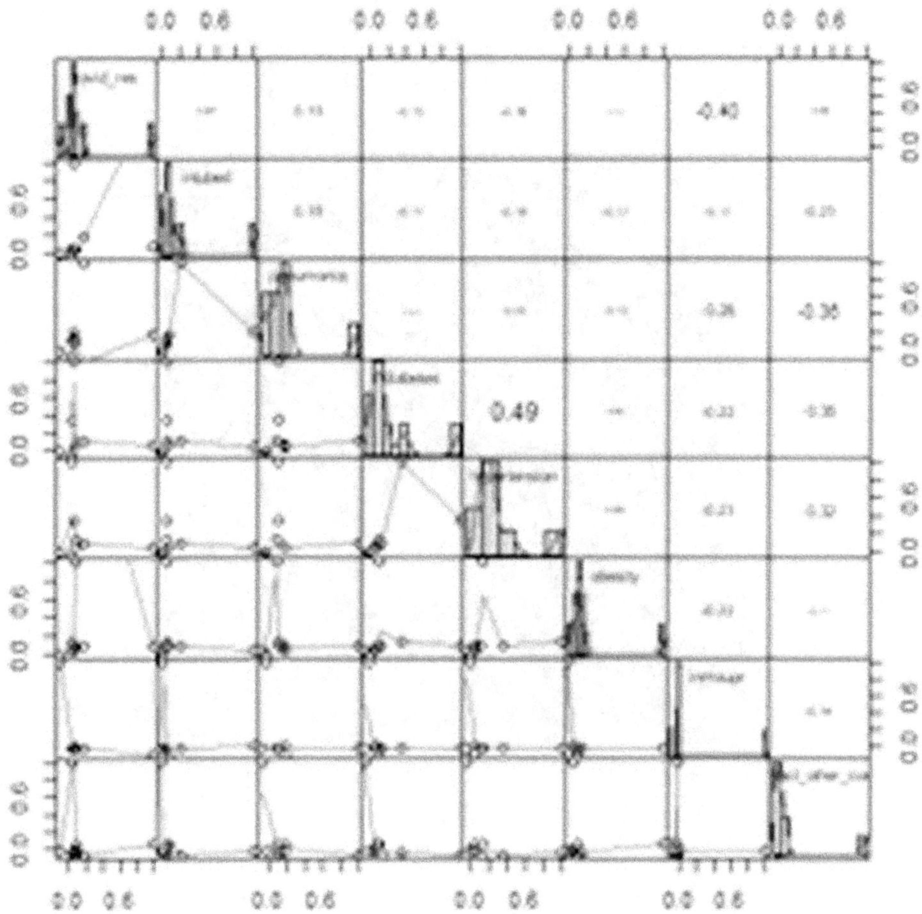

Correlation matrix is studied well based on p-value and correlation to confirm our chosen diseases and their relation for being corona positive. The result matches what we have selected earlier in the paper.

Correlation Plot is used to graphically present the correlation matrix, as it is important part of data analytics. In the plot blue dots and its intensity shows the strength of correlation whereas pink/red color dots the negative correlation details.

Looking at the graph again we can say that sex, asthama, inmsupr etc are not correlated to covid_res. This plot directly helps us to analyse the disease responsible for corona detection.

Figure 4. Correlation plot of all diseases

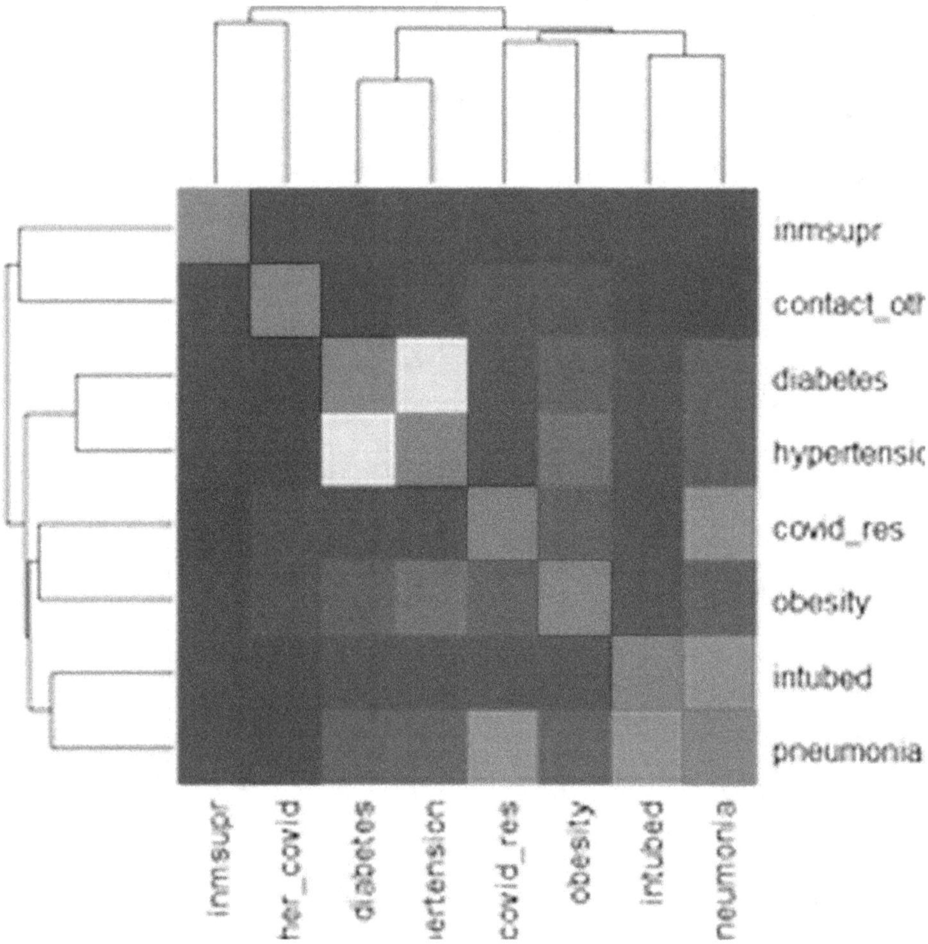

Figure 5. Correlation plot with insignificant value eliminated

```
Confusion Matrix and Statistics

          Reference
Prediction     1     2
         1 26130 15293
         2     0     0

               Accuracy : 0.6308
                 95% CI : (0.6261, 0.6355)
    No Information Rate : 0.6308
    P-Value [Acc > NIR] : 0.5022

                  Kappa : 0

 Mcnemar's Test P-Value : <2e-16

            Sensitivity : 1.0000
            Specificity : 0.0000
         Pos Pred Value : 0.6308
         Neg Pred Value :    NaN
             Prevalence : 0.6308
         Detection Rate : 0.6308
   Detection Prevalence : 1.0000
      Balanced Accuracy : 0.5000

       'Positive' Class : 1
```

Next step in data visualization and analysis if covid result is to find the insignificant diseases that are not very relevant for our prediction and analysis. The graph plotted shows very clearly that insignificant data are left blank. The plot shows the combination of correlogram and significance test. Here we have considered to plot values of correlation that is greater than 0.01 and are eliminated with a space.

Scatter Plot graph in our case is used to plot the chart of correlation matrix. The plot below many useful information-like diagonals is used to visualize the distribution of every variable. It also helps to understand the bivariate scatter plots with fitted line along with correlation and significance level with the help of stars as seen in upper part of the plot. To understand significance better we consider the p-values (0-> "***",0.00-> "**", 0.05-> "*", 0.1 -> "." And 1-> "".

Figure 6. Scatter plot of all variables considered

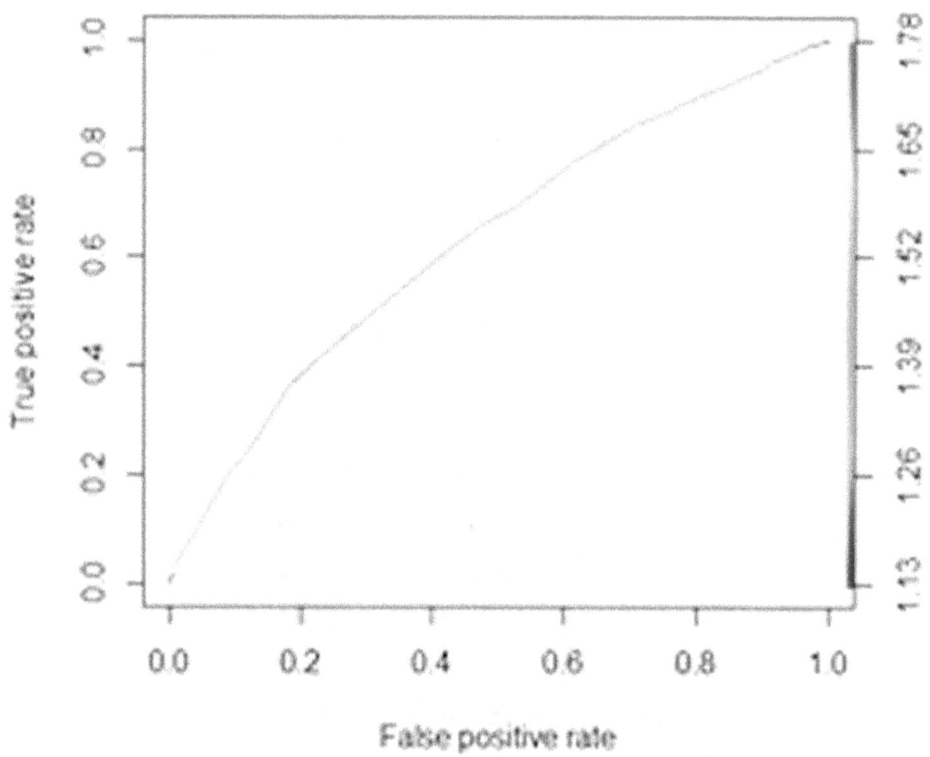

Heat map is considered as another important visualization to show the distribution and correlation of variables with each other. Lighter shade depicts the low level of co relation whereas darker shade means high level of correlation.

Figure 7. Heat map of all variables.

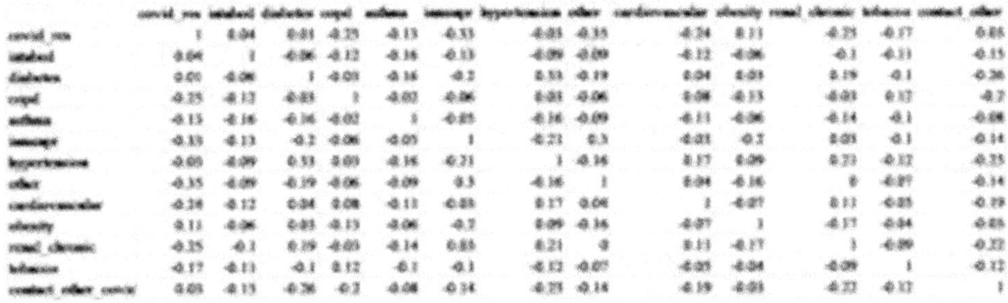

From the analysis of co- relation considering correlation coefficient, significant and insignificant variables, visualization using scatter plot and heat map, we found that the following disease are more likely to cause covid infection in the patient -Pneumonia, Diabetes, hypertension, obesity, intubed, inmsupr and contact with infected person.

Next step after data filtering and confirming the pre diseases in the patient we move toward analysing the diseases that are causing the infection or are responsible to attract corona virus and make a patient ill.

The table below shows the correlation between covid_res and other variables that have been retrieved after filtering or pre-processing and analysing the data.

Figure 8. P values details of 8 variables.

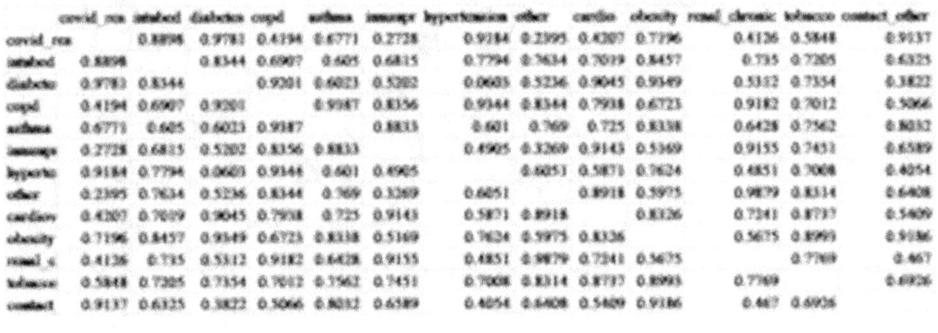

This data analysis not only helps to understand the covid situation but also the relation of one disease with other like hypertension and are related, diabetes and hypertension are related and so on. The below plot is showing the correlation and p values matrix for selected 7 diseases that can cause covid. It clearly shows that covid_res is affected by hypertension, intubed, pneumonia, diabetes, obesity and contact with others whereas inmsurp is not so likely to cause the new infection. Also, we found that pneumonia is most likely to cause covid followed by other diseases.

The plot below shows the insignificant data with blank space and we can find that inmsupr has negative relation with covid_res and other diseases. So, it is the most irrelevant disease pre condition in the patient to cause any other disease along with covid.

Figure 9. Correlation plot of related variables

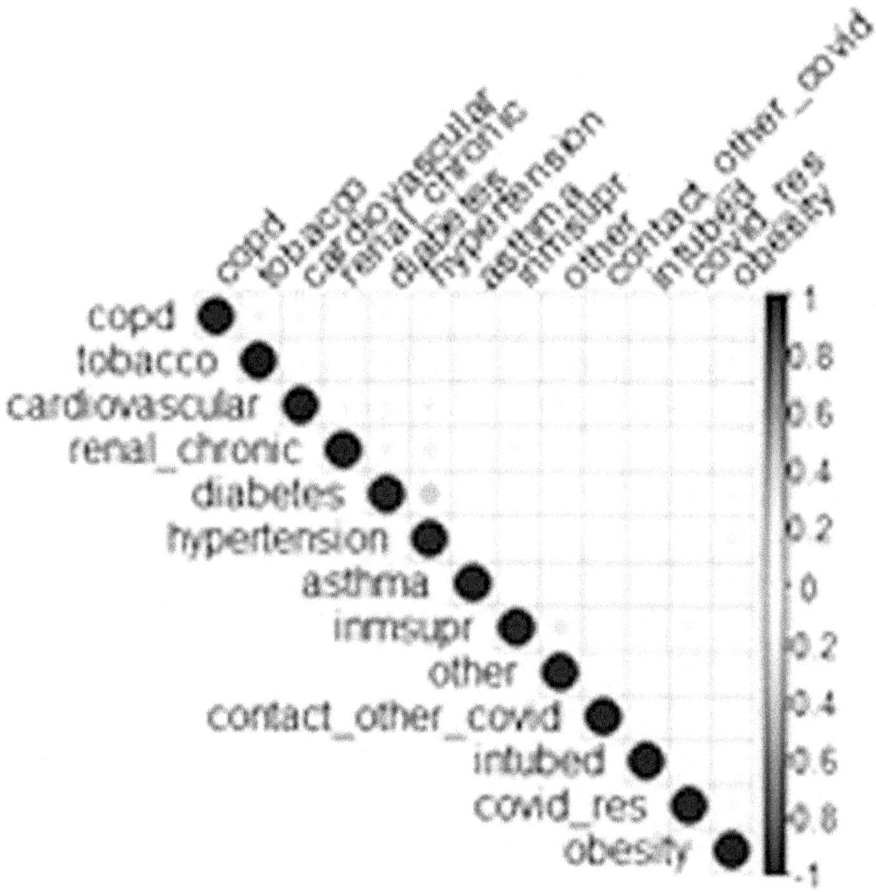

Figure 10. Insignificant plot of related variables

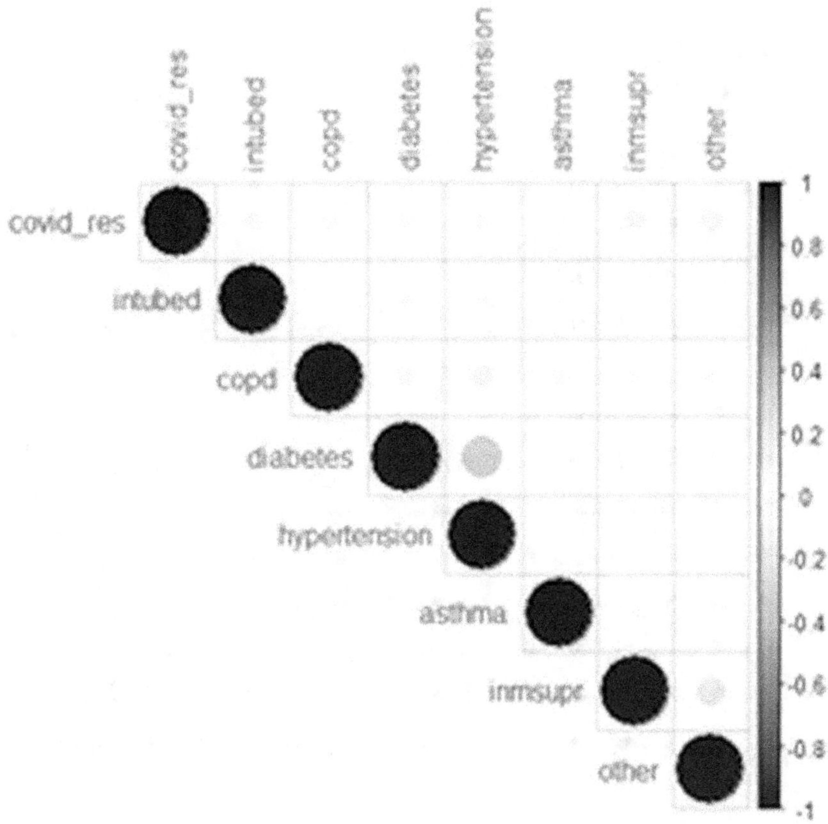

Figure 11. Scatter plot matrix of related 8 variables

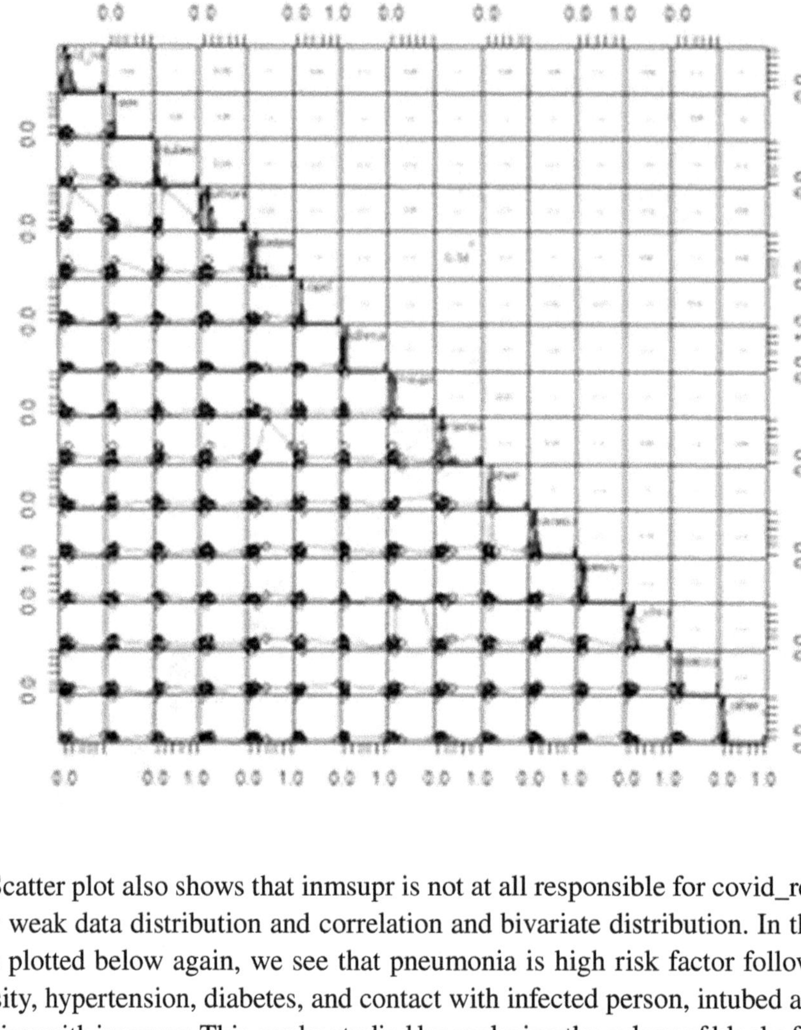

Scatter plot also shows that inmsupr is not at all responsible for covid_res with very weak data distribution and correlation and bivariate distribution. In the heat map plotted below again, we see that pneumonia is high risk factor followed by obesity, hypertension, diabetes, and contact with infected person, intubed and low relation with inmsupr. This can be studied by analysing the colour of blocks formed. Darkest block has less significance and light have most significance.

Figure 12. Heatmap showing related variables

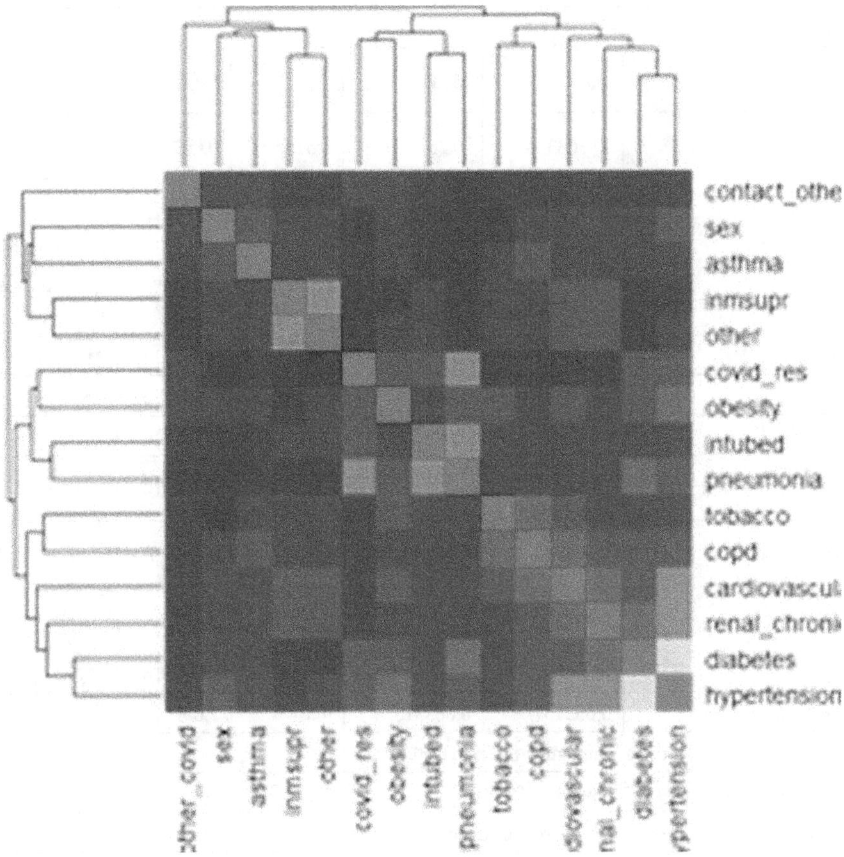

In the next section of the paper after analysing the diseases and their relation now we form a model for predicting and checking the accuracy of model. Logistic Regression is used when we have values in form of 0, 1 or true or false or yes/no (Muhammad, Islam, Usman, & Ayon, 2020).

Logistic Regression is given by following mathematical formula:

$$\text{Logistic regression } (p) = \ln\left((p)/(1 - p)\right) \qquad (1)$$

The dependent values show as 1 for Yes and 0 for No. Here Logistic Regression is used to form a model and model fit is tested based on ROC and its performance, confusion matrix, Area under Curve and accuracy factor of the model. Logistic Regression is used to estimate the relationship of the variables. We understand the change in value dependent variable because of one or more independent variables.

Table 1. Summary of logistic regression

Deviance Residuals:					
Min	1Q	Median	3Q	Max	
-0.7722	-0.3633	-0.2675	0.5155	0.867	
Coefficients:					
	Estimate	Std. Error	t Value	Pr(>|t|)	Signif. codes: 0 '***' 0.001 '**' 0.01 '*' 0.05 '.' 0.1 ' ' 1
(Intercept)	1.200527	0.033599	35.731	2.00E-16	***
Obesity	0.095883	0.005853	16.382	2.00E-16	***
Pneumonia	0.17331	0.005386	32.176	2.00E-16	***
Diabetes	0.036621	0.005704	6.42	1.38E-10	***
Hypertension	0.007734	0.005622	1.376	0.169	
Intubed	0.037918	0.007103	5.338	9.44E-08	***
Inmsupr	-0.23556	0.012946	-18.196	2.00E-16	***
contact_other_covid	0.052157	0.005315	9.814	2.00E-16	***
(Dispersion parameter for gaussian family taken to be 0.2207208)					
Null deviance: 9647 on 41422 degrees of freedom					
Residual deviance: 9128 on 41415 degrees of freedom					
AIC: 54920					
Number of Fisher Scoring iterations: 2					

Understating the output is equally important as generating it. Deviance Residuals is the measure of a model fit. Next section is coefficients that gives values for standard error, t values and associated p-value. The logistic regression coefficients give the change in the log odds of the outcome for a one-unit increase in the predictor variable. For every one-unit change in obesity, the result of being corona positive is increased by 0.095 and so for all other values. Whereas the value of inmsupr is taken differently, change in its value may affect the result by -0.23.

Now, we check the overall effect of all variables in the model-using chi square and we get the result as

1. Chi Square

$$X2 = 22795.7,\ df = 7,\ P\ (> X2) = 0.0$$

This shows that overall effect of all diseases is statistically significant with p value = 0.0

Next is to perform and evaluate the Prediction of model. Using confusion matrix value, we see that there are no false negative and negative whereas true negative is less than true positive, which again indicates that the result generated or model fit is acceptable. In addition, the model fit accuracy generated is 63%. Which is acceptable and we can say that there are no major diseases so likely to cause covid result as positive.

2. Confusion Matrix

Figure 13. Summary of confusion matrix and statistics

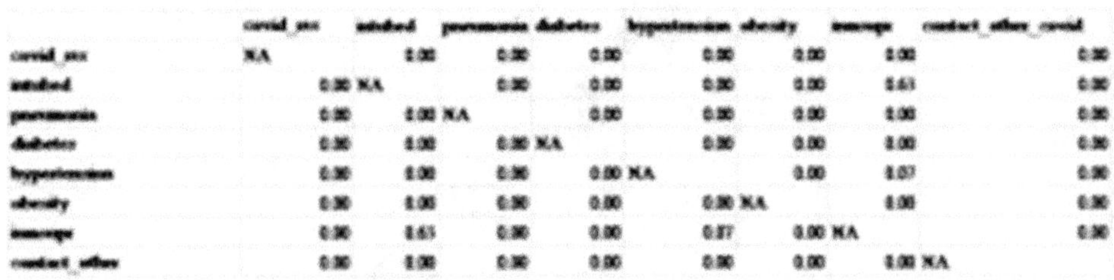

3. ROC Curve is another variable, which checks the model fit and accuracy. From the Curve generated for performance it can be seen that the there is a model fit and accuracy is acceptable. Are Under Curve is 63% which is greater than 50% so again it is acceptable.

Figure 14. ROC showing performance

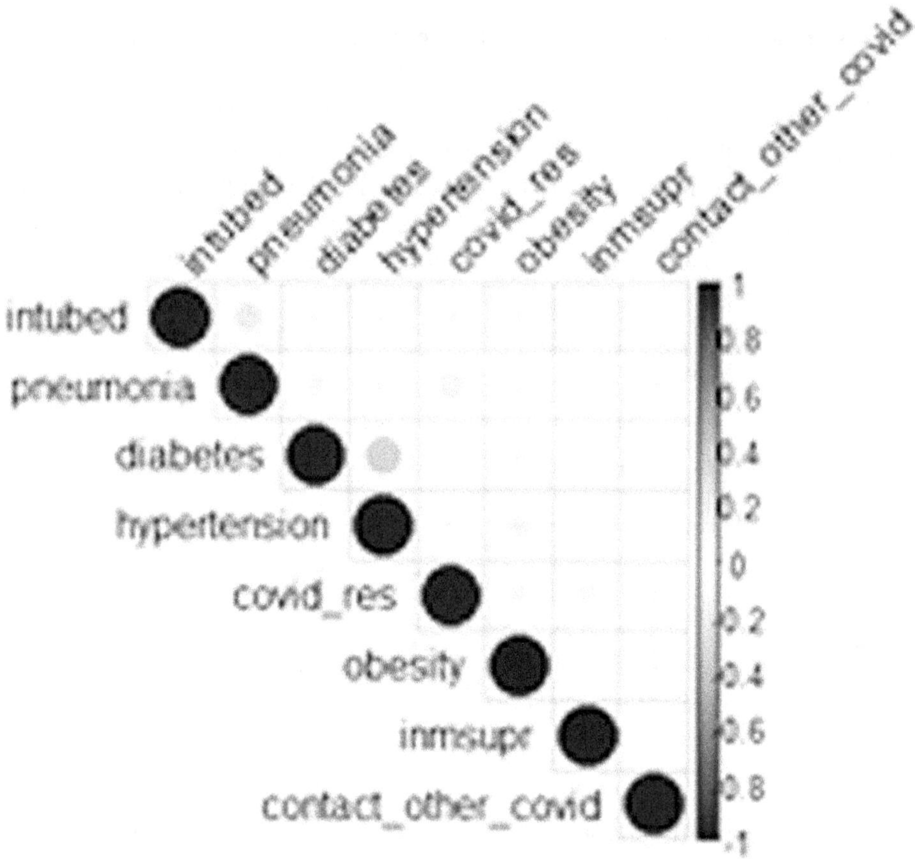

4. Precision And Recall- It is used to verify our model fit based on values generated from confusion matrix. As shown in the table below we can see that the values show the acceptable scores.

Table 2. Precision recall and F1- score

Precision	Recall	F1- Score
0.63	1	0.77

CONCLUSION

Lots of work had been done to predict the symptoms responsible for spread of corona virus across the world. Also, the spread pattern along with best suitable ML Technique/ algorithm is studied across to find the better Accuracy, Precision, Recall, F1 factor etc. Here we have tried to work upon analysing the pre-existing chronic conditions that are responsible for COVID infection. And it is found that Pneumonia has highest severity for getting covid positive followed by obesity, diabetes, hypertension, intubed. Here Logistic Regression Model is used and evaluation is done on the basis of ROC and its performance, confusion matrix, Area under Curve and accuracy factor of the model. All the criteria considered shows significant relation between the variables and Chi Square test shows that all chronical diseases are significant. The prediction is done using logistic regression and based on various metrics like Precision, Recall, F1- Score, AUC we have found that Obesity, Pneumonia, Diabetes, Intubed, Inmsurp and Contact with others are responsible for getting the infection.

REFERENCES

Azarafzaa, M., Azaraf, M., & Akgun, H. (2021). Clustering method for spread pattern analysis of corona-virus (COVID-19) infection in Iran. *Journal of Applied Science, Engineering, Technology, and Education*, *3*(1), 1–6. doi:10.35877/454RI.asci31109

Balabdaoui, F., & Mohr, D. (2020, May 13). Age-stratified model of the COVID-19 epidemic to analyze the impact of relaxing lockdown measures: nowcasting and forecasting for Switzerland. medRxiv. doi:10.1101/2020.05.08.20095059

Chandu, V. (n.d.). *Identification of spatial variations in COVID-19 epidemiological data using KMeans clustering algorithm: A global perspective*. doi:10.13140/RG.2.2.18339.48163

Chatterjee, J. M., Sujath, R., & Hassanien, A. E. (2020, May 30). A machine learning forecasting model for COVID-19 pandemic in India. *Stochastic Environmental Research and Risk Assessment, 34*, 959–972. doi:(0123456789() doi:10.1007/s00477-020-01827-8(0123456789().,-volV)

Davies, N. G., Klepac, P., Liu, Y., Prem, K., Jit, M., Pearson, C. A. B., Quilty, B. J., Kucharski, A. J., Gibbs, H., Clifford, S., Gimma, A., van Zandvoort, K., Munday, J. D., Diamond, C., Edmunds, W. J., Houben, R. M. G. J., Hellewell, J., Russell, T. W., Abbott, S., ... Eggo, R. M. (2020). Age-dependent effects in the transmission and control of COVID-19 epidemics. *Nature Medicine*, *26*(8), 1205–1211. doi:10.103841591-020-0962-9 PMID:32546824

Fleitas, P. E., Paz, J. A., Simoy, M. I., Vargas, C., Cimino, R. O., Krolewiecki, A. J., & Aparicio, J. P. (2020, October). Understanding the value of clinical symptoms of COVID-19. A logistic regression model. *MedRvix*. doi:10.1101/2020.10.07.20207019

Gawriljuk, V. O., Zin, P. K., Foil, D. H., Bernatchez, J., Beck, S., Beutler, N., . . . Ekins, S. (2020). Machine Learning Models Identify Inhibitors of SARS-CoV-2. bioRxiv. doi:10.1101/2020.06.16.154765

Iwendi, C., Bashir, A. K., Peshkar, A., Sujatha, R., Chatterjee, J. M., Pasupuleti, S., Mishra, R., Pillai, S., & Jo, O. (2020, July). COVID-19 Patient Health Prediction Using Boosted Random Forestlgorithm. *Frontiers in Public Health*, *8*, 357. Advance online publication. doi:10.3389/fpubh.2020.00357 PMID:32719767

Khanday, A. M., Rabani, S. T., Khan, K. R., Rouf, N., & Mohi Ud Din, M. (2020, June). Machine learning based approaches for detecting COVID-19 using clinical text data. *International Journal of Information Technology*. doi:10.1007/s41870-020-00495-9

Kumar, J., Harshavardhan, A., Bhukya, H., & Prasad, A. K. (n.d.). *Advanced machine learning-based analytics on COVID-19 data using generative adversarial networks. In Materials Today: Proceedings*. Elsevier. doi:10.1016/j.matpr.2020.10.053

Muhammad, L. J., Islam, M., Sharif, U. S., & Ayon, I. S. (2020). *Predictive Data Mining Models for Novel Coronavirus (COVID-19)*. Infected Patients Recovery. Research Square., doi:10.21203/rs.3.rs-33247/v1

Muhammad, L. J., Islam, M., Usman, S. S., & Ayon, S. I. (2020). Predictive Data Mining Models for Novel Coronavirus (COVID-19) Infected Patients' Recovery. *SN Computer Science*, *1*(4), 206. doi:10.100742979-020-00216-w PMID:33063049

Pourhomayoun, M., & Shakibu, M. (2020). Predicting Mortality Risk in Patients with COVID-19 Using Artificial Intelligence to Help Medical Decision-Making. medRxiv. doi:10.1101/2020.03.30.20047308

Prakash, K. B., Imambi, S. S., Mohammed, I., Kumar, T. P., & Pavan, Y. N. (2020). Analysis, Prediction and Evaluation of COVID-19 Datasets using Machine Learning Algorithms. *International Journal of Emerging Trends in Engineering Research*.

Salehi, A. W., Baglat, P., & Gupta, G. (2020, June 12). Review on machine and deep learning models for the detection and prediction of Coronavirus. *Materials Today: Proceedings*, *33*, 3896–3901. Advance online publication. doi:10.1016/j.matpr.2020.06.245 PMID:32837918

Sarmir, S., Islam, M. T., Salwa, M., Hossain, Z., Shah, M., Hasan, N. M., . . . Haque, A. M. (2021, May 31). Duration and Risk Factors of Post-COVID Symptoms Following Recovery Among the Medical Doctors in Bangladesh. *Cureus*. Advance online publication. doi:10.7759/cureus.15351

Singh, N. P., Sonbhadra, S. K., & Agarwal, S. (2020, June 1). COVID-19 Epidemic Analysis using Machine Learning and Deep Learning Algorithms. medRxiv. doi:10.1101/2020.04.08.20057679

Somasekar, J., Kumar, P. P., Sharma, A., & Ramesh, G. (2020, September 14). *Machine learning and image analysis applications in the fight against COVID-19 pandemic: Datasets, research directions, challenges and opportunities.* doi:10.1016/j.matpr.2020.09.352

Sun, K., Chen, J., & Viboud, C. (2020, April). Early epidemiological analysis of the coronavirus disease 2019 outbreak based on crowdsourced data: A population-level observational study. *The Lancet. Digital Health*, *2*(4), e201–e208. doi:10.1016/S2589-7500(20)30026-1 PMID:32309796

Valencia, D. N. (2020, March 24). Brief Review on COVID-19: The 2020. *Cureus*. Advance online publication. doi:10.7759/cureus.7386 PMID:32337113

Chapter 8
College Recommendation System Using K-Means Algorithm

Harshita Chadha
Maharaja Agrasen Institute of Technology, India

Neelam Sharma
Maharaja Agrasen Institute of Technology, India

Shruti Gupta
Maharaja Agrasen Institute of Technology, India

Nitish Pathak
Bhagwan Parshuram Institute of Technology, India

ABSTRACT

Machine learning is one of the most exciting technologies of the 21st century. This is true both because of the complex issues it possesses the capability to solve, as well as its wide range of application areas. One of the most apparent applications of this technology is in the form of recommendation systems. The chapter outlines a clustering-based recommendation system and contrasts the viability of such systems in comparison with traditional recommender methodologies. The utility of such systems is further highlighted by the demonstration of an application. Utilizing massive online data and the foursquare application programming interface, the colleges across the US have been clustered into groups based on the similarity indexes. This segmentation is then used to present a rudimentary recommendation system that can prove to be useful for prospective students looking to make decisions about higher education institutions to find universities similar to the ones they already have in mind.

DOI: 10.4018/978-1-6684-4225-8.ch008

INTRODUCTION

Machine learning is fast becoming one of the most versatile and fast-growing technologies. One of the main reasons behind the boom in machine learning and the data science industry, in general, is the exponential increase in the amount of data that is generated by users of the internet. According to a report (Wurmser, 2009), there were about 3.5 billion smartphone users all over the world in 2020 and each such user spent about 3 hours 10 minutes on their device every day. That amounts to about 11.095 billion minutes of device usage every day, and every such minute spent online generates massive amounts of data. Since data is the main driving fuel behind them, machine learning (ML) and Artificial Intelligence (AI) technologies are adept enough today to utilize it (Verma et al., 2019).

One of the most common ways in which this data is utilized is via recommender systems that can spell out financial gains for large corporations in the long run (Singh et al., 2021). By suggesting relevant products, these systems optimize user experience thus prolonging end-user engagement. However, despite the apparent utility, these recommendation systems suffer from many inherent disadvantages most of which have to do with the rigidity of framework implementations (Khusro et al., 2016).

This article serves to explore the inherent disadvantages in recommendation systems and thus suggests a clustering-based approach that helps overcome these (Mohamed et al., 2019). For the purpose of demonstrating the utility of our suggested approach, an application-based overview of the approach is presented. An example of a clustering-based recommendation system for US college applications is presented in this article. Applying to universities for higher education can be a daunting task. Adding to the stress of the job is the prospect of having to move away and thus scouring to find an institution located in a place where amenities pertaining to one's personal needs are readily available. This article focuses to help resolve this plight by leveraging university location data and the foursquare API. As an end result, a segmented database is created that can serve as an application-specific recommendation system.

COMPARISON OF K-MEANS CLUSTERING AND K-NEAREST NEIGHBOUR

Traditionally, recommender systems are clubbed into two large categories: content-based and collaborative filtering systems. However, in recent years it has been observed that these systems have inherent disadvantages associated mostly with the amount of flexibility pertaining to the nature of data fed. To overcome this disadvantage multiple hybrid systems and alternative approaches have been proposed. In this article, clustering is used as a solution and the difference between classification and clustering is highlighted here to indicate the reason behind this choice. The first step to solve the problem would be to assess the input data. There can be quite a few algorithms to take a pick from. The main difference discussed in this section is between classification and clustering. To segment the data, the following algorithms are considered: K-Nearest Neighbors and K-means Clustering.

K-Nearest Neighbors or KNN is a classification algorithm that belongs to the supervised learning class. It accepts labeled input data and then classifies data based on preset classes. These sets are then recommended to the user based on the previous interest they have shown in a given class (Cunningham et al., 2007). K-Means Clustering, on the other hand, is a clustering algorithm that is encased in unsupervised learning. It operates on unlabeled input data and divides it on the basis of similarity index. Data points that are similar are put into one cluster, and the clusters altogether as dissimilar as possible (Jin et al., 2011; Sathya et al., 2013).

Here, the results were wielded by making use of K-Means Clustering. The major advantage it gives us over using KNN would be that the input data does not need to be labeled which in turn reduces the stages of training and testing to only grouping. KNN uses explicit feedback to recommend colleges to the student, whereas by using K-Means clustering, the user is added to a cluster based on the similarity index and then recommended, the colleges that similar users have

liked as well. This allows the novel user to span different colleges that may have something different to offer and similar users have liked as well. Using KNN, the colleges would be recommended based on the feedback of the user only (Likas et al., 2003).

The project utilizes the K-means clustering algorithm to segment the colleges across the United States based on their similarity to each other. This similarity index is determined based on the 10 most popular venues found around the institution. Thus, once a prospective student can identify one such institution that meets all their needs, the data can help them in finding several others that may match up to the criteria that they are considering and recommend colleges that may fit them based on other similar students.

THE DATA

The data, housing critical information such as the name of the universities in the United States and their specific locations, was sourced from a website called opendatasoft. The data was composed of all post-secondary educations defined by the Integrated Post-Secondary Education System (IPEDS), National Center for

Education Statistics, US Department of Education. The following illustration in figure 1 shows an estimate of the geographical spread of the data acquired.

Figure 1. Geographical spread of opendatasoft college data

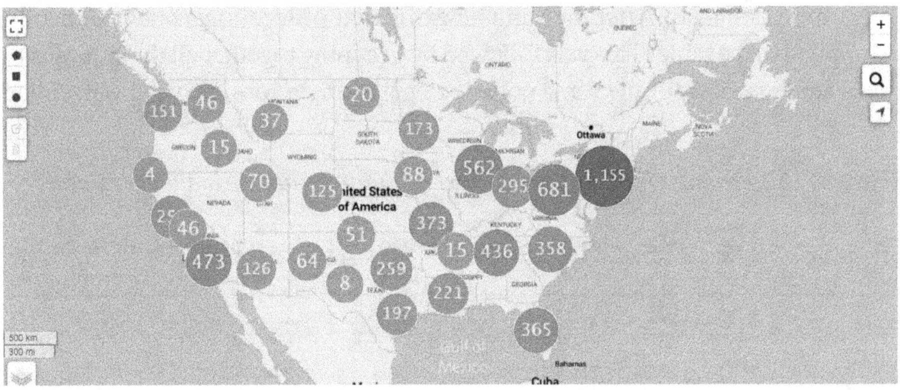

The original dataset constituted about 45 attributes and more than 7500 entries. Keeping in mind the scope of the present applications, a large number of the attributes were irrelevant as a consequence of which the data had to be cleaned and wrangled before it could be used for development purposes. From the large data frame, five main attributes were singled out and extracted to a new more relevant database. These were: the name of the college, its state, the city it was located in, the zip code, and the geographical coordinates. Figure 2 below illustrates the null value-free, partially created database as sourced from (Opendatasoft, 2018).

Figure 2. Partially created data set as obtained by leveraging opendatasoft resources.

	zip-code	Name	City	State	Latitude	Longitude
0	34952	FORTIS INSTITUTE-PORT SAINT LUCIE	PORT SAINT LUCIE	FL	27.191599818	-80.249507429
1	99301	COLUMBIA BASIN COLLEGE	PASCO	WA	46.251436412	-119.118516839
2	25301	WEST VIRGINIA JUNIOR COLLEGE-CHARLESTON	CHARLESTON	WV	38.34799259	-81.634181682
3	82240	EASTERN WYOMING COLLEGE	TORRINGTON	WY	42.079353804	-104.190823112
4	614	UNIVERSITY OF PUERTO RICO-ARECIBO	ARECIBO	PR	18.469199	-66.74114

Leveraging the Foursquare API

The second step in data creation was to leverage the foursquare application programming interface (API) to obtain college-specific venue information. Using the Foursquare API, about 100 venues per college located within a radius of 500 meters were obtained. After the suitable extraction of locations, one hot encoding was applied to the table. This was followed by grouping together all the columns that had separate venues for the same college to get a single row entry per one college.

Figure 3. Data frame with the top 20 most popular venues around every college

	College Name	1st Most Common Venue	2nd Most Common Venue	3rd Most Common Venue	4th Most Common Venue	5th Most Common Venue	6th Most Common Venue	7th Most Common Venue	8th Most Common Venue	9th Most Common Venue	...	11th Most Common Venue	12th Most Common Venue
0	ACADEMY OF ART UNIVERSITY	Coffee Shop	Gym / Fitness Center	Café	Art Museum	Hotel	Sandwich Place	Salad Place	Art Gallery	Boutique	...	Tea Room	Dim Sum Restaurant
1	ACAYDIA SCHOOL OF AESTHETICS	Mexican Restaurant	Asian Restaurant	Chinese Restaurant	Sandwich Place	Snack Place	Bank	Bakery	Rock Club	Indian Restaurant	...	Breakfast Spot	Thai Restaurant
2	ALASKA BIBLE COLLEGE	Pizza Place	Café	Shipping Store	Ice Cream Shop	Coffee Shop	Mediterranean Restaurant	Bookstore	Museum	Sandwich Place	...	Clothing Store	Tourist Information Center
3	ALLEN SCHOOL-BROOKLYN	Pizza Place	Deli / Bodega	Grocery Store	Gym	Bagel Shop	Yoga Studio	Wine Shop	Bakery	Pet Store	...	Middle Eastern Restaurant	Gym / Fitness Center
4	AMERICAN BUSINESS AND TECHNOLOGY UNIVERSITY	Art Gallery	History Museum	Lawyer	Health & Beauty Service	Yoga Studio	Event Space	Dry Cleaner	Dumpling Restaurant	Electronics Store	...	Ethiopian Restaurant	Event Service

With the help of one-hot encoding, a list of the top 20 most common venues by category around each college was chosen and extracted to form the final dataset that shall be fed into the K-means clustering algorithm. This dataset is illustrated in figure 3 above.

PROJECT SPECIFICATIONS

This section deals with the explanation of the application of Machine Learning to the created dataset. The K-means clustering algorithm is used to cluster the colleges based on the top 20 most popular venues present in a 500-meter radius around them. The K-means algorithm groups data unsupervised based on the similarity of the data points to each other. It is a type of partition clustering which essentially means that the algorithm serves to divide the data points into non-overlapping clusters without any internal structure information about any subset. The algorithm tries to minimize intra-cluster distances while at the same time attempting to maximize the inter-cluster distances (Kodinariya et al., 2013). These distances are found either by using the Euclidean distance formula or via the cosine similarity technique. Since the algorithm is susceptible to fall prey to convergence to a local minimum thereby not giving optimum results, it must be run iteratively to land on the global minimum. The biggest problem with using this algorithm, however, is the determination of the optimum K value for the best clustering results. This can be done under cluster accuracy measurements (Yual et al., 2019).

In this case, the average distances between the data points within the clusters are calculated and treated as error parameters. This metric of error is based on the assumption that the less the intracluster distances, the denser the cluster, the superior the clustering obtained. This parameter is termed as the sum of squared errors (SSE) and is used to obtain the optimum value of K. The SSE of a clustering cycle is plotted against the value of k for that cycle and this way, a plot of SSE vs K value is obtained. The elbow point of this plot is treated as the optimum K value for the given data set. For the present dataset, the value of K was chosen to be 4 and the elbow plot can be illustrated in figure 4.

Figure 4. Elbow plot for college clustering with K-means algorithm

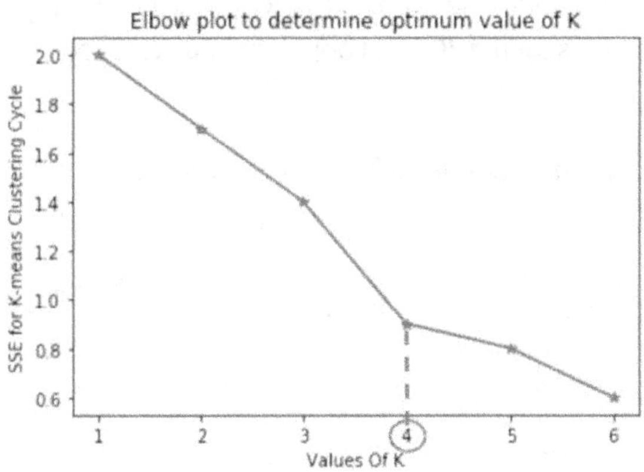

Present your perspective on the issues, controversies, problems, etc., as they relate to theme and arguments supporting your position. Compare and contrast with what has been, or is currently being done as it relates to the chapter's specific topic and the main theme of the book.

Dataflow for College Clustering using K-Means

This section serves to provide a holistic overview of the working of the entire system right from the data manipulation to the rudimentary user interface to find similar college recommendations. The data is first imported into the notebook from the opendatasoft repository after which it is cleaned and wrangled to extract necessary information. The latitude and longitude of each college specified in the cleaned data are hence used to iteratively send a get request to the foursquare API that returns 100 venues for each college located within a 500-meter radius. Following this, a new data frame is created that stores the information of all such venues separately. Further processing is done on the data to combine all the venue entries for every college. This is done via one-hot encoding followed by the group by clause passing the college name as an argument. For the sake of simplicity, using the one-hot encoded values, the list of top 20 most popular venue categories from around each college is selected and a database containing this information is made.

This data is fed into the K-means clustering algorithm that returns a list of labels for each of the entries of the table. These label values are substituted back into the main table so that now a new data frame is created that has as its attribute: the name of the college, the city and state where it is located, its geographical coordinates, followed by 20 attributes for the 20 most popular venue categories around it. The data

Figure 5. Data flow diagram for clustering college data through the K-means algorithm

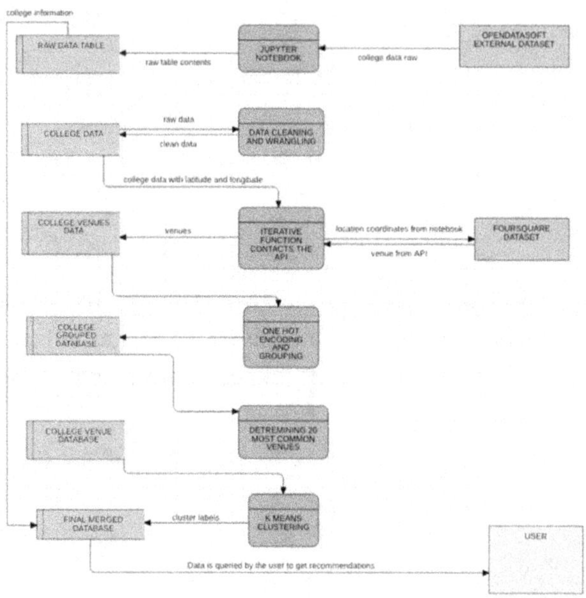

frame can be suitably queried by the user to obtain similar college recommendations. The user can use the table to find the name of the college they are most interested in identifying the cluster label and in turn obtain a list of those most similar to their choice, venue vise, is output. A rudimentary user interface has been created to illustrate this usage and shall be presented in the next section along with the result visualizations. The data flow diagram is also illustrated in figure 5.

RESULTS AND ANALYSIS

Figure 6 illustrates a geographical plot of a portion of the data that is clustered. To maintain clarity of visualization the entirety of the data frame resources was not realized.

Figure 6. Visualization of clustered data points

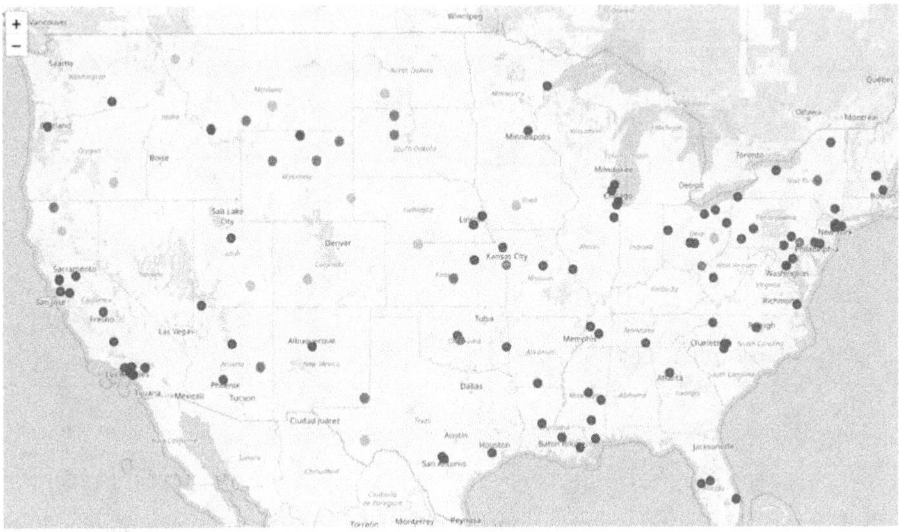

The data points were visualized using python's folium library. The map helps understand the distribution of different types of colleges segmented based on venue categories better. For instance, it can be inferred that the purple markers indicate all those data points that have a K label of 1 assigned to them. A deeper examination reveals that hotels and bars are the most popular venue categories for these colleges. This indicates that they are located near a tourist destination. This information helps to get a deeper insight into the locality without ever having to visit or scour the web for hours to get the relevant information for each college individually.

For utility demonstration, a user interface was put in place that could suitably be query the clustered data to obtain relevant information. An example of one such user interaction is demonstrated in the illustration in figure 7. The user is prompted to enter the name of an institution they are already partial towards. The parser looks for the institution in the database and identifies its cluster label.

Figure 7. Rudimentary user interface showcasing utility.

```
Enter the name of the college you are currently interested in...WEST VIRGINIA JUNIOR COLLEGE-CHARLESTON
The top colleges similar to yours are...
1 .  FORTIS INSTITUTE-PORT SAINT LUCIE

2 .  COLUMBIA BASIN COLLEGE

3 .  YORK TECHNICAL COLLEGE

4 .  DYERSBURG STATE COMMUNITY COLLEGE

5 .  SEWANEE-THE UNIVERSITY OF THE SOUTH

The kind of venues you will find around this institution most popularly are:

1 .  Bar

2 .  Sports Bar

3 .  Bank

4 .  Pizza Place

5 .  American Restaurant

6 .  Café
```

Subsequently, the program outputs a list of the 5 most similar colleges based on cluster segmentation labels. Further, the most common venues around such an institution are also listed out for further information.

CONCLUSION

The project utilizes an unsupervised learning based, restriction-free methodology to utilize data on the internet that is otherwise seemingly of no proper utility and using external tools and machine learning transforms it into useful information. The results obtained, reinforce our initial assumption: the application of powerful and versatile technology to untapped data resources can help obtain exciting results and new insights about the physical phenomenon that can be generalized and used to make helpful predictions. Furthermore, the end product of the undertaking also helps us demonstrate how deviations from traditional recommender methodologies can help overcome the rigidity of their approach and enhance the scope of applicability.

REFERENCES

Cunningham, P., & Delany, S. (2007). k-Nearest neighbour classifiers. *Mult Classif Syst.*

Jin, X., & Han, J. (2011). *K-Means Clustering*. . doi:10.1007/978-0-387-30164-8_425

Khusro, S., Ali, Z., & Ullah, I. (2016). *Recommender Systems: Issues*. Challenges, and Research Opportunities. doi:10.1007/978-981-10-0557-2_112

Kodinariya, T., & Makwana, P. (2013). Review on Determining of Cluster in K-means Clustering. *International Journal of Advance Research in Computer Science and Management Studies., 1*, 90–95.

Likas, A., Vlassis, N., & Verbeek, J. J. (2003). The global k-means clustering algorithm. *Pattern Recognition, 36*(2), 451-461.

Mohamed, M., Khafagy, M., & Ibrahim, M. (2019). *Recommender Systems Challenges and Solutions Survey*. . doi:10.1109/ITCE.2019.8646645

Sathya, R., & Abraham, A. (2013). Comparison of Supervised and Unsupervised Learning Algorithms for Pattern Classification. *International Journal of Advanced Research in Artificial Intelligence., 2*. Advance online publication. doi:10.14569/IJARAI.2013.020206

Singh, P., Pramanik, D., Dey, A. K., & Choudhury, P. (2021). Recommender Systems: An Overview, Research Trends, and Future Directions. *International Journal of Business and Systems Research., 15*(1), 14–52. doi:10.1504/IJBSR.2021.111753

US Colleges and Universities. (n.d.). *Opendatasoft*. https://data.opendatasoft.com/explore/dataset/us-colleges-and-universities%40public/information /

Verma, K., Bhardwaj, S., Arya, R., Mir, S. U. I., Bhushan, M., Kumar, A., & Samant, P. (2019). Latest Tools for Data Mining and Machine Learning. *International Journal of Innovative Technology and Exploring Engineering, 8*(9S), 18-23. doi:10.35940/ijitee.I1003.0789S19

Wurmser. (2019). US Time Spent with Mobile 2019: Smartphones Gain Minutes, but New Challengers Emerge. *Yoram Emarketer, 30*. https://www.emarketer.com/content/us-time-spent-with-mobile-2019

Yuan, C., & Yang, H. (2019). *Research on K-Value Selection Method of K-Means Clustering Algorithm*. doi:10.3390/j2020016

Chapter 9
Emergency Vehicle-Based Vehicle Detection Model

Poonam D. Chakravarty
Rai School of Engineering, India

Jigar Dilsukhray Pandya
Gujarat State Tribal Education Society, India

Akshara Dave
Indus University, India

Yagnesh Rathod
ITI Poshina, India

Sailesh Suryanarayan Iyer
https://orcid.org/0000-0002-0619-9829
Rai University, India

ABSTRACT

In India, the population is large, and it is very crucial to maintain emergency vehicle services properly. Everyday many accidents occur due to lack responsibility of drivers and lack awareness of motor act laws. Here the authors have developed a very unique solution and try to find unsolved issues in more technical way. The system is about tracing follower vehicles of emergency vehicles using RFID (radio frequency identification). Camera set on an emergency vehicle will capture an image of a trespassing vehicle and read the RFID tag. It will trace the details of the vehicle using an RFID reader. There is one front camera which will check the traffic signals, and if red signal already on red state, then it will not consider the follower vehicle, but if there is green signal and vehicle is following the emergency vehicle, then RFID will scan it, trace it. E-memo is generated and sent to the nearest RTO office for penalty.

DOI: 10.4018/978-1-6684-4225-8.ch009

INTRODUCTION

As we all know that there are lots of issues of traffic violence is there in India. Not only general public but various government or emergency vehicles suffer a lot. Some- times due to lack of awareness and lack of responsibility, various accidental activities can also be done. Different road accidents in countries and mainly in India amounted to around 449 thousand in 2019. Each year, about three to five percent of the GDP of the country was invested in road accidents. About 70 percent of accidents involved young Indians. The country has about one percent of the global vehicle population but it accounted for six percent of the world's road traffic accidents (Statista, 2021). About 30% of deaths are caused due to delayed ambulance. Another Indian government data shows more than 50% of heart attack cases reach hospital late, which can constitute unavail- ability of ambulances too but majority of it is due to patients stuck in traffic.(Bhushan, 2021). The authors of one of the paper proposed a model to pass emergency vehicle smoothly. The model suggested having every vehicle RFID tag.(Bhushan,)

The RFID tag is static and placed on the vehicle. For reading RFID Tags the model has max232 and PIC16F877A system. When an ambulance is approaching the junc- tion, the model communicates to the traffic controller in the junction to turn ON the green light until the emergency vehicle passes. This system is also useful to find sto- len vehicles. It also lessens the manpower. It considers only one road of the traffic junction, and it should have longer range RFID readers (EShanthini. E & Sreeja., G 2016)

To protect this, we suggest here our proposed model for emergency vehicles. In our solution, the front and rear cameras are on the vehicle as described in figure.1. All other vehicles have RFID tags. Camera captures RFID tags of following vehicles. As we know that In India, to follow emergency vehicles or not providing way for emer- gency vehicles, there is a fine of 10,000 Rs. It is easy to find the vehicles which are not providing the way, but it is hard to find emergency vehicle followers. Our system recognizes those vehicles, scans their RFIDs, get the information, e-memo will be generated and send it to the vehicle owner.

Literature Survey

In one of the article, the authors discussed own VNPR (Vehicle Number Plate Recog- nition) system. The system detects moving objects, captures the images, extracts the text form the image and save the text on a web page. For implementation, they have used Raspberry pi and Open CV with python programming. This system can be used for traffic violation detection purpose to capture and recognize number plate.(Shobayo et al 2020)

In another paper, the authors have made an application which is useful in accidental situations. The application scans the number plate of the vehicle which had an acci- dent, it recognizes the number and checking the database, it retrieves the correspond- ing data sets from database and notifies the victim's family and nearest emergency center (Anicham et al)

One of the system describes Smart traffic signal technology using IoT. It will be useful for emergency vehicles to reach time to time. The steps in the model are: (1) The web camera captures the traffic image on the road (2) Conversion from RGB to grayscale (3) Image recognition (4) Find presence and absence of vehicles in road images (5) Object counts and the traffic lights glows automatically (N.J. et al, 2021)

One of the system is used for basic recognition of number plate and read license id from number plate of vehicle image. It is used for mostly parking area and traffic line signals. Which will transfer digital data into plain text data. This system mainly using morphological operation in different way. Which includes image improvement, gray scale transformation, bilateral filtering edge detection and segmentation. We will also use OCR for text recognition. For more functionalities we can add face acknowledgement which can be used for criminal databases. This system has make vehicle management system very easy. It reduce number of error is number plate detection. It reduce manpower. Very fast and affordable, reduce maintenance cost also (Abdul Qawy A et al, 2018)

A one of the paper proposed a system which will clear the flow of traffic signal by converting red signal into green. This system also using RFID technology. It is also helpful for stolen vehicle by using this technology. It also send automatic message to

near police station. This system can also help to detect emergency vehicle. Major issue of traffic, especially for urban area can be solved by using this system. Traffic challenges can be overcome .Emergency vehicle like fire brigade, ambulance and police car facing many problems and as a results we have many loss of human lives. We can add ZIGBEE idea for remote sensor and can also utilize road divider idea (Bastani Zadeh R et al,2018). A paper define the system which is uses smart IR sensor to detect moving object.

There are cameras to extract text from the images. Here Python is used for character recognition and also some libraries used to work with raspberry Pi. All data can be received and able to deliver clear text on web pages for database. It is able to detect fast moving vehicles at different times. It also helps to reduce traffic problem at very cheapest cost for smart city (Bhandari R et al, 2019)

In a paper discussed that when any emergency vehicle faces any traffic it is difficult to control manually by traffic officer. So at that time there can be a system which can capture the image using front camera of ambulance when driver push emergency button and it will be stored in database using raspberry pie3.This image will be sent to control room of RTO office. So they can be able to trace the image of number plate. After that respective trespassers can be put into the penalty payments as per rules. This system helps emergency vehicles to save human lives (Ma Y et al,2019)

One of the paper discussed about the emergency vehicle detection and traffic reduction during heavy traffic using RFID and IOT. To solve the traffic problem this RFID tags counts the number of amount times the image scanned in fixed duration. If the no of images scanned passes to 10 then the green signal will increase to 15 sec- onds as per the system setup. If the counter of images fall bellows to 5 then green light will reset to 5 seconds. So image acquisition done by camera and then send for RGB to grayscale for processing. Then differentiate absence and presence of vehicle in road images. The program will run based on count of the images and control the traffic signals. So road safety, traffic problem, pollution can be reduced (Xu X et al,2018)

There is a paper which focus system which is vehicle counting under complex transportation environment. In this situation boundary and edges not detected easily by system. So it will use optical flow method for moving object detection and shadow detection algorithm. It will also use particle filter algorithm for accurate image. This system is limited to daytime with good and poor illumination(K. Tejas et al,2019)

There is a paper which deals with the effective use of wireless technology and high speed micro controller for smooth and clear flow of traffic. This system is implement- ed by RFID reader to detect RFID tag which is placed in vehicle. When any vehicle comes near to traffic area, the reader trace it. Only the codes are matched information send to control room and traffic signal can be managed. It is also helps to detect stolen vehicle. It is also provide longitudinal and latitudinal information of vehicle which is stored in the server using IOT module (Subhadeep Patra,David Van Hamme,2020)

One of the paper define there is an android application which is used for vehicle detection system. In Emergency time this application will automatically recognize the number from the vehicle number plate and send the SMS or notification to the some groups or contacts. So relatives of can get information of person (Swarup Kulkarni, Dr. Roshani Ade,2017) After considering these types of many systems, we have found a model.

Proposed Model

Sometimes it observed the emergency vehicle accidents in public domain or in heavy traffic routes of many countries or mainly in a country part of India where the popula- tion crunch is high mostly. The part of 11% accidents from major all accidents are happened with Emergency vehicles like Ambulances, Fire Fighters and Police cars and vans. The proposed technique is highly working on the base concept of image detection & image processing (includes image identification).

The footage of CCTV camera which install in an emergency vehicle can help to evaluate the whole task having the camera setup on both of the side front and back side of the vehicle. The front camera used for detecting an upcoming traffic signal or traffic lights as well the back camera is used for detect the vehicle & RFID tag or barcode tag which stick on the front of the vehicle. The main concept behind this proposed technique is to pre- vent the road accidents with emergency vehicles by civilian vehicles. In some of the special case the civilian people are driving their vehicles in very rush behind the emergency vehicle to overcome the traffic situation. As acting for vehicle driving by people like this cause the sever accident of civilian vehicle and emergency vehicle. So the proposed system will capture the vehicle images which are aggressively driven behind the EVs. The captured image will look for the RFID tag of vehicle and get the vehicle and owner details by scan the Tag or Code. The owner will get the fine deduction or E-Memo for violating the rule or against the carelessness.

Figure 1. Vehicle detection system while moving EVs.

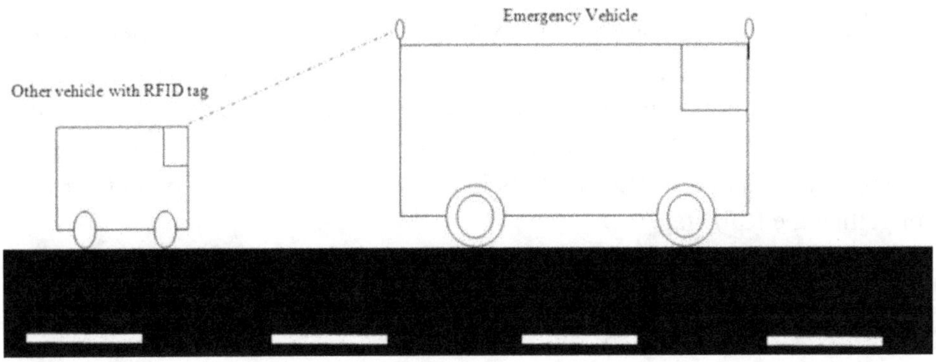

The successor vehicle has the RFID tag having the vehicle and owner full details like owner name, residential address, vehicle company, vehicle name, date of purchase, owner reference etc.

Vehicle Detection Scenarios

The EV has two cameras for capturing images for 2 different circumstances which mentioned as below.

Scenario 1: Successor vehicle RFID read while EV moving on the road.
Scenario 2: Successor vehicle RFID read while EV STOP signal detect.

Scenario 3: EV stuck in heavy traffic

Scenario 1: Successor Vehicle RFID Read While EV Moving on the Road

As fig. 1 shows process stage of image capturing which mainly tracing for vehicle RFID tag from EV's rear camera. If any EV running on the road in own speed and other vehicles are giving the way to EV in case any of the vehicle which coming for chase captured by the EV's rear camera. The mechanism of capturing image will be executed after 2 successful warnings. The signal receiver of each vehicle received the signal and notifies the drier by blinking the small notification light on vehicle dash- board. After the 1st blinks the sensor will check for the distance of EV and vehicle. If the distance measured the increment will not going to take any step. If any vehicle still chasing will get the 2nd time notification by blinking of notification light and again sensor measure the distance. If the distance measured the increment will not going to take any step else if the distance decreases at 3rd attempt will issue the fine e- slip or e-memo to the driver in registered number and at registered address.

Scenario 2: Successor Vehicle RFID Read While EV STOP Signal Detect

This scenario explained by fig. 2. If EV is in front of the RED traffic light and any other vehicle get come closer and sensor detect the very less distance, at that time no any notification or e-memo send to the successor vehicle just due to no signaling al- low to move forward. In this case the front camera of EV is also trying to detect the signal light status. If RED STOP light on the rear camera will not try to process for getting the RFID tag.

Figure 2. EV in front of traffic signal

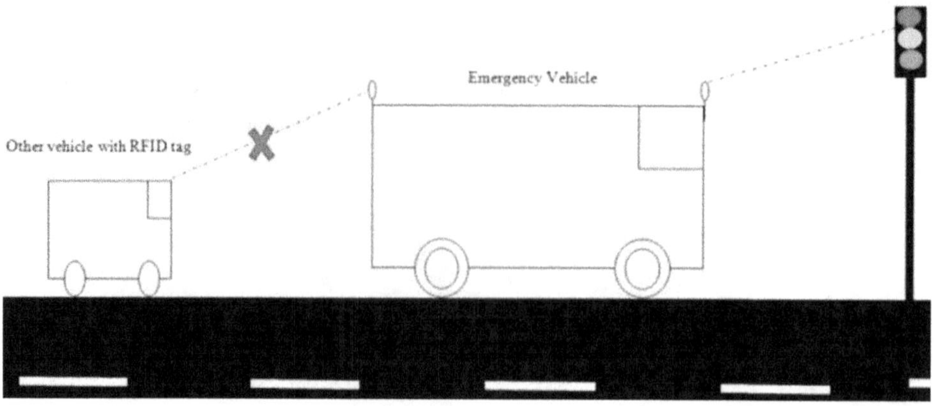

Scenario 3: EV Stuck in Heavy Traffic Situation

There are some situations for traffic jam wills not going to involve the image capturing of successor vehicle while detecting other vehicles in front of EVs as mentioned in fig. 3.

Figure 3. EV in front of vehicle and also followed by successor Vehicle

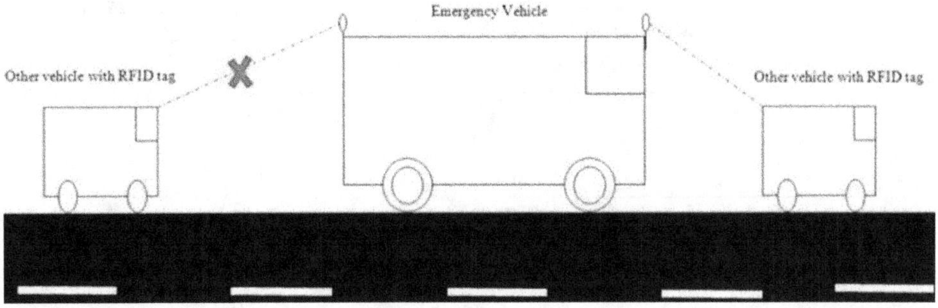

Working Architectural flow of Proposed System

With reference of 3 different possibilities the generation of working flow for proposed model is based on some conditional aspects. As explained in Fig. 4 the flow starts from measure the distance between any successor's vehicle and any EV (ambulance, fire fighter or police van). The first flow condition check for *"if any successor vehicle is with safe distance to emergency vehicle or not"?* In case, the acknowledgment is for yes then no need to go for further process but if the acknowledgment received no then need to check for other 2 conditions like *"If the EV is successor of any other vehicle? AND If any traffic light is blinking red for signal off"?* In this case the ac- knowledgment is yes then no need to go for further process else initiated the further process for execution. As the next process step the RFID Tag read by the EV's scan- ner and try to trace the detail and send the notification to the vehicle. After receives notification successor vehicle has to discontinue the chasing EV.If the vehicle con- tinue chasing again after notify him the system send another notification to the vehicle by checking the condition *"If sent Notification Twice?"* In case of continues chasing after receives the notification twice the digital E-Memo assigns to the successor vehi- cle. The Owner details and vehicle details are already stores in the RFID Tag at the time of vehicle registration. So it will be easy to trace the details and send E-Memo to appropriate culprit. The appropriate sensors are worked there for measure the distance between EV and other vehicle, sending the notification from EV and receive the noti-fication to successor vehicle. The notification will get in form of lighting source and beep notice.

Figure 4. Working architecture of proposed system

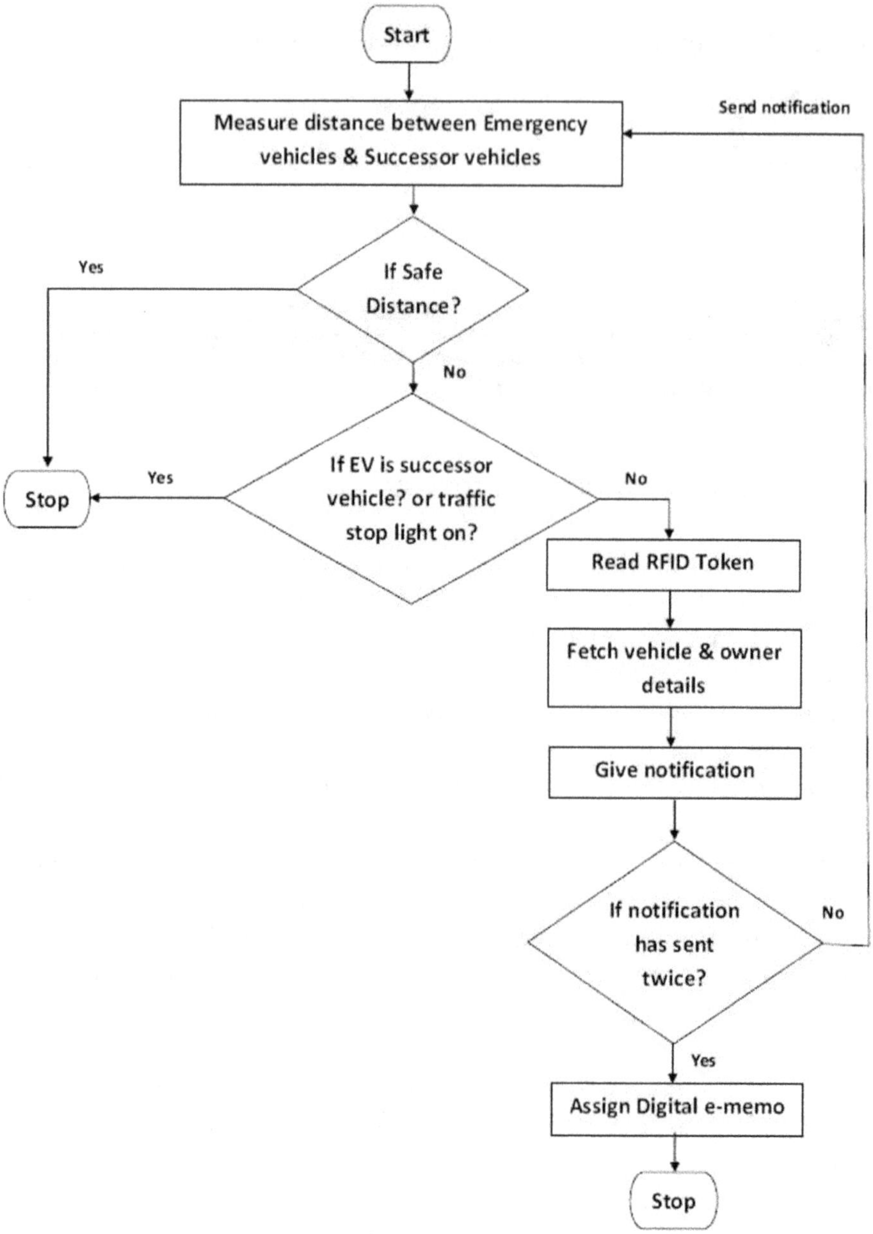

CONCLUSIONS AND FUTURE OPPORTUNITY

Our System suggests the way to a unique problem of emergency vehicle followers. We can use RFID tags, Image processing techniques to minimize the problem. In future, we will implement this system in some specific areas using FastTags (kind of RFID tag used in collecting tolls in India nowadays). Our system will have many benefits like emergency vehicles can reach timely to the destination, it minimizes the accidents, etc. After implementation, we also can further add several other functional- ities like alarm notification to vehicles, E-SMS Memo, etc. (EShanthini, 2016)(Shobayo, 2020)(Anicham et al.,)(N, J., S, V., P, K., S, S., & S, N., 2021) (AbdulQawy, A, 2018)(Bastani Zadeh et al., 2018)(Bhandari, et al., 2019)(Ma et al., 2019)(Xu et al., 2018)(Tejas et al., 2019)(Patra, 2020)(Kulkarni, 2017)

REFERENCES

https://www.statista.com/statistics/746954/india-number-of-road-accidents/

AbdulQawy, A., Elkhouly, R., & Sallam, E. (2018). Approaching rutted road-segment alert using smartphone. *13th International Conference on Computer Engineering and Systems (ICCES)*, 341–346.

Anicham, Diviya, Kaviya, & Gajalakshmi. (n.d.). An Emergency Notification Android Application Using Number Plate Recognition. *International Journal of Advanced Networking & Applications*, 246-250.

Bastani Zadeh, R., Ghatee, M., & Eftekhari, H. R. (2018). Three-phases smartphone-based warning sys- tem to protect vulnerable road users under fuzzy conditions. *IEEE Transactions on Intelligent Transportation Systems*, *19*(7), 2086–2098. doi:10.1109/TITS.2017.2743709

Bhandari, R., Raman, B., & Padmanabhan, V. (2019). Fullstop: A camera-assisted system for characterizing unsafe bus stopping. *IEEE Trans. Mob. Comput.*

BhushanS. (2021). https://medium.com/@bkshashi9/ambulance-stuck-in-traffic-scary-right-7b539ce518b9

N, J., S, V., P, K., S, S., & S, N. (2021, June). IOT and Computer Vision based Digital Signaling and Security for Indian Traffic Signal. *International Research Journal of Engineering and Technology*, *8*(6), 2336–2339.

Kulkarni & Ade. (2017). Intelligent Traffic Control System Implementation for Traffic Violation Control, Congestion Control and Stolen Vehicle Detection. IJES, 5(2).

Ma, Y., Zhang, Z., Chen, S., Yu, Y., & Tang, K. (2019). A comparative study of aggressive driving behav- ior recognition algorithms based on vehicle motion data. *IEEE Access: Practical Innovations, Open Solutions*, *7*, 8028–8038. doi:10.1109/ACCESS.2018.2889751

Patra & Van Hamme. (2020). *Detecting Vehicles' Relative Position on Two-Lane Highways Through a Smartphone-Based Video Overtaking Aid Application.* Academic Press.

Shanthini & Sreeja. (2016). Improved Traffic Control Systems For Emergency Vehicle Clearance And Stolen Vehicle Detection. *International Research Journal of Engineering and Technology*, *3*(3), 630–635.

Shobayo, O. (2020). Development of Smart Plate Number Recognition. *Applied Computational Intelligence and Soft Computing*.

Tejas, K., Reddy, K. A., Reddy, D. P., Bharath, K., Karthik, R., & Kumar, M. R. (2019). *Efficient license plate recognition system with smarter Interpretation through IoT. In Soft Computing for Problem Solving*. Springer.

Xu, X., Yu, J., Chen, Y., Zhu, Y., Qian, S., & Li, M. (2018). Leveraging audio signals for early recognition of inattentive driving with smartphones. *IEEE Transactions on Mobile Computing*, *17*(7), 1553–1567. doi:10.1109/TMC.2017.2772253

ADDITIONAL READING

Ravi Kumar, J. M. S. V. (2021). Automatic Vehicle Number Plate Recognition System Using Machine Learning. *Materials Science and Engineering*.

Song, H., Liang, H., Li, H., Dai, Z., & Yun, X. (2019). *Vision-based vehicle detection and counting system using deep learning in highway scenes*. Song European Transport Research Review. doi:10.118612544-019-0390-4

Sun, W., Sun, M., Zhang, X., & Li, M. (2020). *Moving Vehicle Detection and Tracking Based on Optical Flow Method and Immune Particle Filter under Complex Transportation Environments*. Hindawi Complexity Volume. doi:10.1155/2020/3805320

Chapter 10
Evaluation and Visualization of Blockchain-Based Healthcare

Riten Bhagra
VIT University, Bhopal, India

Gaurav Choudhary
School of Computing Science and Engineering, VIT University, Bhopal, India

ABSTRACT

Blockchain is a distributed ledger apparatus that utilizes timestamped blocks to manage and verify various transactions between two parties in a network with the help of cryptographic algorithms. As compared to popular opinion, blockchain not only serves as a backbone to cryptocurrencies but it can also serve as a pioneering solution to various disciplinary fields such as healthcare, logistics, and education. The approach taken by this chapter to tackle the various problems faced by the healthcare sector utilizes blockchain technology. It avails smart contracts as well as the decentralized nature of the blockchain and smart tokens to ensure smooth interactions between the patients and the physicians. Protection and maintaining the integrity of healthcare data without overlooking any security threats are the primary objectives of this chapter. This chapter discusses a model that uses a cloud-based decentralized model for the storage of patient data.

INTRODUCTION

Blockchain is especially an encouraging and revolutionary technology because it helps to reduce the risk, find out fraud, and brings transparency in a scalable way for uncountable uses. Blockchain technology is a system that contains transactional

DOI: 10.4018/978-1-6684-4225-8.ch010

records, also referred to as the block of the public in several databases, known as the "chain," in a network that is connected through peer-to-peer nodes. Typically, this storage is also identified as a 'digital ledger.' Every transaction in this ledger is approved by the digital signature of the owner, which authenticates the transaction and protects it from any sort of external interference. Hence, the information contained by the digital ledger is highly secured. In other words, the digital ledger is just like an excel spreadsheet that is shared among various computers in a network, in which the storage of transactional records is based on actual purchases. The interesting thing about blockchain is that the data can be seen by anyone, but still, they cannot corrupt it (Jiang,2018)

Without the need for a central authority, blockchain technology paves a path for decentralized and distributed systems with the use of cryptographic principles, transactions are both secure and trustworthy. Thanks to the success of cryptocurrencies, blockchain technology have become quite fashionable in recent years and has penetrated various domains. Blockchain has a variety of built-in features like distributed ledger, decentralized storage, authentication, security, and immutability. Apart from this, it has done wonders in fields like Healthcare. Blockchain is a novel network that eliminates the need for a single, centralized authority while still allowing for safe and pseudo-anonymous transactions and agreements to be made directly between parties. The development of Decentralized Apps can work together with blockchain and provide for on-chain storage. Software applications and infrastructure systems, such as electronic medical records (EMRs), should be able to collaborate easily, share data, and use the data through government institutions and app providers in an interoperable healthcare setting (Abdullah Al Omar,2017; De Aguiar,2020)

However, today's healthcare researchers and professionals face inconsistent evidence, delayed communications, and gapped patient workflows because of vendor-specific and overlapping health networks, making it difficult to deliver customized care (Gaurav Choudhary, 2020;Ilsun You,2018). The lack of a trusted connection that can connect these disparate health systems to form an end-to-end reachable network is a major issue. The healthcare sector has been under constant data breaches. The healthcare data that the various organizations store is very critical because it contains individual's personally identifiable information (Philip Virgil Astillo,2017; Vojislav B Miˇsi´c,2019). This data is considered to be a treasure of sensitive information by criminals. Data of such a magnitude can cause serious damages to an individual's financial and personal reputation. This information can be used by cybercriminals for huge financial payoffs.

Blockchain which is known for its security can provide a huge relief to the healthcare sector. Due to the properties mentioned above, blockchain has emerged as a promising means of providing this trustworthy connection.

Figure 1. How blockchain meets the various healthcare requirements. (Thomas McGhin,2019)

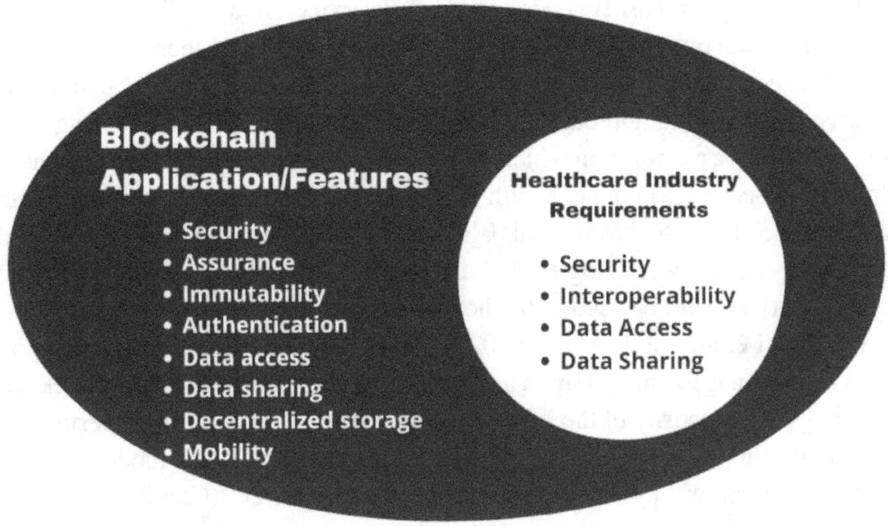

RELATED WORK

Chen et al. (Chen HS,2015) proposed a new patient-centered blockchain model so that patients, as well as the hospitals, can access the medical records easily. The records would be stored on Ethereum Swarm, a distributed cloud storage platform. To solve this problem of data control and ownership, they decided to employ multi-signature contracts. Multisignature contracts require the patient as well as the hospital, to sign their private keys on a transaction so that they can be authenticated. So, the patient cannot edit the record without proper permission from the required authorities, but he/she has control over who can access his/her records. A potential concern is the large number of transactions that a user must pay for the computation, which would lead to huge costs.

Li et al.(Hongyu Li,2018) have researched a novel data preservation system (DPS) for storing patient's data while respecting the user's privacy and maintaining the integrity of stored data. It offers a huge advantage over the general data electronic DPS since it can deal with situations in which data is lost or tampered with. This proposed DPS was implemented in the Ethereum platform. The cost of reserving the patient's data was very less and while retrieving the records the response time was less than 1 sec. However, if a small amount of data needs to be stored there would be some usable space wasted. Also, this method cannot recognize image content.

Mettler et al. (M. Mettler,2016) explain how blockchain technology offers countless services in the health care sector. It helps in the eradication of the trusted third party which was earlier needed for market services. It made direct transactions possible between two parties. Gem, a US-based start-up uses Ethereum technology to build direct patient-physician relationships. healthbank, a Swiss start-up, takes a relatively interesting approach for handling data transactions and sharing health data. They offer a platform on which a user can store as well as manage their health information securely. Patient- health data (e.g., blood pressure, sleeping cycles) can be fetched from the health bank Blockchain. However, it just tells us about the existing blockchain technologies that are present and do not implement any new technology that could be used in the healthcare sector.

Theodouli et al. (A. Theodouli,2020) presented a healthcare data sharing and permission management system in a secure and auditable way by using Blockchain technology. The security of the system is enhanced as it has additional permissions to maintain the integrity of the data which is accessible. Their model works on three layers mainly: Layer 1(Web/Cloud platforms), Layer 2(Cloud Middleware), Layer 3(Blockchain network). Smart contracts embedded in the model are Registry Contract(contains a registry of users), Patient Data Contract(contains hashed Patient Healthcare data), Permissions Contract(monitors the Permission management of Patient Data). This has added value in fields such as Patient Data Integrity, Patient anonymity, Audit, and Accountability. However, it lacks the process of sharing healthcare data for clinical and research purposes.

Zhang et al. (Peng Zhang,2022) proposed a model that can help in sharing and scaling clinical data securely through blockchain technology. This model adopts a public key cryptographic system and the FHIRChain design addresses five key problems: system modularity, user identification, and authentication, data exchange is secure, data access permissions, and data formats are consistent. It can further address other healthcare issues, such as coordinating stakeholders across the healthcare industry and provide patients easy and secure access to their medical data. So, it addresses data integrity, interoperability, access control in a well-defined manner. However, this model lacks shared decision making which can be improved by using blockchain configurations in a testbed environment like the blockchain template provided by AWS. Ekblaw A et al. (Ariel Ekblaw,2016) proposed a decentralized platform called MedRec that helps patients to access his/her medical history using a record management system. This platform uses permission blockchain to manage data sharing, authentication, and other security attributes in the healthcare domain. This blockchain design when integrated with the existing data storage provider enables interoperability as it retrieves patient medical records. Medical researchers were given many incentives to contribute to the mining of the blockchain as they collected aggregated healthcare metadata. However, the design aspect of the MedRec system needs to be improved. Along with this, more security tests of the system should be carried out to check whether it is secure or not. Krawiec et al. (Krawiec R,2018) put forward several existing points in health information systems and the corresponding opportunities that are provided by blockchain technologies. The authors have also discussed that how blockchain can be used in health IT systems so that health organizations. patients and health providers can collaborate.

Peter et al. (Peter,2016) proposed a model that analyses and assembles various concepts in blockchain-related technologies. it focuses on blockchain solving common interoperability problems in healthcare. These methods suggest that blockchain can protect electronic health records, create trust and enable conditional patient privacy. Blockchain technology will help to rebuild trust in healthcare.

Dubovitskaya et al. (Dubovitskaya,2018) has proposed a permission-based blockchain framework that majorly focuses on sharing and managing medical records for cancer patients. This design uses a membership service, which helps in authenticating users using a username password. Patient identity was created by combining information (like date of birth, first names, zip code), and then it was encrypted for security. The Medical data files are uploaded to a secure cloud server and their access is managed by the blockchain logic.

Kuo et al. (Tsung-Ting Kuo, 2017) presented a review paper that contained many blockchain applications in healthcare systems, such as modern healthcare data ledger, improved medical record, and benefits for each described application. Some major challenges for healthcare blockchain technology are condentiality and 51 percent attacks. The authors have also published a few implementations techniques that may solve the challenges: (a) encryption of sensitive data and storing confidential data off the chain to protect the confidentiality, (b) keeping ongoing transactions on the chain to increase network scalability, and (c) virtual private networks can prevent the 51% attacks.

Gropper A (Adrian Gropper,2016; Fan K, Wang,2018)introduced a unique "HIE of One" system. It promotes the use and creation of blockchain identities for physicians and also raises concerns for the various patient matching challenges in the IT health systems. Patients need to install a digital wallet on digital devices(like a mobile phone) to create unique blockchain-based IDs. Then, this is communicated with the rest of the network. This system only uses the blockchain-based ID, thus we can securely manage access to patient data. The major benefits of this technology are (1) re-decentralization of a trusted relationship between the patients and doctors (2) Doctors and patients can use different self-sovereign technologies to support their economic, security, and privacy interests.

Fan et al. (Fan K,2018) proposes a model that successfully resolves the problem of sharing and storing large-scale data in an EMR system. Patients have easy access to the EMRs of different hospitals thus avoiding the previous medical data being segregated into different databases. Data collaboration and sharing can help hospitals get an early understanding of patients' medical history before the patient's consultation. The authors propose an efficient privacy-preserving and sharing the scheme guarantee users' privacy contained in his data by utilizing the combination of encryption technology and access control protocol. The model can quickly return the location of encrypted information, improving the efficiency of the system.

Ismail et al. (L. Ismail,2019) proposed a lightweight blockchain architecture for healthcare data management. This model has low communication and computation overhead. The authors have replaced the consuming mining consensus protocol of the Bitcoin network with an energy-efficient consensus protocol. Model's architecture divides the nodes into clusters where each cluster maintains the ledger which is more efficient than the Bitcoin network. Their architecture uses an HBCM which generates the blocks and orders the transactions. This approach solves the issue of forking which is a major issue in Bitcoin which if it presents in the healthcare domain can lead to life-threatening situations. They have also analysed the effectiveness of their architecture by examining different models which exist in Bitcoin. However, it lacks integration of IoT into the proposed architecture for smart healthcare blockchain solutions. The model hasn't been implemented in real life yet, so it still possesses some security, privacy, and performance risks.

Griggs et al. (Griggs KN,2018) addresses the various security problems in data transactions transfer in an IoT healthcare network. The authors proposed utilizing blockchain-based smart contracts which would perform real-time analysis as well help in keeping transaction metadata logs for medical sensors. This model utilizes a permission-based blockchain to execute smart contracts. These contracts would evaluate information collected by a patient's IoT devices and bases on the information collected it would trigger alerts for the patient and the concerning physicians. The Smart Contracts are encoded in Solidity. However, it does not address the security of IoT devices. The largest challenge for this model is to maintain security at every individual node. On a large scale, key management becomes a major issue because multiple smart devices are broadcasting their transactions to many nodes which are waiting to verify the next block.

Hathaliya et al. (J. Hathaliya,2019) has provided insights into the blockchain-based RPM system for healthcare systems. The model contains the blockchain-based system architecture which overcomes the security and privacy issues of the traditional system. The architecture combines the decentralized AI and blockchain which helps in analytics, diagnosis, validation of medical reports, and decision making. However, it does not address the various security challenges to the RPM-based system.

BLOCKCHAIN AND HEALTHCARE

A blockchain is a continuous chain of blocks that are time-stamped and linked using cryptographic headers. The chain constantly grows and new blocks get added to its end. Each new block holds a hash value to the context of the previous block. The shareholders, also referred to as nodes of the blockchain are organized in a peer-to-peer network. Each node within the network has 2 keys: a public key that is used for encrypting the messages and a private key that is used for decrypting the messages sent to a node. This method is known as asymmetric cryptography which helps in ensuring robustness, consistency, and transparency within a blockchain environment.

Blockchain architecture can be divided into three layers:

The infrastructure contains all the necessary hardware components required for a blockchain to execute. It also includes nodes(network participants) that can perform various tasks such as initiating transactions, validating transactions and blocks, generating blocks, and maintaining a copy of the ledge. It also contains the storage component used for storing the ledger of transaction records in the network. The platform layer contains the facilities for invoking Remote Procedure Calls (RPCs), REST API for communication between client and blockchain network. The distributed computing layer maintains fault tolerance, local access to data, immutability, privacy, authenticity, and security of the transaction data.

Transactions declared within the blockchain network which is regarded valid by the other nodes present in the network are arranged into time-stamped blocks by specific nodes. The consensus protocol marks what kind of miners are to be chosen for the task and what kind of data should be included in each block. The blocks are then broadcasted to the network, where the validation nodes verify that received block contains valid transactions and that it references the previous block in the chain by using the corresponding hash. If both requirements are fulfilled, the nodes add the block to the blockchain whereas if requirements are not fulfilled, the block is dropped. The various layers of blockchain is shown in Fig 2.

- Public Blockchain: In a public blockchain, all the transactions are fully transparent allowing anyone to take a look at the transactions. Anyone can participate whether as a user, miner, developer, or community member. Since it is open-source, therefore it has no restrictions. Simply an internet connection is required to get access to the network for validating blocks and sending transactions. Participants are rewarded in the form of tokens that are associated with the public blockchain.

- Private Blockchain: Transactions are private, and only ecosystem participants who have been granted permission to enter the network have access to them. To access the networks, participants must give their consent. Enterprises who want to collaborate and exchange data but do not want their confidential business data to be available on a public blockchain should use private blockchains. The organizations that operate them have a lot of influence over the members and governance structures. When it comes to a private blockchain, a token may or may not be linked with it. Some examples include Hyperledger, Ripple.
- Hybrid Blockchain: Hybrid blockchain blends the protection and openness of a public blockchain with the privacy advantages of a permissioned and private blockchain. This gives companies a lot of leeway in terms of deciding what data they want to make available and accessible and what data they want to keep private. For example, Dragonchain.

Figure 2. Various layers of blockchain.

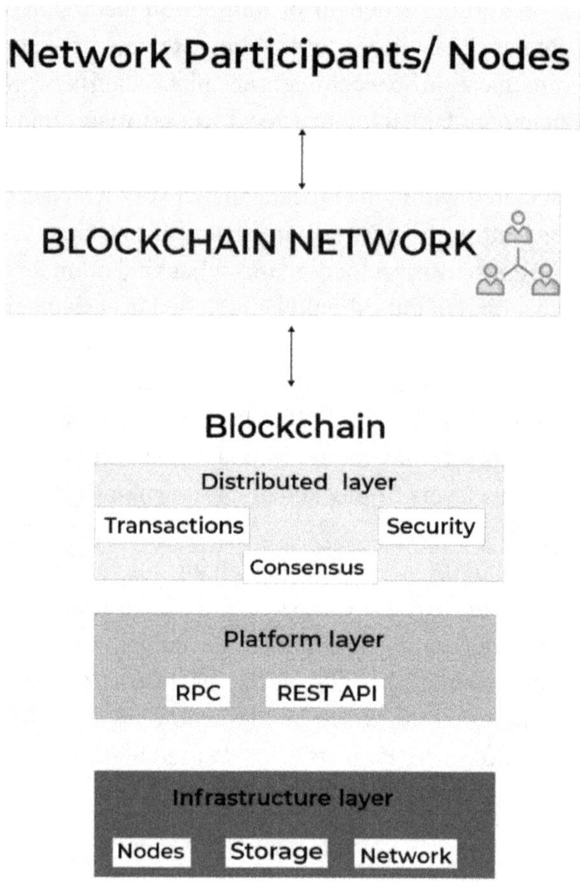

Need for Blockchain in Healthcare

Blockchain has a huge potential in healthcare. Blockchain technology can be used to create a patient-centric approach. It would help in enhancing the reliability and security of patient's data since every patient would have control over his/her record. Efficient exchange of medical records between different healthcare institutions would be facilitated by the use of blockchain technology. The storage of patient's confidential medical records is also a pretty big concern. Blockchain would provide protection against all types of cyberattacks and is also very robust against failures and provides various methods of access control. It provides a good framework for healthcare data. Further application of blockchain relates to the supply chain in pharma and developing measures against counterfeit drugs. While the development of new drugs incur substantial costs related to trials to evaluate the safety and efficacy of the drug, the use of smart contracts allow to facilitate the procedure of the informed consent as well as improve identity management and data quality Providing access to patients for managing their own identify also allow integration of the informed consent procedure while ensuring the privacy of individual health data. A simple blockchain healthcare transaction scenario is shown in Fig 3.

Figure 3. A simple blockchain healthcare transaction scenario.

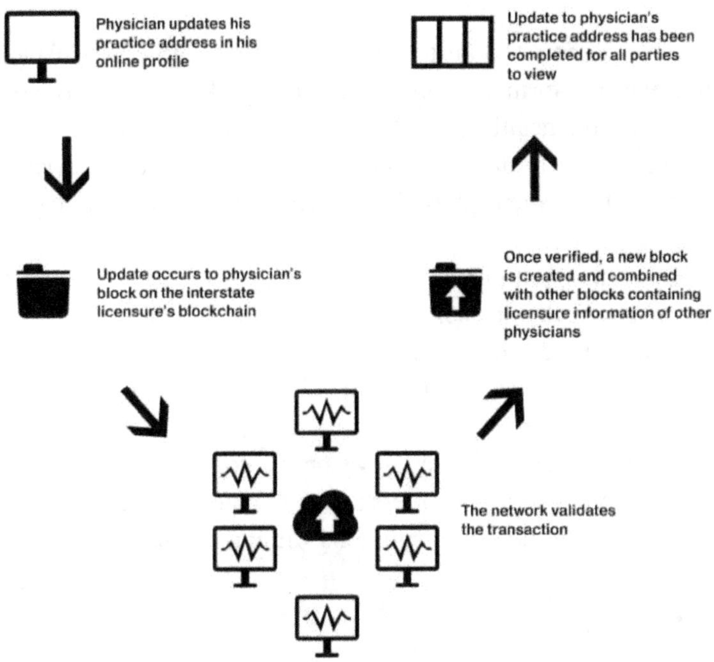

The Healthcare industry is one of the data-intensive industries. A large volume of data is generated, circulated, stored, and accessed on a daily basis. The decentralized nature of blockchain, its openness, and permission-less characteristics offer a unique solution for healthcare. Blockchain technologies greatly improve the flow of data within and among companies and improve the accessibility and security of patient information. Existing electronic health record (EHR) management systems still suffer from many problems during data management(collection, storage, and distribution of data) such as data manipulation, delayed communication, and trust less cooperation. A problem facing healthcare record systems throughout the world is how to share the medical data with more stakeholders for various purposes without sacrificing data privacy and integrity.

To overcome the current scenario, this paper will continue to devise a model that can provide a secure way of handling sensitive healthcare data along with a secure and impeccable transaction model by using blockchain. The patients and as well as physicians can have a flawless experience without worrying about any security threats. The model makes use of the decentralized and distributed nature of blockchain to make sure that the confidentiality, integrity, and availability of the user are not compromised.

BLOCKCHAIN BASED HEALTHCARE SOLUTIONS

After a deep analysis of blockchain technology and how can we use it in the healthcare industry, IndHealth platform was introduced. It is a blockchain-based platform whose main aim is to take the healthcare industry to a whole new level. It offers reliable, easily accessible, and affordable consulting services to a patient from a physician of their choosing. The patient-physician system has the following features:

- Book physician appointments.
- Validity check of the physician's license.
- Secure data storage and access control.
- Safe and anonymous payment module.

The physician-physician system allows online consultations between different physicians to go smoothly. It also offers a feature that facilitates sharing of patient records among different medical institutions without any hassle.

IndHealth

The IndHealth solution works like this. Suppose Bob has a medical query and needs someone to guide him on how to tackle the medical problem. Bob can use the Indhealth web portal to find the required physician for him based on various parameters like prices, availability, ratings, and specialization provided the physician is registered on the IndHealth platform.

When Bob finds his ideal physician, he can directly 1contact the physician through the IndHealth web platform. The physician can then obtain access to Bob's medical records anonymously through a decentralized, secure data bank and then advises Bob through the IndHealth application. The physician/doctors' license can be verified through an immutable database that is regularly updated and is connected with a licensing authority. In return for their service, the doctors' will get compensated for their valuable comments through the IH(IndHealth) token. The simple working of IndHealth is shown in Fig 4.

Figure 4. The working of IndHealth.

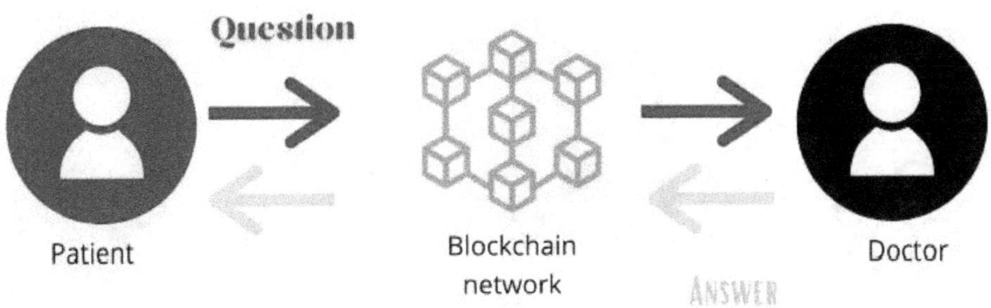

IndHealth Solution Stack

The various applications of blockchain technology can be integrated into our IndHealth solution and various third-party applications. The main protocol stack complies with the various healthcare regulations and has leading technological solutions. The IndHealth platform is currently protocol agnostic. The token functionality helps in the proper functioning of all subsystems and delivers secure interactions. The Indhealth tokens accelerates the following functions:

- Physician recognition and license validity of various physicians/doctors.
- Safe and secure payments.

- Secure and decentralized data storage.

Payment Module

The relationship between a physician and a patient will be defined by a smart contract. The various payment terms and conditions will be available to all the participants of the IndHealth platform. A small fee would be added to every transaction to maintain the service quality of the network. The blockchain will automatically process all transactions which are being carried out on the IndHealth network.

To achieve anonymous communication between a doctor and a patient and prevent direct payments we used tokenization. Data access and communication are already fully anonymized, but direct payments could compromise this. Tokenization helps in achieving pseudo- anonymity between the doctor and the patient. We will use Bolt(Blind Oddchain Lightweight Transactions32) to achieve full transaction anonymity. Other solutions like KYC and AML can be implemented in the future to ensure overall anonymity and regulatory compliance.

Figure 5. Working of the payment module.

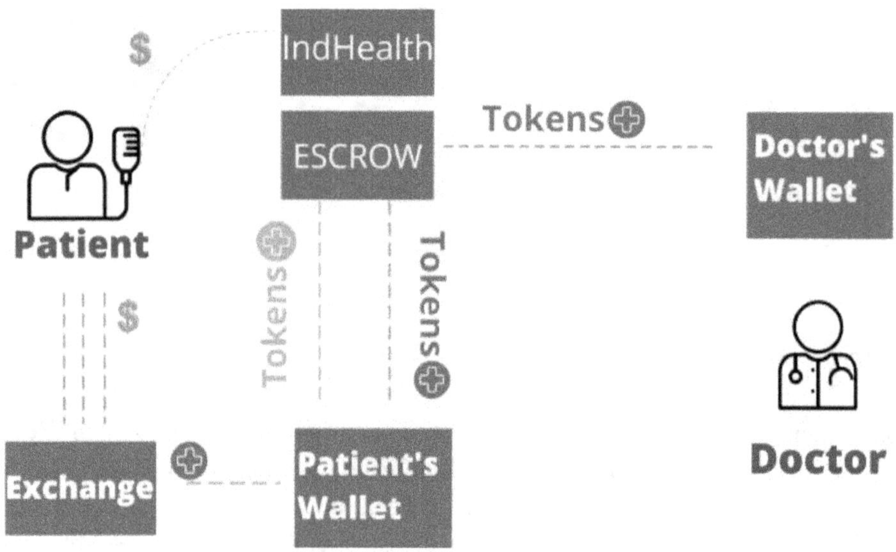

Figure 6. License validation.

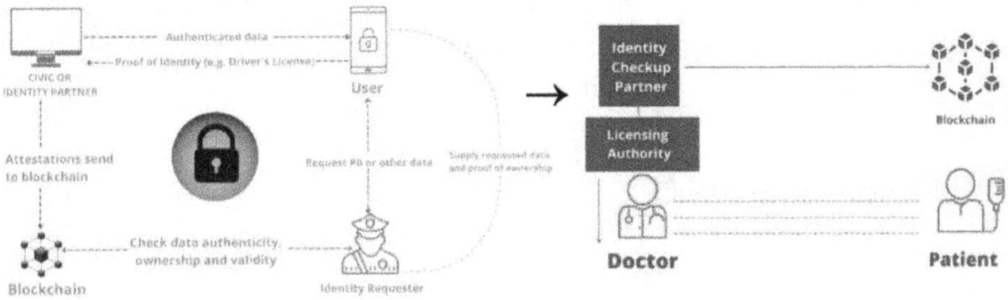

Identity Management

The ability to identify a physician and check his/her license authenticity while maintaining patient confidentiality is one of the key features of the IndHealth platform. The licensing of physicians is heavily regulated. National bodies grant the license and keep track of its validity, depending on the legislature. CIVIC or a similar approach can be used to link an individual person to a certificate, hash it to the blockchain, and allow rigorous checks across the IndHealth Platform. It would enable the physician (or another provider) to apply his name, as well as credentials such as medical school and residency completion, to a blockchain authority once. Every state board of medical examiners or other approving authority, in a perfect scenario, would have access to this digital ledger and, with the candidate's consent, would be able to check the documents that have already been compiled in a digital format.

Decentralized Storage of Data

The centralized storage techniques provide an efficient way of storing and distributing data. But while dealing with sensitive data like health records, the centralized database is not optimal. Using a decentralized database, we can add another layer of security to our system. Data storage is usually off the chain and thus can be deleted. The distributed ledger will be used for running the decentralized storage system and it can also help in hashing the data for immutability. We will use Storj which is a decentralized cloud storage platform.

Storj

Storj is an open-source decentralized storage system that can be used to implement our solution. Whenever a file is stored on Storj, it is encrypted, split into pieces, and then it is stored on various diverse nodes. This helps us to minimize the dangers associated with data breaches. The default encryption is pretty standard and is applied to every file. Your data is only shared with those with whom you want to share it with. The data is divided into a number of shards.

Figure 7. Working of Storj

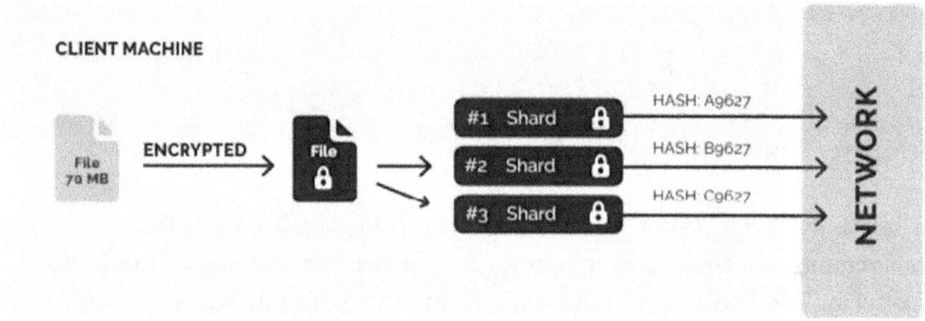

A shard is a portion of an encrypted file that is stored on cloud storage. Before being broken into a number of shards(also called sharding), the files are encrypted client-side. Sharding provides privacy, security, performance, and availability. The encryption key is only shared with the owner of the data so, they have complete control over their data. In the following example, it uses AES256-CTR to encrypt the data but any encryption system can be implemented to protect the data.

FUTURE RESEARCH DIRECTIONS

The Healthcare industry is an extensive data-driven industry. It is very important to properly manage the medical data because proper data management can lead to a better overwhelming experience for the patient as well as the physician. In short, it provides efficient healthcare. Blockchain technology can improve the management of healthcare data because it provides security, assurance, immutability, authentication, decentralization, and mobility. It would help in enhancing the reliability and security of patient's data. It can solve the basic healthcare industry requirements such as security, interoperability, data sharing between different medical institutions, and control over vital data. Recently, blockchain technology is focused more upon cryptocurrencies like Bitcoin, Ethereum, etc.

This research has laid the basic foundation for using blockchain technology with healthcare. Much work remains to be completed and many possibilities can be used to improve this research. Firstly, the future work will include the real-life implementation of the above model in order to evaluate its performance, security, and privacy. The weaknesses and limitations of the above research have suggested the following areas as recommendations for future research.

- Use of IoT devices: The Internet of Things (IoT) is at the peak in the 21st century. The IoT devices can be used to monitor the real-time health of a patient. The wearable devices like fitness bands can be used to monitor blood pressure, doctor appointments, exercise, etc. Any unusual disturbances or changes in the normal lifestyle of a person should be immediately reported to the concerned doctor on the IndHealth platform. Knitting IoT and Blockchain is a very tedious task, primarily because of plenty of setbacks in communication, for a Proof of Work consensus to be established, its massive resource requirements must be met, and lastly because of the computational complexities that it brings along.
- Data accessibility and updating patient data: In the future, we can add a module in which updation of patient data can only place with the prior approval of the patient and only by the authorized doctor. The patient can clearly observe which individual is accessing his medical data. Then he/she can permit the authorized person to append his medical data. Thus, the main focus of this work would be to add, update and retrieve healthcare data only with the patient's approval without comprising the integrity of the blockchain network.
- Drug supply: Drug counterfeiting has been a distressing issue afflicting the healthcare industry. According to the Organization of Health Research Funding 10-30 percent of the drugs are counterfeited or cloned versions that are sold in the markets of the developing countries. The fake medicines/ drugs can be very dangerous and critical to the patient's health because of their composition. They are highly different from the original medicines. The IndHealth platform in the future can directly help the patient to get the medicine from a verified drug supplier and keep a track of the drug as it enters the different drug supply chain phases.

CONCLUSION

It is important to note that blockchain can revolutionize the healthcare industry. The proposed model can replace the traditional healthcare system. The various healthcare organizations can utilize the model to provide the patients a seamless experience. The patients can finally control who has the access to their confidential medical data. Health practitioners can register themselves on the IndHealth platform without any hassle. Their practicing licenses can be validated easily through the blockchain authority. It would enable the physician to apply his name, as well as credentials such as medical school and residency completion, to a blockchain authority once. The model is still to be implemented in real life and thus we cannot assess the various privacy, security, and ethical risks associated with the platform. The model would definitely prevent any leakage of confidential medical information while adhering to the CIA triad.

This chapter tries to devise a model that can solve the existing healthcare industry problems and provide a seamless experience for the doctors as well as the patients. The IndHealth platform uses tokenization to achieve pseudo-anonymity between the patient and physician. Identifying a physician and validating his/her medical license while maintaining patient confidentiality is one of the key features of the model. The model uses a decentralized storage platform called Storj in order to store confidential data. If the healthcare organization is under any cyberattack, the decentralized storage would prevent any data breaches, thus protecting the sensitive data. The payment transaction between the patient and the physician would take place seamlessly. Blockchain is a novel network that eliminates the need for a single, centralized authority while still allowing for safe and pseudo-anonymous transactions and agreements to be made directly between parties. The patient will pay the physician in the form of IndHealth tokens which can be later redeemed for money. The patient data is first encrypted before storing them on the cloud storage platform to add a layer of protection against data breaches. To conclude, this model could solve the challenges currently faced by the healthcare industry while maintaining the CIA(confidentiality, integrity, availability) security triad.

REFERENCES

Al Omar, A., Rahman, M. S., Basu, A., & Kiyomoto, S. (2017). Medibchain: A blockchain based privacy preserving platform for healthcare data. *International conference on security, privacy and anonymity in computation, communication and storage*, 534–543.

Astillo, P. V., Choudhary, G., Duguma, D. G., Kim, J., & You, I. (2021). Trmaps: Trust management in specification-based misbehavior detection system for imd-enabled artificial pancreas system. *IEEE Journal of Biomedical and Health Informatics*.

Chen, H. S., Jarrell, J. T., Carpenter, K. A., Cohen, D. S., & Huang, X. (2019). Blockchain in Healthcare: A Patient-Centered Model. *Biomed J Sci Tech Res.*, *20*(3), 15017–15022.

Choudhary, G., Astillo, P. V., You, I., Yim, K., Chen, R., & Cho, J.-H. (2020). Lightweight misbehavior detection management of embedded iot devices in medical cyber physical systems. *IEEE eTransactions on Network and Service Management, 17*(4), 2496–2510.

De Aguiar. (2020). A survey of blockchain-based strategies for healthcare. *ACM Computing Surveys, 53*(2), 1–27.

Dubovitskaya, A., Xu, Z., Ryu, S., Schumacher, M., & Wang, F. (2018). Secure and Trustable Electronic Medical Records Sharing using Blockchain. *AMIA Annu Symp Proc.,* 650-659.

Ekblaw, Azaria, Halamka, & Lippman. (2016). *A case study for blockchain in healthcare: " medrec " prototype for electronic health records and medical research data*. Academic Press.

Fan, K., Wang, S., Ren, Y., Li, H., & Yang, Y. (2018). MedBlock: Efficient, and Secure Medical Data Sharing Via Blockchain. *J Med Syst., 42*(8), 136.

Griggs, K.N., Ossipova, O., Kohlios, C.P., Baccarini, A.N., Howson, E.A., & Hayajneh, T. (2018). Healthcare blockchain system using smart contracts for secure automated remote patient monitoring. *J Med Syst., 42*(7), 130.

Gropper, A. (2016). Powering the physician-patient relationship with hie of one blockchain health it. In *ONC/NIST use of Blockchain for healthcare and research workshop*. ONC/NIST.

Hathaliya, J., Sharma, P., Tanwar, S., & Gupta, R. (2019). Blockchain-based remote patient monitoring in healthcare 4.0. *2019 IEEE 9th International Conference on Advanced Computing (IACC)*, 87–91.

Ismail, L., Materwala, H., & Zeadally, S. (2019). Lightweight blockchain for healthcare. *IEEE Access: Practical Innovations, Open Solutions, 7*, 149935–149951.

Jiang, S., Cao, J., Wu, H., Yang, Y., Ma, M., & He, J. (2018). Blochie: a blockchain-based platform for healthcare information exchange. 2018 IEEE international conference on smart computing (smartcomp), 49–56.

Krawiec, R., Housman, D., White, M., Filipova, M., Quarre, F., & Barr, D. (2018). *Blockchain: Opportunities for health care*. Proc.

Kuo, Kim, & Ohno-Machado. (2017). Blockchain distributed ledger technologies for biomedical and health care applications. *Journal of the American Medical Informatics Association, 24*(6), 1211–1220.

Li, Zhu, Shen, Gao, Tao, & Liu. (2018). Blockchain-based data preservation system for medical data. *Journal of Medical Systems*, *42*, 6.

Nichol & Brandt. (2016). *Co-Creation of Trust for Healthcare: The Cryptocitizen Framework for Interoperability with Blockchain*. Academic Press.

Theodouli, A., Arakliotis, S., Moschou, K., Votis, K., & Tzovaras, D. (2018). On the design of a blockchain-based system to facilitate healthcare data sharing. *2018 17th IEEE International Conference On Trust, Security And Privacy In Computing And Communications/12th IEEE International Conference On Big Data Science And Engineering (TrustCom/BigDataSE)*, 1374–1379.

Vojislav, Miˇsi'c, Miˇsi'c, & Chang. (2019). Towards a blockchain-based healthcare information system. *2019 IEEE/CIC International Conference on Communications in China (ICCC)*, 13–18.

You, Yim, Sharma, Choudhary, Chen, & Cho. (2018). On iot misbehavior detection in cyber physical systems. *2018 IEEE 23rd Pacific Rim International Symposium on Dependable Computing (PRDC)*, 189–190.

Zhang, P. (2018). Fhirchain: Applying blockchain to securely and scalably share clinical data. *Computational and Structural Biotechnology Journal*, *16*, 267–278.

Chapter 11
Implications Between Environmental Management and Green Transformational Leadership in Greening Organizations

José G. Vargas-Hernández
https://orcid.org/0000-0003-0938-4197
Posgraduate and Research Department, Tecnológico Mario Molina Unidad Zapopan, Mexico

ABSTRACT

This chapter analyzes the implications that the organizational environmental management and green transformational leadership have between them and with the green organizational innovation aimed to achieve sustainable development. It departs from the assumption that green organizations have implemented environmental management and green innovation enhanced by green transformational leadership to achieve organizational sustainable competitive advantage. The method employed is the analytic-reflective based on the review of the theoretical and empirical literature. It is concluded that the environmental management issues guided by a green transformational leadership are related to organizational sustainable development encouraging organizations to act in favor of greening organizations, green consumers, and achieving a sustainable competitive advantage with the Sustainable Development Goals (SDGs).

DOI: 10.4018/978-1-6684-4225-8.ch011

INTRODUCTION

The implementation of the organizational greening scheme is a strong commitment to employee engagement and raises awareness of environmental initiatives to improve environmental sustainability performance. There is a research gap on green practices in industries and on the relationship between environmental management and organizational performance (Poon et.al., 2000, Amores et. al., 2014). Environmental awareness and protection are integrated into green management. The organizational model greening involves intertwined phases of awareness of change through rational schemes and reinforcing identity formation, establishing a management agenda, and recognizing concerns of environmentally sustainable organizational models and addressing concerns.

Green human resources have the main objective to promote functions and actions awareness of organizational environmental management, motivating and developing a sense of pride of employees for participation on a green program. Green human development and training activities create awareness of environmental management value supporting the organizations to establish an eco-friendly management aimed to enhance sustainability

Organizations perform green and environmental management innovation. The environmental awareness and behavior of leaders encourage to their followers, according to the social learning theory, green innovation practices beneficial to the organization performance. Drawing on the social learning theory, organizational environmental leadership influences organizational performance as mediating role of green innovation practices and activities. Green innovation action has a mediating effect of environmental leadership on a firm's financial performance.

Organizational green innovation action has mediation effects of environmental leadership on the organizational environmental performance. Organizational environmental leadership has various environmental benefits influencing green organization identity, playing a crucial role in promoting organizational sustainable development and improving green innovation performance (Greenwood et al. 2012; Pan and Tian 2017). The mediating effect of green innovation actions on environmental leadership and performance have an impact on organizational performance. In achieving these concerns, environmental leadership behaviors motivate followers to adopt green practices and achieve environmental and sustainable development

The engagement of stakeholder´s in bringing about an organizational greening reform requires organizational psychological and social capital for a transformative ability towards creating and developing organizational greening model aimed to combat the negative environmental issues and promote environmental sustainability (Çop, Olorunsola, Alola, 2021). Green engagement and responsibility of the workforce is the result of alignment and integration of green activities into the environmental management as organizational strategy (Hanna, Rocky Newman, & Johnson, 2000; Robertson & Barling, 2013) employee training and developing green leadership (Egri & Herman, 2000; Robertson & Barling, 2013, Teixeira et al., 2012),) and development of a supportive culture (Kitazawa & Sarkis, 2000; Stone, 2000).

Organizational model greening achievements result of commercialization of green innovations although they are far away from the desired environmental performance (Lampikoski, 2012; Anderson, 2009). Green innovation practices are relevant to organizational performance and is essential the acceptance of green organizations by leadership which emphasizes environmental sustainability (Daily et al., 2009). Organizational environmental leadership enhances green innovation practices aimed to boost the overall organizational performance.

Environmental knowledge learning has a moderating effect on environmental leadership and green innovation practice. Green innovation practices bridge organizational environmental leadership and organizational performance. Organizational environmental leadership influence new technology, designing and production, quality management, supplier management products, green image, market share, and so forth.

ENVIRONMENTAL MANAGEMENT

Organizations implement environmental management and green innovation to achieve sustainable competitive advantage (Hart, 1995; Porter & Van der Linde, 1995). Organizational green innovation is an overall plan and identifiable process of organizational environmental management practices proposed and designed by top management (Eisenhardt & Martin, 2000).

Design, development, and implementation of organizational green innovation strategy meet the requirements of stakeholders and avoid environmental costs (Marshall and Mayer 1992) and environmental management create potential benefits for the organization. Environmental management is relevant for organizations to build green product design and development processes. Environmental management factors must be incorporated in the design and development of new green products to cope environmental challenges (Dangelico, Pujari, 2010; Rubik, Frankl, Pietroni, Scheer, 2007; Pujari, Wright, Peattie, 2003; Albino, Balice, Dangelico, 2009).

Organizational and business communities increase awareness and significance of green environmental management methods and techniques. The proactive environmental management is an emerging trend (González-Benito & González-Benito, 2006). Environmental management system of the firms and green teamwork are necessary to ensure system success (Jabbour 2011).

The green team involved in green management practices is defined as people working together aimed to foster environmental learning, resolve environmental sustainability problems and environmental management conflicts (Jabbour, 2013b; Jabbour, 2011). Environmental management may promote an environmental market approach. Sustainability is a multidisciplinary concept explained in other terms such as sustainable development, green business, environmental management, social responsibility, and so forth (Kumar and Christodoulopoulou 2014)

Environmental organizational social responsibility and green management performance are concepts that can be integrated in a framework of environmental management (Chang, Yeh, & Li, 2020; Chuang & Huang, 2015; Chuang & Huang, 2018; Kao, Liu, Huang, & Chang, 2010; Lyon & Maxwell, 2008; Post, Rahman, & Rubow, 2011; Rahman & Post, 2012). Environmental management comprises organizational social responsibility forming an integrative framework with green organizational image and competitiveness. Green corporate image environmental management depends on the industry in which the organization operates (Frooman, 1999).

Green organizational identity is part of the environmental management framework related to environmental protection and maintenance constructed by organizational members to make their behaviors meaningful (Chen, 2011). Green organizational identity has a positive effect on organizational members who jointly construct to make meaningful their behaviors withing an environmental management framework Green organizational identity is the basis of the organizational environmental management framework related to the protection and maintenance of the environment constructed by members to make meaningful behaviors (Chen, 2011).

Environmental management and corporate image contribute to enhance green reputation, green competitiveness, and green organizational performance (Barney 1991; Li, Ngniatedema, and Chen (2017). Environmental management enhance the ecological legitimacy of the firm by exerting initiatives to create and develop organizational green image (Amores-Salvadó, Martín-de Castro, & Navas-López, 2014).

Environmental management supports green innovation practices with the improvement of environmental benefits and efficiency (Chen et al., 2006; Pan & Tian, 2017; Zhang et al., 2015). Green innovation practices are concerned with environmental management policies and procedures and the best green practices (Sroufe & Gopalakrishna-Remani, 2018). Green productivity applies environmental management technologies, tools, and techniques to reduce the environmental impact of organizational activities. Green productivity is the improvement of both environmental performance and productivity of organizations, communities and service providers aimed to achieve higher business performance through and in conjunction with environmental management (Ottman, et al, 2006, Beamon, 1999).

Green human resources management integrates environmental management (Amrutha and Geetha 2020). Green human resource management has different roles in environmental management based on organizational experience, environmental competencies and green capabilities that have an impact on green engagement (Renwick et al. 2013). Green human resources management identifies organizations and corporations that develop green human resources policies, implement common processes, and provide some green initiatives on environmental management and sustainable development including awareness on environmental affairs and the socio-economic well-being of both employees and the organizations. The green human resources management is concerned with the environmental management of economic development (Renwick, Redman, & Maguire, 2013, p. 3).

Green human resources have the main objective to promote functions and actions awareness of organizational environmental management, motivating and developing a sense of pride of employees for participation on a green program. Green human resources management practices have multifaceted effects and potential impacts on environmental management issues involving policies and practices requiring constant monitoring aligned with sustainability environment, economic and social balance (Yusliza, Ramayah, & Othman, 2015, p. 1).

The strength of green human resource policies is related to the adoption intensity of environmental management systems and policies of different companies (Bohdanowicz, Zientara, & Novotna, 2011). Human capital role in implementing environmental management is a mature field known as green human resource management.

The recruitment practices of green employees supported by environmental management ensures the entrants possess the environmental organizational culture and can maintain environmental organizational values (Wehrmeyer 1996). The green organizational culture may support the environmental management. Green human development and training creates environmental awareness about the environmental management value based on methods to use and conserve renewable energy, and reduce waste materials (Zoogah, 2011).

Training and human development practices should focus on development of attitudes, knowledge and skills related to environmental management (Zoogah, 2011, p. 17). There is a positive relationship between environmental management and training (Jabbour 2015). Green knowledge sharing behavior can be applied among members of the organization and must intensify the green product property ownership to engage them on organizational performance through meeting the environmental demands and the environmental management concept (Chang, Wang, Lin, 2020).

The best ideas about green environmental management green issues, eco-friendly practices, etc., come from the workers in specific areas and different sources (Casler, Gundlach, Persons, & Zivnuska, 2010). Participation of employees in organizational environmental programs improves with compensation and green rewards to take up initiatives and practices more committed to achieve efficacy of environmental management programs (Forman and Jorgensen 2001; Berrone and Gomez-Mejia 2009).

Green human resource management has a relationship and a relevant role with environmental strategy and environmental management activities. Green human capital management is instrumental to the fulfillment of organizational environmental management objectives (Hersey, 1998). Environmental management is critical to organizational strategy and public policies affect the development of the green economy conditioning the sustainable environmental technology and the green sector (Hamdouch, Depret, 2010; Starik, Throop, Doody, Joyce, 1996).

ORGANIZATIONAL GREEN TRANSFORMATIONAL LEADERSHIP

Green transformational leadership increases the overall organizational performance hinged on the green resilience of the team and influences the green work engagement in the overall organizational sustainability and green performance (Ramus & Steger, 2000). The leadership styles as contextual factors have an influence on the green outcomes of followers. Albort-Morant, Henseler, Cepeda-Carrión, Leal-Rodríguez, 2018; Chen, Chang, 2013).

The leadership style behaviors signal the team green initiatives to support the organizational green innovation, motivating the collective voluntary followers attempts to innovate green processes and products (Robertson, Barling, 2013; Chen, Chang, 2013; Akgün, Keskin, Byrne, 2009; Adner, Levinthal, 2001; Chang, Chen, 2013). Green leadership can arouse green teams to involve them in green innovative processes and products.

Leadership styles displaying green behaviors voluntarily such as green transformational leadership, and empowering leadership motivate employees to pursue the green teams' interests (Chen, Chang, Lin, 2014; Srivastava, Bartol, Locke, 2006). The leadership style behaviors signal the team green initiatives to support the organizational green innovation, motivating the collective voluntary followers attempts to innovate green processes and products (Robertson, Barling, 2013; Chen, Chang, 2013; Akgün, Keskin, Byrne, 2009; Adner, Levinthal, 2001; Chang, Chen, 2013).

There is no empirical research on transformational leadership or environmental sustainability within organizations (Barling et al., 2010). Research in environmental issues at multiple levels lacks the ability to depict the influences of green leadership on the green team in green innovation. There is a positive effect between green leadership linked to team green innovation, green process and product innovation at the team level and contributes to environmental concerns and issues. Team green efficacy mechanism links the relations between green leadership and team green process and product innovation.

Green leadership promotes the collective interests by organizing and encouraging building teams to give higher value to all the members and contribute to the organizational green development (Ayoko, Chua, 2014). Green leadership give support to the team members to build confidence through environmental motivation signaling shared values on green issues to encourage green outcomes (Epitropaki, Kark, Mainemelis, Lord, 2017).

Green leadership may exert influence on team outcomes to facilitate the environmental practices and improve green innovation to solve environmental problems (Marquis, Jackson, Li, 2015; Anderson, Poto˘cnik, Zhou, 2014; Jabbour, Santos, Fonseca, Nagano, 2013). The environmental oriented leadership behavior is identified as a critical component of organizational green innovation which may benefit the voluntary workplace green behavior affecting the organizational green processes and product innovation. Organizational green leadership has benefits on green innovation (Jabbour, Santos, Fonseca, Nagano, 2013; Pearce, Ensley, 2004). Organizational green leadership engagement into the green activities voluntarily to create and develop a collective green environmentally oriented behavior (Kim, Kim, Han, Jackson, Ployhart, 2017).

Leadership green behavior is relevant to identify and predict green innovation (Robertson, J.L.; Barling, J. (2013). Afsar, Badir, Kiani, 2016; Flannery, May 1994; Goldstein, Cialdini, Griskevicius, 2008; Egri, Herman, 2000; Fernández, Junquera, Ordiz, 2006). The environmental oriented behaviors of leaders are indirect towards environmentally specific outcomes through motivations, beliefs, and satisfactions (Robertson, Barling, 2013; Kim, Kim, Han, Jackson, Ployhart, 2017; Whitmarsh, O'Neill, 2010). Green prosocial motivation has a mediating role on the effect of leadership on teams' green innovation leading to psychological mechanisms supporting the impact of green leadership on green team behavior.

Leadership voluntary workplace green behaviors signal motivation extending to team green to explain the indirect influence on team green innovation (Kim, Kim, Han, K.; Jackson, Ployhart, 2017; Bandura, 2000). Leaders should exert behaviors to develop the collective efficacy of the team capabilities to realize green motivations and participate in green activities. Green leaders demonstrate green behaviors voluntarily to develop collective confidence in green teams to develop green innovative processes and to produce green innovative products Robertson, J.L.; Barling, 2013).

Teams engaged in green activities driven by leadership green oriented behavior and environmental sustainability passion can build towards green innovation and increase environmental behavior. Leadership specific green voluntary behavior may exert effects on team support to achieve environmental sustainability and green innovative outcomes (Robertson, Barling, 2013).

Leaders act as a role model to activate similar activities in followers Robertson, Barling, 2013). Leadership voluntary workplace green behavior is characterized by motivation based on green activities and the willingness of the team members to work together through communications and interactions resulting in capabilities to engage and accomplish green tasks and on the formation of green endeavors (Kim, Kim, Han, Jackson, Ployhart, 2017; Liu, Chen, Tao, 2015).

The role model perspective (Robertson, Barling, 2013) and the social cognitive theory (Bandura, 1986). sustain the influence of the leadership voluntary workplace green behaviors on the team green processes and products innovation and find a positive relationship and a mediation effect of team green efficacy on the leadership voluntary workplace green behaviors team green innovation relationship. Organizational green leadership based on social cognitive theory, uses motivational environmentally specific motivation processes as a mechanism to contribute to team green innovation (Bandura, 1986, 2000).

Leadership promoting follower's efficacy beliefs about green capabilities, according to the social cognitive theory of Bandura (1986), improve the green environment (Chen, Chang, Lin, 2014; Liu, Chen, Tao, 2015; Mahroum, Al-Saleh, 2013). Leadership organizational green innovation is a mediating role of individual psychological intervening mechanisms of spiritual, intrinsic motivation and environmental passion states, for example (Robertson, Barling, 2013; Afsar, Badir, Kiani, 2016). Other potential psychological mechanisms may influence of leadership voluntary workplace green behaviors at the team level on green innovation (Mittal, Dhar, 2016).

The role model perspective (Robertson, and Barling, 2013) and social cognitive theory (Bandura, 1986) support that the team green is influences by the leaders of voluntary workplace green behaviors and the team green innovation in the dimensions of green process innovation and green product innovation (Albort-Morant, Henseler, Cepeda-Carrión, Leal-Rodríguez, 2018). Pursuing organizational green process and product innovation is for the best interest of team members stimulated by green leadership boosting the use of team knowledge and the abilities to produce green outcomes (Chen, Chang, 2013; Edmondson, Nembhard, 2009). Organizational environmental knowledge learning has a moderating effect on environmental leadership on organizational green innovation actions. Green innovation actions have a mediating role on the relationship between environmental leadership and performance.

Environmental knowledge learning has an impact on environmental leadership and this on green innovation. Organizational environmental knowledge learning has a moderating effect on environmental leadership on organizational green innovation strategy. The level of environmental knowledge learning helps the members of the organization to understand the environmental vision of leaders and facilitate green innovation practices. Organizational environmental knowledge learning involves the design and implementation of organizational green innovation strategy and actions that have an impact on the style of environmental leadership. Leadership voluntary behavior and attitudes should boost abilities of the team members to work together in green innovative processes and products.

Leaders enact green behaviors of organizing and leading voluntarily to show their commitment to organizational green innovative outcomes (Kim, Kim, Han, Jackson, Ployhart, 2017). Leaders' voluntary workplace green behaviors can promote green innovative outcomes and activate similar practices and complete tasks in green innovation in their followers (Robertson, Barling, 2013; Kim, Kim, Han, Jackson, Ployhart, 2017; Elenkov, Manev, 2005). Leadership behavior in organizations is identified to predict and promote green innovation outcomes. (Robertson, J.L.; Barling, J. (2013). Afsar, Badir, Kiani, 2016; Flannery, May 1994; Goldstein, Cialdini, Griskevicius, 2008; Egri, Herman, 2000; Fernández, Junquera, Ordiz, 2006).

The leaders' voluntary workplace green behavior is linked to motivational mechanisms and forms of team green process and product innovation. The motivations of voluntary workplace green behaviors at the level of team green efficacy explains the indirect influence of leaders on team green innovation Kim, Kim, Han, Jackson, Ployhart, 2017; Bandura, 2000). The mechanism of environmentally specific motivation links the styles of green leadership behaviors to team green innovation (Kim, Kim, Han, Jackson, Ployhart, 2017). Leaders' environmental behaviors towards environmentally specific outcomes are indirect through motivations, beliefs, and satisfaction (Robertson, Barling, 2013; Kim, Kim, Han, Jackson, Ployhart, 2017; Whitmarsh, O'Neill, 2010).

Green behaviors of leaders boost green innovative outcomes, thus leaders in voluntary workplace green discretionary behaviors in activities that benefit the sustainability of an ecofriendly organization engage followers in green innovation practices, not controlled by a system and policy of environmental protection Kim, Kim, Han, Jackson, Ployhart, 2017) Leadership voluntary workplace green behavior enact eco-friendly behaviors improves the organizational environmental performance (Chaudhary, 2020). Leaders' voluntary workplace green behavior effects on team green efficacy leads to team performance improvement of green process and product innovation (Jabbour, Santos, Fonseca, Nagano, 2013; Chen, Farh, Campbell-Bush, Wu, Wu, 2013)

Leaders of voluntary workplace green behaviors can create the basis for environmental practices and actions within the green team aimed to enhance their capabilities of achieving green innovation which may result in team green process and product innovation. The green innovation literature discusses a share construct supporting the managerial implications between the organizational leadership managing the green team´s innovation (Wu, Wu, Yuan, 2019; Chaudhary, 2020).

Leaders in voluntary workplace green behaviors achieve green innovation emphasized by the development of managerial abilities. Organizations should select potential leaders and provide them coaching, training and development programs to improve managerial and leadership abilities. Organizational green performance is improved with green transformational leadership and environmental training (Chen & Chang, 2013; Jabbour, 2013b). Organizational leadership should be aware of training to improve the abilities to design, develop and implement green actions aimed to increase the environmental performance. Green team mediates between green leadership and green team process and product innovation.

The green team structure nurtures innovation affected by leadership focusing on the relationship between voluntary workplace green behaviors and green innovation (Jabbour, Santos, Fonseca, Nagano, 2013; Chen, Farh, Campbell-Bush, Wu, Wu, 2013; Lim, Ployhart, 2004). Organizational green leadership inspire the green teams to implement environmental criteria to develop manufacturing innovative processes of green innovative products (Akgün, Keskin, Byrne, 2009). Green leadership fostering green innovation should cultivate the confidence of the team members in engaging green innovative activities. Selected candidates to the green team can motivate each other to produce green innovative processes and products.

Association between voluntary workplace green behaviors and team innovation, the leadership becomes very relevant (Anderson, Potoˇcnik, Zhou, 2014). The leaders of voluntary workplace green behavior affect the team´s green process and product innovation and stimulates the team-level green innovation with implications on the leadership development for sustainable environmental innovation. Leaders of voluntary workplace green behaviors have influence on team green innovation and the organizational green innovative outcomes. Leadership green behavior is relevant to identify and predict green innovation (Robertson, J.L.; Barling, J. (2013). Afsar, Badir, Kiani, 2016; Flannery, May 1994; Goldstein, Cialdini, Griskevicius, 2008; Egri, Herman, 2000; Fernández, Junquera, Ordiz, 2006).

The team green innovation dimension reveals the knowledge of leadership voluntary workplace green behavior determining the team green innovation behaviors (Dangelico, 2016). Leadership focusing on team has relevant consequences (Lim, Ployhart, 2004) and there is a significant relationship between the leader's voluntary workplace green behavior and team green innovation. Green leadership voluntary workplace green behaviors enact green teams to facilitate green innovative processes and products. The characteristics of motivation establish the process of leadership impacts on green behaviors (Norton, Parker, Zacher, Ashkanasy, 2015).

Implications

There is a green relationship between the green leadership behavior and the green behavior of individual subordinates. Individuals and group leaders in the workplaces are in a relationship between conscientiousness, moral reflectiveness, and voluntary environmental behavior (Kim et al. 2016). Organizational leaders consider a moral mandate and a legal requirement the greening of the organizational model to maintain the right and legitimacy to operate (Hart and Milstein 2003).

The leadership environmental norms, their environmentally transformational leadership and the pro environmental behaviors influence the organizational greening practices and activities in the workplace. Pro-environmental leadership behaviors that enact environmental leader behaviors are relevant factors in the process of greening the organizations. Environmental behaviors meeting the environmental standards is related to environmental leadership and organizational performance considering green innovation practices as a mediating role and environmental knowledge learning as a moderating role. Organizational environmental leadership is crucial for the design and implementation of green innovation practices.

Environmental knowledge learning moderates' organizational environmental leadership on green innovation practices. Environmental knowledge learning has a moderating effect on the relationship between organizational environmental leadership and green innovation practices. Organizational environmental leadership contributes to environmental and financial performance (Liu et al., 2018). Organizational green innovation strategy has a mediating effect of environmental leadership on organizational financial performance.

Leadership pro-environmental behavior facilitates the organizational sustainable innovation processes and products by encouraging followers to initiate green practices, green information, and green knowledge. Chen, Chang, 2013; Bossink, 2007). Green leadership promotes information exchange on organizational environmental concerns and issues, eco-friendly supplies, recycling activities, waste management, etc., enabling to experience green behaviors during the engagements in innovation processes and products (Kim, Kim, Han, Jackson, Ployhart, 2017)

Green transformational leadership has positive effects on green work engagement as the mediating role and green team resilience. Green transformational leadership inspire organizational members to achieve environmental and sustainable development goals exceeding environmental performance (Chen and Chang 2013). The environmental engagement and the leadership and pro-environmental behaviors at the workplace predict the pro-environmental behaviors o employees at the workplace.

The green psychological climate is a mechanism of ethical leadership that influence organizational citizen behavior ethics and has relevant implications on their relationship boundaries and knowledge drawn from social role theory and (Eagly, 1987; Eagly and Wood, 2012) and from socialization theory (Schahn and Holzer, 1990). The perception of psychological green climate affects the pro-environmental behaviors mediating the link between ethical leadership and organizational environmental citizenship behavior (Dumont et al., 2017; Norton et al., 2017; Zhou et al., 2018).

Green organizational psychological climate has a mediating role in the relationship between ethical leadership and organizational citizen behavior ethics, contributing in enhance knowledge in green organizational psychological climate (Norton et al., 2017; Zhou et al., 2018), and OCBE (Robertson and Barling, 2017; Saeed et al., 2018; Testa et al., 2018; Woosnam et al., 2019), ethical leadership (Brown et al., 2005; Ahn et al., 2018; Bavik et al., 2018; Byun et al., 2018), and outcomes of organizational citizens behavior ethics.

There is a direct relationship between ethical leadership and organizational citizen behavior ethics regarding green organizational psychological climate as a mediating variable. Green organizational psychological climate mediates the relationship between ethical leadership and organizational citizens behavior ethics (Robertson and Barling, 2013, 2017; Norton et al., 2017). Ethical leadership on green organizational psychological climate and organizational citizen behavior ethics have effects on pro-environmental standards and behaviors moderated by gender. A theoretical framework developed by Khan, Jianguo, Ali, Saleem and Usman (2019) links organizational environmental citizenship behavior and ethical leadership in a relationship mediated by a green psychological climate.

A framework linking gender and ethical leadership, green psychology climate and organizational citizen ethics enhance the effects of environmental social learning processes. The structural research framework model suggests that gender is mediating green organizational psychological climate in the relationship between ethical leadership and organizational citizen behavior ethics.

Gender should be considered in the analysis of pro-environmental behaviors knowing that moderates the relationship between ethical leadership and organizational citizenship behavior ethics trough psychological green climate (Eisler et al., 2003; Xiao and McCright, 2015; Kennedy and Kmec, 2018; Vicente-Molina et al., 2018). Green organizational psychological climate mediates the relationship between ethical leadership and organizational citizen behavior ethics.

The theories of broaden-and-build, job demand–resource, and conservation of resource are a framework for green transformational leadership, green work engagement as the mediating role to green team resilience. Green transaction leadership has a mediation effect on green work engagement (Chen, Chang, & Lin, 2014). Environmentally transformational leadership is linked to subordinates' environmental behavior at the workplace. Transformational leadership is associated with workers' engagement with the environmental behaviors and safety (Barling, Weber, & Kelloway, 1996; Dvir, Eden, Avolio, & Shamir, 2002; Mullen & Kelloway, 2009).

Implications

The transformational leadership behaviors focus on pro-environmental initiatives of transformational leadership that affect employees' pro-environmental engagement and behaviors. Green organizational transformational leadership is oriented towards developing a green vision and shaping pro-environmental behaviors affecting the engagement of employees in pro-environmental green behaviors (Robertson and Barling, 2017; Wang et al., 2018). Green transformational leadership shapes the green-related organizational vision towards development towards green-related issues and provides feedback on the ecological issues and concerns. Green transformational leadership should articulate the organizational green vision and organizations green norms and regulations providing the followers with the required resources to achieve their objectives (Mittal and Dhar, 2016).

Research on organizational green transformation and operationalization of environmental sustainability in multilayered industry settings including resources and non-resource intensive organizations model greening. The construct of green transformational leadership supports the finding that perception that transforms the green team resilience and resistance of negative organizational practices (Chen and Chang 2013; Podsakoff, Mackenzie, and Bommer 1996). Environmental norms influence leaders' environmental transformational leadership and the workplace environmental behaviors, and the leaders' role in the employees' engagement for the environmentally friendly behaviors.

Green transformational leadership should focus on the effects of other leadership variables in promoting environmental behaviors in the workplace and human resource practices in the organizational greening (Jackson, Renwick, Jabbour, & Muller-Camen, 2011; Jackson & Seo, 2010). Environmentally transformational leadership predicts environmental norms and pro-environmental behaviors which in turn predict at the workplace the environmental engagement of employees.

Green transformational leadership has effects on green team resilience, although this relationship has not been well studied. Green transformational leadership positively influences green team resilience. There is empirical evidence of the relations of leader–team level green innovation Norton, Parker, Zacher, Ashkanasy, 2015; Kim, Kim, Han, Jackson, Ployhart, 2017) Green transformational leadership provide models stimulate task-engaging behaviors, (Bass & Stogdill, 1990). Green transformational leadership and dynamic capabilities improve green product development performance (Chen & Chang, 2013).

Green servant leadership signals followers the green practices and activities that can support green innovative outcomes. Green servant leadership has a positive influence on innovation and can predict ecological outcomes (Cai, Lysova, Khapova, Bossink, 2018; Luu, 2019). Green innovation strategies, practices and actions have a mediating role between organizational environmental leadership and organizational performance.

CONCLUSION

Organizational sustainable environmental management may overcome environmental challenges, strengthen corporate image, and achieve green consumption goals. Environmental management issues are related to organizational sustainable development encouraging organizations to act in favor of green consumers and achieving sustainable development goals (SDGs)

Organizational leadership capability environmentally oriented is a motivation for developing and implementing green innovation practices and activities. Organizational environmental leadership under certain circumstances motivate green innovation practices and organizational performance. Organizational green innovation strategy has a mediating effect of environmental leadership on organizational environmental performance. Organizational green innovation practices bridge environmental leadership and organizational performance. Green innovation practices bridge environmental leadership and organizational sustainable development.

Green performance management as the culmination of corporate strategy affected by the environmental management, is the process of enhancement professional skills of employees to achieve organizational objectives. Green human resources have the promise to all organizational stakeholders, researchers, and academics to support green initiatives and have an influence on the formulation and implementation of environmental management strategies. Organizational environmental leadership need to meet the expectations, demands and pressures of the stakeholders by implementing, as an antecedent of green innovation, practices and activities aimed to producing green products to meet the needs and demand of consumers and improve environmental benefits.

Managers and employees should be knowledgeable about environmental leadership styles adopt green innovation strategies, practices and operations and organizational performance. Organizational environmental knowledge learning supports environmental leaders to improve green technologies, process, and products with an impact on organizational performance. Environmental knowledge learning is a condition for environmental leadership that has an impact on green innovation practices

REFERENCES

Adner, R., & Levinthal, D. (2001). Demand heterogeneity and technology evolution: Implications for product and process innovation. *Management Science*, *2001*(47), 611–628. doi:10.1287/mnsc.47.5.611.10482

Afsar, B., Badir, Y., & Kiani, U. S. (2016). Linking spiritual leadership and employee pro-environmental behavior: The influence of workplace spirituality, intrinsic motivation, and environmental passion. *Journal of Environmental Psychology, 2016*(45), 79–88. doi:10.1016/j.jenvp.2015.11.011

Ahn, J., Lee, S., & Yun, S. (2018). Leaders' core self-evaluation, ethical leadership, and employees' job performance: The moderating role of employees' exchange ideology. *Journal of Business Ethics, 148*(2), 457–470. doi:10.100710551-016-3030-0

Akgün, A. E., Keskin, H., & Byrne, J. (2009). Organizational emotional capability, product and process innovation, and firm performance: An empirical analysis. *Journal of Engineering and Technology Management, 2009*(26), 103–130. doi:10.1016/j.jengtecman.2009.06.008

Albino, V., Balice, A., & Dangelico, R. M. (2009). Environmental strategies and green product development: An overview on sustainability driven companies. *Business Strategy and the Environment, 2009*(18), 83–96. doi:10.1002/bse.638

Albort-Morant, G., Henseler, J., Cepeda-Carrión, G., & Leal-Rodríguez, A. L. (2018). Potential and realized absorptive capacity as complementary drivers of green product and process innovation performance. *Sustainability, 2018*(10), 381. doi:10.3390u10020381

Amores-Salvadó, J., Martín-de Castro, G., & Navas-López, J. E. (2014). Green corporate image: Moderating the connection between environmental product innovation and firm performance. *Journal of Cleaner Production, 83*, 356–365. doi:10.1016/j.jclepro.2014.07.059

Amrutha, V. N., & Geetha, S. N. (2020). A systematic review on green human resource management: Implications for social sustainability. *Journal of Cleaner Production, 247*, 119131. doi:10.1016/j.jclepro.2019.119131

Anderson, N., Potočnik, K., & Zhou, J. (2014). Innovation and creativity in organizations: A state-of-the-science review, prospective commentary, and guiding framework. *Journal of Management, 40*(5), 1297–1333. doi:10.1177/0149206314527128

Anderson, R. (2009). *Confessions of a Radical Revolutionist*. St Martin's Press.

Bandura, A. (1986). The explanatory and predictive scope of self-eficacy theory. *Journal of Social and Clinical Psychology, 1986*(4), 359–373. doi:10.1521/jscp.1986.4.3.359

Bandura, A. (2000). Exercise of human agency through collective efficacy. *Curr. Dir. Psychol., 2000*(9), 75–78. doi:10.1111/1467-8721.00064

Barling, J., Christie, A., & Hoption, C. (2010). Leadership. In S. Zedeck (Ed.), *APA handbook of industrial and organizational psychology* (Vol. 1). American Psychological Association.

Barling, J., Weber, T., & Kelloway, K. E. (1996). Effects of transformational leadership training on attitudinal and financial outcomes: A field experiment. *The Journal of Applied Psychology, 81*(6), 827–832. doi:10.1037/0021-9010.81.6.827

Barney, J. (1991). Firm resources and sustained competitive advantage. *Journal of Management, 17*(1), 99–120. doi:10.1177/014920639101700108

Bass, B. M., & Stogdill, R. M. (1990). *Bass & Stogdill's handbook of leadership: Theory, research, and managerial applications*. Simon and Schuster, Free Press.

Bavik, Y. L., Tang, P. M., Shao, R., & Lam, L. W. (2018). Ethical leadership and employee knowledge sharing: Exploring dual-mediation paths. *The Leadership Quarterly, 29*(2), 322–332. doi:10.1016/j.leaqua.2017.05.006

Beamon, B. M. (1999). Designing the Green Supply Chain. *Logistics Information Management, 12*(4), 332–342. doi:10.1108/09576059910284159

Berrone, P., & Gomez-Mejia, L. R. (2009). Environmental performance and executive compensation: An integrated agency-institutional perspective. *Academy of Management Journal, 52*(1), 103–126. doi:10.5465/amj.2009.36461950

Bohdanowicz, P., Zientara, P., & Novotna, E. (2011). International hotel chains and environmental protection: An analysis of Hilton's we care! programme (Europe, 2006–2008). *Journal of Sustainable Tourism, 19*(7), 797–816. doi:10.1080/09669582.2010.549566

Bossink, B. A. (2007). Leadership for sustainable innovation. *International Journal of Technology Management & Sustainable Development, 2007*(6), 135–149. doi:10.1386/ijtm.6.2.135_1

Brown, M. E., Treviño, L. K., & Harrison, D. A. (2005). Ethical leadership: A social learning perspective for construct development and testing. *Organizational Behavior and Human Decision Processes, 97*(2), 117–134. doi:10.1016/j.obhdp.2005.03.002

Byun, G., Karau, S. J., Dai, Y., & Lee, S. (2018). A three-level examination of the cascading effects of ethical leadership on employee outcomes: A moderated mediation analysis. *Journal of Business Research, 88*, 44–53. doi:10.1016/j.jbusres.2018.03.004

Cai, W., Lysova, E. I., Khapova, S. N., & Bossink, B. A. (2018). Servant leadership and innovative work behavior in Chinese high-tech firms: A moderated mediation model of meaningful work and job autonomy. *Frontiers in Psychology, 2018*(9), 1767. doi:10.3389/fpsyg.2018.01767 PMID:30327623

Casler, A., Gundlach, M. J., Persons, B., & Zivnuska, S. (2010). Sierra Nevada Brewing Company's thirty-year journey toward sustainability. *People & Strategy, 33*, 44–51.

Chang, C. H., & Chen, Y. S. (2013). Green organizational identity and green innovation. *Management Decision, 2013*(51), 1056–1070. doi:10.1108/MD-09-2011-0314

Chang, T. W., Chen, Y. S., Yeh, Y. L., & Li, H. X. (2020). Sustainable consumption models for customers: Investigating the significant antecedents of green purchase behavior from the perspective of information asymmetry. *Journal of Environmental Planning and Management, 2020*, 1–21.

Chang, T. W., Yeh, Y. L., & Li, H. X. (2020). How to shape an organization's sustainable green management performance: The mediation effect of environmental corporate social responsibility. *Sustainability, 12*(21), 9198. doi:10.3390u12219198

Chaudhary, R. (2020). Green human resource management and employee green behavior: An empirical analysis. *Corporate Social Responsibility and Environmental Management, 2020*(27), 630–641. doi:10.1002/csr.1827

Chen, G., Farh, J. L., Campbell-Bush, E. M., Wu, Z., & Wu, X. (2013). Teams as innovative systems: Multilevel motivational antecedents of innovation in R&D teams. *The Journal of Applied Psychology, 2013*(98), 1018–1027. doi:10.1037/a0032663 PMID:23565898

Chen, Y. S. (2011). Green organizational identity: Sources and consequence. *Management Decision, 2011*(49), 384–404. doi:10.1108/00251741111120761

Chen, Y. S., & Chang, C. H. (2013). The determinants of green product development performance: Green dynamic capabilities, green transformational leadership, and green creativity. *Journal of Business Ethics, 116*(1), 107–119. doi:10.100710551-012-1452-x

Chen, Y. S., Chang, C. H., & Lin, Y. H. (2014). Green transformational leadership and green performance: The mediation effects of green mindfulness and green self-efficacy. *Sustainability, 6*(10), 6604–6621. doi:10.3390u6106604

Chen, Y. S., Lai, S. B., & Wen, C. T. (2006). The influence of green innovation performance on corporate advantage in Taiwan. *Journal of Business Ethics*, *67*(4), 331–339. doi:10.100710551-006-9025-5

Chuang, S. P., & Huang, S. J. (2015). Effects of business greening and green IT capital on business competitiveness. *Journal of Business Ethics*, *128*(1), 221–231. doi:10.100710551-014-2094-y

Chuang, S. P., & Huang, S. J. (2018). The effect of environmental corporate social responsibility on environmental performance and business competitiveness: The mediation of green information technology capital. *Journal of Business Ethics*, *150*(4), 991–1009. doi:10.100710551-016-3167-x

Çop, S., Olorunsola, V. O., & Alola, U. V. (2021). Achieving environmental sustainability through green transformational leadership policy: Can green team resilience help? *Bus Strat Env.*, *30*, 671–682. https://onlinelibrary.wiley.com/doi/epdf/10.1002/bse.2646 doi:10.1002/bse.2646

Daily, B. F., Bishop, J. W., & Govindarajulu, N. (2009). A conceptual model for organizational citizenship behavior directed toward the environment. *Business Strategy and the Environment*, *13*(4), 209–222.

Dangelico, R. M. (2016). Green product innovation: Where we are and where we are going. *Business Strategy and the Environment*, *2016*(25), 560–576. doi:10.1002/bse.1886

Dangelico, R. M., & Pujari, D. (2010). Mainstreaming green product innovation: Why and how companies integrate environmental sustainability. *Journal of Business Ethics*, *2010*(95), 471–486. doi:10.100710551-010-0434-0

Dumont, J., Shen, J., & Deng, X. (2017). Effects of green HRM practices on employee workplace green behavior: The role of psychological green climate and employee green values. *Human Resource Management*, *56*(4), 613–627. doi:10.1002/hrm.21792

Dvir, T., Eden, D., Avolio, B. J., & Shamir, B. (2002). Impact of transformational leadership on follower development and performance: A field experiment. *Academy of Management Journal*, *45*, 735–744.

Eagly, A. H. (1987). *Sex differences in social behavior: A social role interpretation*. Erlbaum.

Eagly, A. H., & Wood, W. (2012). Social role theory. In P. A. M. Van Lange, A. W. Kruglanski, & E. T. Higgins (Eds.), *Handbook of theories of social psychology* (pp. 458–476). Sage. doi:10.4135/9781446249222.n49

Edmondson, A. C., & Nembhard, I. M. (2009). Product development and learning in project teams: The challenges are the benefits. *Journal of Product Innovation Management*, *2009*(26), 123–138. doi:10.1111/j.1540-5885.2009.00341.x

Egri, C. P., & Herman, S. (2000). Leadership in the North American environmental sector: Values, leadership styles, and contexts of environmental leaders and their organizations. *Academy of Management Journal*, *2000*(43), 571–604.

Eisenhardt, K. M., & Martin, J. A. (2000). Dynamic capabilities: What are they? *Strategic Management Journal*, *32*(10-11), 1105–1121. doi:10.1002/1097-0266(200010/11)21:10/11<1105::AID-SMJ133>3.0.CO;2-E

Eisler, I., Le Grange, D., & Asen, K. (2003). Family interventions. In J. Treasure, U. Schmidt, & E. van Furth (Eds.), *Handbook of eating disorders: Theory, treatment and research* (pp. 291–310). Wiley.

Elenkov, D. S., & Manev, I. M. (2005). Top management leadership and influence on innovation: The role of sociocultural context. *Journal of Management*, *2005*(31), 381–402. doi:10.1177/0149206304272151

Epitropaki, O., Kark, R., Mainemelis, C., & Lord, R. G. (2017). Leadership and followership identity processes: A multilevel review. *The Leadership Quarterly*, *2017*(28), 104–129. doi:10.1016/j.leaqua.2016.10.003

Fernández, E., Junquera, B., & Ordiz, M. (2006). Managers' profile in environmental strategy: A review of the literature. *Corporate Social Responsibility and Environmental Management*, *2006*(13), 261–274. doi:10.1002/csr.109

Flannery, B. L., & May, D. R. (1994). Prominent factors influencing environmental activities: Application of the environmental leadership model (ELM). *The Leadership Quarterly*, *1994*(5), 201–221. doi:10.1016/1048-9843(94)90012-4

Forman, M., & Jorgensen, S. (2001). The social shaping of participation of employees in environmental work within enterprises—Experiences from a Danish context. *Technology Analysis and Strategic Management*, *13*, 71–90. doi:10.1080/09537320120040455

Frooman, J. (1999). Stakeholder influence strategies. *Academy of Management Review*, *24*(2), 191–205. doi:10.5465/amr.1999.1893928

Goldstein, N. J., Cialdini, R. B., & Griskevicius, V. (2008). A room with a viewpoint: Using social norms to motivate environmental conservation in hotels. *The Journal of Consumer Research*, *2008*(35), 472–482. doi:10.1086/586910

González-Benito, J., & González-Benito, O. (2006). A review of determinant factors of environmental proactivity. *Business Strategy and the Environment, 15*(2), 87–102. doi:10.1002/bse.450

Greenwood, L., Rosenbeck, J., & Scott, J. (2012). The role of the environmental manager in advancing environmental sustainability and social responsibility in the organization. *Journal of Environmental Sustainability, 2*(2), 59–75.

Hamdouch, A., & Depret, M. H. (2010). Policy integration strategy and the development of the 'green economy': Foundations and implementation patterns. *Journal of Environmental Planning and Management, 2010*(53), 473–490. doi:10.1080/09640561003703889

Hanna, M. D., Rocky Newman, W. R., & Johnson, P. (2000). Linking operational and environmental improvement through employee involvement. *International Journal of Operations & Production Management, 20*(2), 148–165. doi:10.1108/01443570010304233

Hart, S. L. (1995). A natural-resource-based view of the firm. *Academy of Management Review, 20*(4), 986–1014. doi:10.5465/amr.1995.9512280033

Hart, S.L., & Milstein, M.B. (2003). Creating sustainable value. *Acad. Manag. Exec., 17*(2), 56-69.

Hersey, K. (1998). A close look at ISO 14000. *Professional Safety, 43*, 26–29.

Jabbour, Ch., & Chiappetta, J. (2015). Environmental training and environmental management maturity of Brazilian companies with ISO14001: Empirical evidence. *Journal of Cleaner Production, 96*, 331–338. doi:10.1016/j.jclepro.2013.10.039

Jabbour, Ch. J. Ch. (2011). How green are HRM practices, organizational culture, learning and teamwork? A Brazilian study. *Industrial and Commercial Training, 43*(2), 98–105. doi:10.1108/00197851111108926

Jabbour, C. J. C. (2013b). Environmental training and environmental maturity of Brazilian companies with ISO14001: Empirical evidence. *Journal of Cleaner Production, 43*(2), 1–8.

Jabbour, C. J. C., Santos, F. C. A., Fonseca, S. A., & Nagano, M. S. (2013). Green teams: Understanding their roles in the environmental management of companies located in Brazil. *Journal of Cleaner Production, 2013*(46), 58–66. doi:10.1016/j.jclepro.2012.09.018

Jackson, S. E., Renwick, D. W. S., Jabbour, C. J. C., & Muller-Camen, M. (2011). State-of-the-art and future directions for green human resource management: Introduction to the special issue. *German Journal of Research in Human Resource Management*, *25*(2), 99–116. doi:10.1177/239700221102500203

Jackson, S. E., & Seo, J. (2010). The greening of strategic HRM scholarship. *Organizational Management Journal*, *7*(4), 278–290. doi:10.1057/omj.2010.37

Kao, M. R., Liu, C. Y., Huang, Y. C., & Chang, N. J. (2010). Research of the relationship among business green management, environmental performance, and competitive advantage. *Journal of Management & Systems*, *17*(2), 255–278.

Kennedy, E. H., & Kmec, J. (2018). Reinterpreting the gender gap in household pro-environmental behaviour. *Environmental Sociology*, *4*(3), 299–310. doi:10.1080/23251042.2018.1436891

Khan, M. A. S., Jianguo, D., Ali, M., Saleem, S., & Usman, M. (2019). Interrelations Between Ethical Leadership, Green Psychological Climate, and Organizational Environmental Citizenship Behavior: A Moderated Mediation Model. *Frontiers in Psychology*, *10*, 1977. doi:10.3389/fpsyg.2019.01977 PMID:31555173

Kim, A., Kim, Y., Han, K., Jackson, S. E., & Ployhart, R. E. (2016). Multilevel influences on voluntary workplace green behavior: Individual differences, leader behavior, and coworker advocacy. *Journal of Management*, *43*(5), 1335–1358. doi:10.1177/0149206314547386

Kim, A., Kim, Y., Han, K., Jackson, S. E., & Ployhart, R. E. (2017). Multilevel influences on voluntary workplace green behavior: Individual differences, leader behavior, and coworker advocacy. *Journal of Management*, *2017*(43), 1335–1358. doi:10.1177/0149206314547386

Kitazawa, S., & Sarkis, J. (2000). The relationship between ISO 14001 and continuous source reduction programs. *International Journal of Operations & Production Management*, *20*(2), 225–248. doi:10.1108/01443570010304279

Kumar, V., & Christodoulopoulou, A. (2014). Sustainability and branding: An integrated perspective. *Industrial Marketing Management*, *43*(1), 6–15. doi:10.1016/j.indmarman.2013.06.008

Lampikoski, T., (2012). Green, innovative, and profitable: a case study of managerial capabilities at Interface Inc. *Tech. Inn. Manag. Rev.*, *2*(11), 4-12.

Li, S., Ngniatedema, T., & Chen, F. (2017). Understanding the impact of green initiatives and green performance on financial performance in the US. *Business Strategy and the Environment*, *26*(6), 776–790. doi:10.1002/bse.1948

Lim, B. C., & Ployhart, R. E. (2004). Transformational leadership: Relations to the five-factor model and team performance in typical and maximum contexts. *The Journal of Applied Psychology*, *2004*(89), 610–621. doi:10.1037/0021-9010.89.4.610 PMID:15327348

Liu, J., Chen, J., & Tao, Y. (2015). Innovation Performance in New Product Development Teams in China's Technology Ventures: The Role of Behavioral Integration Dimensions and Collective Efficacy. *Journal of Product Innovation Management*, *2015*(32), 29–44. doi:10.1111/jpim.12177

Liu, J., Yuan, Y., & Zhang, J. (2018). Environmental leadership in organizations: A literature review and prospects. *Economic Management Journal*, *40*(10), 193–208.

Luu, T. T. (2019). Green human resource practices and organizational citizenship behavior for the environment: The roles of collective green crafting and environmentally specific servant leadership. *Journal of Sustainable Tourism*, *2019*(27), 1167–1196. doi:10.1080/09669582.2019.1601731

Lyon, T. P., & Maxwell, J. W. (2008). Corporate social responsibility and the environment: A theoretical perspective. *Review of Environmental Economics and Policy*, *2*(2), 240–260. doi:10.1093/reep/ren004

Marquis, C., Jackson, S. E., & Li, Y. (2015). Building sustainable organizations in China. *Management and Organization Review*, *2015*(11), 427–440. doi:10.1017/mor.2015.37

Marshall, M. E., & Mayer, D. W. (1992). Environmental training: It's good business. *Business Horizons*, *35*(2), 54–58. doi:10.1016/S0007-6813(05)80197-1

Mittal, S., & Dhar, R. L. (2016). Effect of green transformational leadership on green creativity: A study of tourist hotels. *Tourism Management*, *57*, 118–127. doi:10.1016/j.tourman.2016.05.007

Mullen, J. E., & Kelloway, E. K. (2009). Safety leadership: A longitudinal study of the effects of transformational leadership on safety outcomes. *Journal of Occupational and Organizational Psychology*, *82*(2), 253–272. doi:10.1348/096317908X325313

Norton, T. A., Parker, S. L., Zacher, H., & Ashkanasy, N. M. (2015). Employee green behavior: A theoretical framework, multilevel review, and future research agenda. *Organization & Environment*, *2015*(28), 103–125. doi:10.1177/1086026615575773

Norton, T. A., Zacher, H., Parker, S. L., & Ashkanasy, N. M. (2017). Bridging the gap between green behavioral intentions and employee green behavior: The role of green psychological climate. *Journal of Organizational Behavior, 38*(7), 996–1015. doi:10.1002/job.2178

Ottman, J. A., & ... . (2006). Avoiding Green Marketing Myopia. *Environment, 2006*, 48.

Pan, C., & Tian, H. (2017). Environmental leadership, green organizational identity, and corporate green innovation performance. *Chinese Journal of Management, 14*(6), 832–841.

Pearce, C. L., & Ensley, M. D. A. (2004). reciprocal and longitudinal investigation of the innovation process: The central role of shared vision in product and process innovation teams (PPITs). *Journal of Organizational Behavior, 2004*(25), 259–278. doi:10.1002/job.235

Podsakoff, P. M., MacKenzie, S. B., & Bommer, W. H. (1996). Transformational leader behaviors and substitutes for leadership as determinants of employee satisfaction, commitment, trust, and organizational citizenship behaviors. *Journal of Management, 22*(2), 259–298. doi:10.1177/014920639602200204

Poon, J. (2000). Enhancing corporate image in organisations. *Management Research News, 23*(5/6), 28–54. doi:10.1108/01409170010782037

Porter, M., & Van der Linde, C. (1995). Green and competitive: Ending the stalemate. *Harvard Business Review, 73*(5), 120–134.

Post, C., Rahman, N., & Rubow, E. (2011). Green governance: Boards of directors' composition and environmental corporate social responsibility. *Business & Society, 50*(1), 189–223. doi:10.1177/0007650310394642

Pujari, D., Wright, G., & Peattie, K. (2003). Green and competitive: Influences on environmental new product development performance. *Journal of Business Research, 2003*(56), 657–671. doi:10.1016/S0148-2963(01)00310-1

Rahman, N., & Post, C. (2012). Measurement issues in environmental corporate social responsibility (ECSR): Toward a transparent, reliable, and construct valid instrument. *Journal of Business Ethics, 105*(3), 307–319. doi:10.100710551-011-0967-x

Renwick, D. W. S., Redman, T., & Maguire, S. (2013). Green Human Resource Management: A review and research agenda. *International Journal of Management Reviews, 15*(1), 1–14. doi:10.1111/j.1468-2370.2011.00328.x

Robertson, J. L., & Barling, J. (2013). Greening organizations through leaders' influence on employees' pro-environmental behaviors. *Journal of Organizational Behavior, 34*(2), 176–194. doi:10.1002/job.1820

Robertson, J. L., & Barling, J. (2017). Toward a new measure of organizational environmental citizenship behavior. *Journal of Business Research, 75*, 57–66. doi:10.1016/j.jbusres.2017.02.007

Rubik, F., Frankl, P., Pietroni, L., & Scheer, D. (2007). Eco-labelling and consumers: Towards a re-focus and integrated approaches. *International Journal of Technology Management & Sustainable Development, 2007*(2), 175–191. doi:10.1504/IJISD.2007.016932

Saeed, B. B., Afsar, B., Hafeez, S., Khan, I., Tahir, M., & Afridi, M. A. (2018). Promoting employee's proenvironmental behavior through green human resource management practices. *Corporate Social Responsibility and Environmental Management, 26*(2), 424–438. doi:10.1002/csr.1694

Schahn, J., & Holzer, E. (1990). Studies of individual environmental concern: The role of knowledge, gender, and background variables. *Environment and Behavior, 22*(6), 767–786. doi:10.1177/0013916590226003

Srivastava, A., Bartol, K. M., & Locke, E. A. (2006). Empowering leadership in management teams: Effects on knowledge sharing, efficacy, and performance. *Academy of Management Journal, 2006*(49), 1239–1251. doi:10.5465/amj.2006.23478718

Sroufe, R., & Gopalakrishna-Remani, V. (2018). Management, social sustainability, reputation, and financial performance relationships: An empirical examination of US firms. *Organization & Environment, 32*(3), 331–362. doi:10.1177/1086026618756611

Starik, M., Throop, G. M., Doody, J. R., & Joyce, M. E. (1996). Growing an environmental strategy. *Business Strategy and the Environment, 1996*(5), 12–21. doi:10.1002/(SICI)1099-0836(199603)5:1<12::AID-BSE37>3.0.CO;2-X

Stone, L. J. (2000). When case studies are not enough: The influence of corporate culture and employee attitudes on the success of cleaner production initiatives. *Journal of Cleaner Production, 8*(5), 353–359. doi:10.1016/S0959-6526(00)00037-8

Teixeira, A. A., Jabbour, C. J. C, & Jabbour, A. B. L. (2012). *Relationship between green management and environmental training in companies located in Brazil.* Academic Press.

Testa, F., Corsini, F., Gusmerotti, N. M., & Iraldo, F. (2018). Predictors of organizational citizenship behavior in relation to environmental and health & safety issues. *International Journal of Human Resource Management*. Advance online publication. doi:10.1080/09585192.2017.1423099

Vicente-Molina, M. A., Fernández-Sainz, A., & Izagirre-Olaizola, J. (2018). Does gender make a difference in pro-environmental behavior? The case of the Basque Country university students. *Journal of Cleaner Production*, *176*, 89–98. doi:10.1016/j.jclepro.2017.12.079

Wehrmeyer, W. (1996). *Greening people: Human resources and environmental management*. Greenleaf.

Whitmarsh, L., & O'Neill, S. (2010). Green identity, green living? The role of pro-environmental self-identity in determining consistency across diverse pro-environmental behaviours. *Journal of Environmental Psychology*, *2010*(30), 305–314. doi:10.1016/j.jenvp.2010.01.003

Woosnam, K. M., Strzelecka, M., Nisbett, G. S., & Keith, S. J. (2019). Examining millennials' global citizenship attitudes and behavioral intentions to engage in environmental volunteering. *Sustainability*, *11*(8), 2324. doi:10.3390u11082324

Wu, P. J., Wu, T. J., & Yuan, K. S. (2019). "Green" Information Promotes Employees' Voluntary Green Behavior via Work Values and Perceived Accountability. *Sustainability*, *2019*(11), 6335. doi:10.3390u11226335

Xiao, C., & McCright, A. M. (2015). Gender differences in environmental concern: Revisiting the institutional trust hypothesis in the USA. *Environment and Behavior*, *47*(1), 17–37. doi:10.1177/0013916513491571

Yusliza, M. Y., Ramayah, T., & Othman, N.-Z. (2015). While examining adoption factors, HR role and attitude towards using e-HRM is the start-off in determining the successfulness of green HRM? *Journal of Advanced Management Science*, *3*, 337–343. Retrieved November 5, 2014, from www.joams.com

Zhang, B., Wang, Z., & Lai, K. H. (2015). Mediating effect of managers' environmental concern: Bridge between external pressures and firms' practices of energy conservation in China. *Journal of Environmental Psychology*, *43*, 203–215. doi:10.1016/j.jenvp.2015.07.002

Zhou, S., Zhang, D., Lyu, C., & Zhang, H. (2018). Does seeing "mind acts upon mind" affect green psychological climate and green product development performance? The role of matching between green transformational leadership and individual green values. *Sustainability*, *10*(9), 3206. doi:10.3390u10093206

Zoogah, D. (2011). The dynamics of Green HRM behaviors: A cognitive social information processing approach. *Zeitschrift fur Personalforschung*, *25*(2), 117–139. doi:10.1177/239700221102500204

Chapter 12
Incorporating Blockchain in QoS-Based Sematic Web Services Selection: An Overview

Sunita Jalal
College of Technology, Pantnagar, India

Chetan Singh Negi
College of Technology, Pantnagar, India

Rakesh Kumar Yadav
IFTM University, India

Reeta Verma
College of Technology, Pantnagar, India

ABSTRACT

Web service technology and cloud computing offer promising solutions for developing distributed business applications published and invoked over the internet. Web service discovery finds the most relevant web services satisfying the functional needs of users. Semantic description of web services offers more accurate matching results between user query and web services descriptions. User's quality of service (QoS)-based criteria select the best web service from a set of functionally similar web services. The risk of privacy leakage of users is possible as service providers collect user personal data without any agreement. It isn't easy to assess the trustworthiness of values of QoS attributes provided by different service providers. Blockchain is an emerging technique that can overcome privacy leakage and identifying web services with fake values of QoS attributes. This chapter gives an overview of semantic web services, discovery and QoS-based selection of semantic web services, and the use of blockchain technology in the selection of web services.

DOI: 10.4018/978-1-6684-4225-8.ch012

INTRODUCTION

Service Oriented Architecture (SOA) is well known architecture style for software development. SOA based applications are implemented using web services. As defined by the W3C (World Wide Web Consortium), Web Services are software components designed to support interoperable machine-to-machine interaction over a network such as the Internet. Another way to define a web service is that it is a server running on a computer workstation that listens for web requests over a particular port on a network and serves various functionalities or data. Web Services interact with one another and other systems to fulfil requests and carry out workflows and complex transactions. Web services have mechanisms to provide services over various environments and allow the clients to adapt rapidly to changes in their environments. Web services have significant uses in the automation of a task and efficiency improvement. Using the concept of web services, machines are able to interact with each other with no human intervention.

Web Services are usually described syntactically using input and output parameters. These parameters are known as the functional requirements of web services. Other than these functional parameters, web services are described using some of their QoS parameters. The QoS (Quality of Service) parameters are the performance measures of the web service that describe a web service more than just a formal description of the job it performs.

Basic web service architecture is shown in Figure 1. It contains three components: Service Provider, Service Registry, and Service Requester.

Figure 1. Basic web service architecture

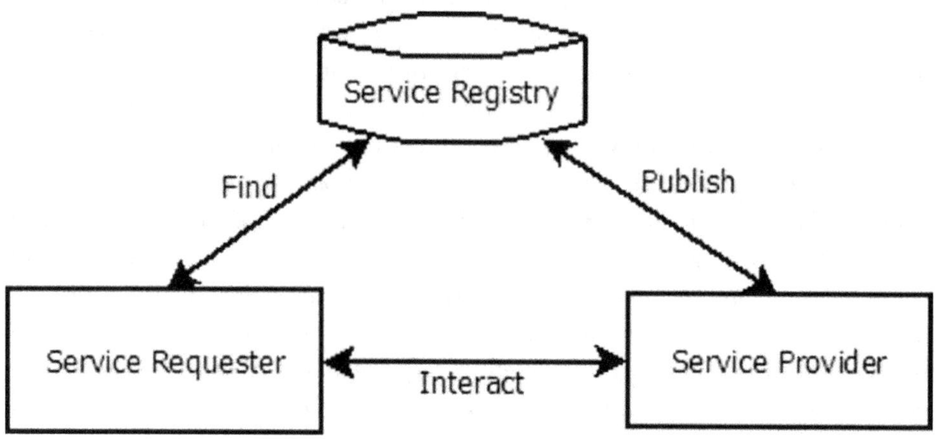

The service provider hosts a network-associable module (web service). It defines a service description for the web service and publishes it to a service registry. The service requester uses a find operation to retrieve the service description from the service registry. It uses the service description to bind with the service provider and invoke with the web service implementation. Service registry stores information about service providers and their services. Service requesters find service descriptions and obtain binding information for services during the discovery process.

Web services composition is a way to develop a business application by using existing value-added web services for different functions of the business application. The relevant web services are identified through discovery. Web service discovery is the process of finding relevant web services that satisfy the functional requirements of user request. Web service discovery can be based on either syntactic approach i.e. keyword-based search or semantic approach. UDDI registry and online web directories support keyword-based search. Semantic approach based web service discovery provides better results as it considers semantic information associated with web services using domain ontologies and WordNet based ontology. QoS-based web service selection focuses on selecting the best web services from the set of functionally similar web services based on the user's QoS constraints.

Blockchain is a fairly new technology that covers multiple computer science applications like cryptographic hash, distributed consensus mechanisms and asymmetric cryptography-based digital signature. Blockchain can be used in web services selection for removing unreliable web services. This chapter explains the discovery and QoS based selection of semantic web services and use of blockchain technology in web services selection.

WEB SERVICE DESCRIPTION

A web service can be described syntactically and semantically.

Web Service Description Language (WSDL)

A web service is described syntactically using standards WSDL 1.0 and WSDL 2.0. Description document contains interface description, protocol binding for invocation, location of service using endpoints. Figure 2 shows WSDL elements and their description.

The <types> element contains XML schema definition of the data types used in web service.

The <message> element defines input and output messages exchanged between service and service consumer. The <portType> element describes operations of web service with input and output messages of the operations. The <binding> element describes invocation details to call an operation of the web service. The <service> element defines the ports for referencing actual location of the web service.

Figure 2. WSDL elements

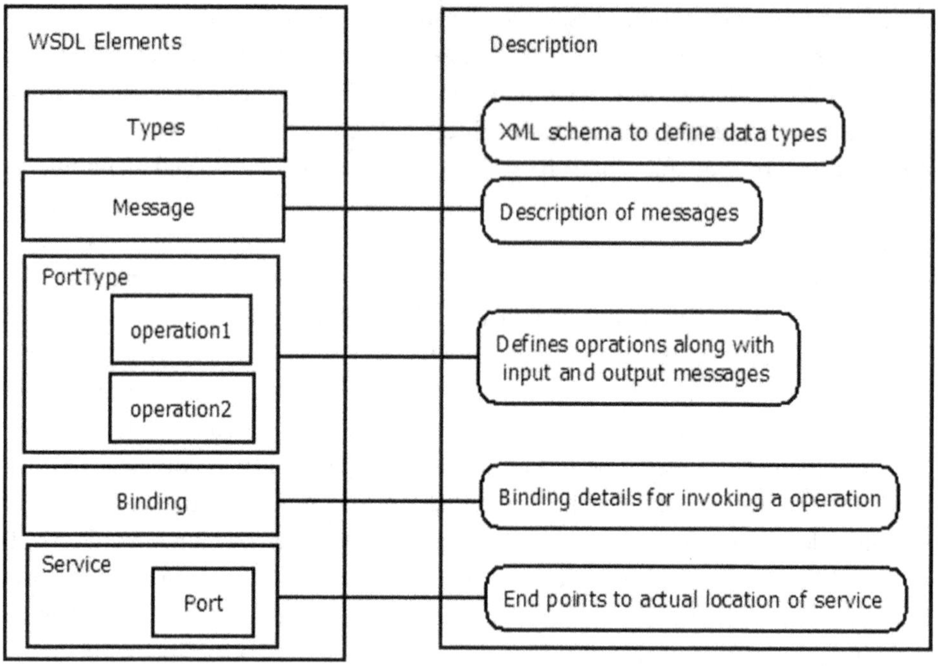

Semantic Description of Web Services

The Semantic Web is intended to construct a web of meaning. Semantic web technology supports data interpretation by machine. It uses ontology as a data model to define data semantically. Ontology is an effective means to specify the concepts of a domain formally. Semantic web services facilitate automation of discovery, selection and composition of web services.

Ontology has four key elements: concepts, instances, relations and axioms.

- Concept-A concept represents a group of objects that share common characteristics. The concepts of a domain are arranged in a hierarchical structure where a higher-level concept is a super-class, and its subordinate concept is a sub-class.
- Instance-An instance is an object of a concept class.

- Relation-A relation describes the relationships between two concepts in a specified domain.
- Axiom- An axiom imposes constraints on instances using logic-based languages to verify the consistency of the ontology.

Some of the widely used approaches for semantic web services are WSMO, OWL-S, WSDL-S, and SAWSDL.

Web Service Modeling Ontology (WSMO)

WSMO (De Bruijn et al., 2005) defines conceptual model and axiomatization for semantic web services. WSMO group includes Web Service Modeling Language (WSML) and Web Service Modeling Execution(WSMX). WSML is a formal language for defining ontology and rules for the semantic web, and WSMX provides an execution environment and implementation for WSMO.

The main elements of WSMO are ontologies, web services, goals, and mediators.

- Ontologies provide conceptual semantics related to a domain used by other WSMO components.
- The web services component describes functional and non-functional aspects of a web service.
- The goals component is used to specify the objectives related to the user's functional and non-functional requirements that a web service should satisfy.
- The mediators component connects heterogeneous elements and offers mediation features for handling mismatches.

Mediation is supported at data level, protocol level, and process level. At data level, mediators resolve mismatches between heterogeneous data sources. At protocol level, mediators handle mismatches between heterogeneous communication protocols. Process Level mediators provide mediation for heterogeneous business processes. WSMO defines four types of mediators: OO Mediators, GG Mediators, WG Mediators, and WW mediators. OO Mediators connect heterogeneous ontologies with mediation. GG Mediators link Goals where source goal is refined into target goal. WG Mediators connect web services to various goals. WW Mediators links web services by handling mismatches between them.

The functional aspect of the web service is shown by its capability. The capability is defined using a sub-class of class web service. It is shown in Figure 3.

Figure 3. Class capability

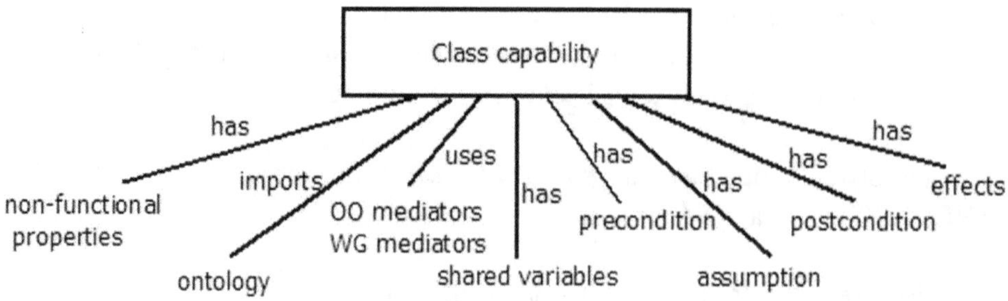

A capability is described by defining the precondition, postcondition, assumption, and effect. Precondition defines conditions that must be met to execute a web service. Assumption expresses the state of the domain before the execution of Web service. Postcondition specifies conditions over the output of web service. Effects specify the state of the domain after completing the execution of the web service. Shared variables define common variables used by precondition, assumption, postcondition, and effects expressions. Ontologies are imported using ontology mediators. Interface description of a web service defines the realization of the web service's functionality from both a choreography and orchestration view. Choreography defines the sequence of exchange of messages among multiple web services. Orchestration defines the execution of the composition of web services, and information flow between web services is described using a workflow model.

OWL-S

OWL-S (Martin et al., 2004) describes semantic web services within OWL based framework. Ontology for service is presented in Figure 4.

Figure 4. OWL-S ontology

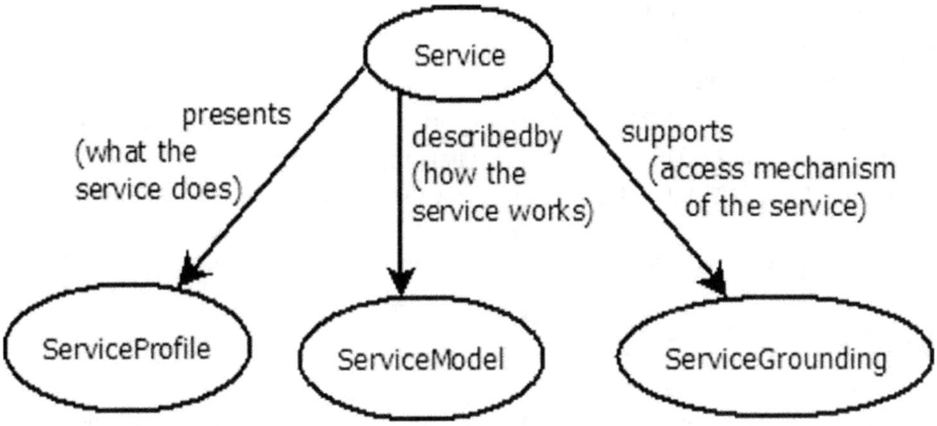

A published service instance presents a service profile, is described by service model, and supports service grounding description for communication protocol. The service profile expresses the functions and features of a web service. The execution state of service is determined by precondition and effects. The profile class has following properties: hasParameter, hasInput, hasOutput, hasPrecondition, hasResult. Service model describes how a service performs its function. It defines data flow and control flow of service. The ProcessModel is a sub-class of class ServiceModel. A process can be atomic, simple, and composite. Atomic process has no subprocesses and directly invocable. It executes in a single step. Simple process is not invocable and provides abstract view of an atomic process and composite process. The service grounding explains access mechanism of the service. It describes binding protocol, formats of messages.

Example of Input, Output, Precondition and Effects

<profile:**hasInput**> <process:Input rdf:ID="User">..</process:Input> </profile:hasInput>

<profile: **hasInput**> <process:Input rdf:ID="Train">..</process:Input> </profile:hasInput>

<profile:**hasOuptut**> <process:Output rdf:ID="Ticket">.. </profile:hasOuptut> </process:Output>

<process:**hasPrecondition**>

<expr:expressionBody rdf:datatype="http:www.w3.org/2001/XMLSchema#string">

<pred name=http://127.0.0.1/TrainTransport_Ontology.owl#isAvailable>

```
<param>?http://127.0.0.1/BookTrain.owl#Train</param>
</pred> </expr:expressionBody>
</process:hasPrecondition>
<process:hasEffect>
<pred name=http://127.0.0.1/TrainTransport_Ontology.owl#BookedFor">
<param>?http://127.0.0.1/BookTrain.owl#Train</param>
<param>?http://127.0.0.1/BookTrain.owl#Customer</param>
</pred>
</process:hasEffect>
```

WSDL-S

Types of semantics associated with web services are data, functional, non-functional, and execution. Data semantics provide semantic definition to data present in input and output messages. Functional semantics provide semantic description to a capability of web service. Non-functional semantics define semantics for non-functional properties of web service. Execution semantics define semantics for data and control flow of web services in a composition. WSDL-S (Akkiraju et al., 2005) offers a method to annotate the data and functionalities of Web services presented in WSDL with semantic concepts expressed in a given domain model. The domain model can have one or more ontologies. Semantic annotation with the web services elements allows automatic discovery and composition of web services. Semantic annotation with different elements is done using modelReference and schemaMapping attributes. WSDL-S also provides semantic annotation on precondition and effects of the function.

Example of Semantic Annotations in WSDL-S Document

```
<xs:element name= “processOrderResponse” type=”xs:string
sem:modelRyference=”OrderOntology#OrderConfirmation”
sem:schemMappmg=” http://127.0.0.1/wsdl-s/example.xq”/>
</xs:schema>
</types>
<interface name:=”OrderInterface”>
<operation name=”processOrder” pattern=wsdl:in-out
modelReference = “rq:#RequestQuote” >
<input messageLabel = “processOrderRequest”
element=”tns;processOrderRequest’7>
<output messageLabel =”processOrderResponse”
element=”processOrderResponse”/>
```

```
<!—Precondition and effect are added as extensible elements on an operation>
<sem:precondition name="ExistingAccountPrecond"
sem:modelReference="OrderOntology#AccountExists">
<sem:effcct name="ltemReservedEffect"
<sem:modelReference="OrderOntology#ItemReserved"/>
</operation>
</interface>
```

Semantic Annotations to WSDL (SAWSDL)

SAWSDL (Kopecký et al., 2007) defines a set of extension to WSDL by providing semantic annotations to different parts of WSDL. Semantic annotation is achieved through modelReference, liftingSchemaMapping, loweringSchemaMapping.

A model reference is provided with each element within WSDL and XML schema. However, SAWSDL describes its meaning only for <wsdl:interface>, <wsdl:operation>, <wsdl:fault>, <xs:element>, <xs:simpleType>, <xs:attribute>, and <xs:complexType>.

Model references adhere the service to the client requirements; however the semantic model may still differ from the input/output structure. The lifting schema mappings enhance data from XML schema to a semantic model and lowering schema mappings shrink data from a semantic model into an XML schema structure.

SEMANTIC WEB SERVICE DISCOVERY

Semantic web service discovery uses domain ontology to represent concepts used in the description of web service. Matchmaking algorithms are utilized to find matching degree between parameters of user request and services advertised in service registry. Exact, Plug in, Subsume, and Fail are potential degree of match. (Paolucci et al., 2002)

- Exact match: The exact match indicates that concepts in the user request and advertised web service are equivalent.
- Plug-in match: The degree of match between the user request and advertised web service is plug-in if concepts defined in user request are a subset of concepts described in the advertised web service.
- Subsume match: The degree of match between the user request and advertised web service is subsume if concepts defined in the user query are a superset of concepts described in the advertised web service

- Fail: When there is no subsumption relation between the advertised web service and the user request, it indicates no matching between the user query and the advertised service.

Klusch et al. (2006) developed OWLS-MX semantic matchmaker and used logic based reasoning with information retrieval techniques for discovery of semantic web services. Semantic matchmaker considers exact, subsumes, subsumed-by, plug in and nearest neighbour degree of matches. In **(Meditskos et al., 2009)**, authors proposed discovery approach based on ontology created using OWL-S service profile ontology. In **(Averbakh et al., 2009)**, the authors used user feedback to enhance the quality of semantic web service discovery. Resource Description Framework (RDF) triple is used to store user feedback.

Skoutas et al. (2007) presented a ranking mechanism to evaluate the semantic similarity between a web service and user query. Domain ontology is used to present the concepts of web services. Degree of match between advertised service and user query is computed using precision and recall measures. **Jiao et al. (2017)** implemented web services classification by fuzzy C-means method. The comprehensive semantic similarity comprises latent semantic similarity and ontology semantic similarity. Singular vector decomposition is used to compute the latent semantic similarity. Ontology semantic similarity is determined using Lowest common ancestor (LCA) of concepts.

In **(Chen et al., 2017)**, the authors proposed a semantic similarity measure incorporating multiple conceptual relationships. The is-a, has-a and antinomy conceptual relationships among concepts in WordNet and other ontologies are used. Semantic similarity comprises weighted combination of service textual description similarity and interface signature similarity.

WordNet is lexical database implemented by Princeton University. It is represented as a semantic network that comprises synsets and relations. Synsets are interconnected with various semantic relations. Synset contains the group of words they have similar meaning. Wordnet describes word senses representing several different meaning of a lemma. Wordnet includes following relations between synsets:

- Synonymy – It is used to represent relation between two words having similar meanings.
- Antonymy – The relation between two words is antonymy if they are opposite in meanings.
- Hyponymy/Hypernymy – It is a IS-A relation. For example rose is hyponym of flower and flower is hypernym of rose.
- Polysemy- It shows that a word has more than one sense. For example word 'bank' can be a financial institution or a river bank.

- Meronymy/Holonymy- It is part-whole relation. Compressor is meronym of air-conditioner. Air-conditioner is holonym of compressor.

Sim-path(C1, C2) = 1/ (pathlen(C1,C2))

WordNet can calculate the similarity between nouns or between verbs. The path-length similarity method computes similarity between two concepts by finding the length of shortest path between two concepts. It is defined as

The similarity between two concepts can be based on maximum senses between two concepts. The similarity between two concepts can also be computed using the shared amount of information of the lowest common subsume of the two concepts.

In **(Xue et al., 2015)**, the authors proposed semantic matching for restful web services using WordNet ontology. RESTful web services are described using Web Application Description Language (WADL). WADL is an xml based file that describes resources provided by RESTful web services. Each resource element contains method element which describes request with input parameters and response. Response describes either representation of output or any fault information. A bipartite graph model is used to find similarities between parameters of methods of services. Matching between resource-nodes of services is computed using a generalized vector space model. In **(Sangers et al., 2013),** the authors proposed the discovery framework for WSMO semantic web services. Web services are discovered by matching between user query keywords and semantic web service descriptions. Web services descriptions are parsed using NLP techniques. Different semantic web services can use different ontologies, and NLP techniques provide support to handle ambiguity problems between these ontologies. The discovery engine defines seven different stages of information about the web service. The importance of each stage to the matching process is distinct, and it is expressed in weight. The sum of weights of all the stages is equal to 1. The weights are determined using Genetic Algorithm (GA). Each individual in GA is represented using a chromosome that comprises nine variables for weights. The seven weights express the importance of web service information, and two weights for sense similarity and lexical similarity. Fitness function is based on the average precision values for every user query.

Kaouan et al. (2020) presented k-means clustering based approach for the construction of ontology network from corpus of semantic web service descriptions. The center-objects of clusters denote concepts in the web service ontology. The constructed ontology network is used to store and discover semantic web services. The similarity between two web services in clustering is computed using Jaccard similarity. **Lu (2021)** described a novel method for web service discovery that considers both I/O interface similarity and contextual similarity. The HASA, ISA, and ANTONYMY relations are used between two concepts. WordNet 3.0 lexical database is used to compute semantic similarity between two concepts. The shortest path between concepts C1 and C2 in the service ontology is defined as the sum of edges in path to the lowest common ancestor concept. The semantic distance between C1 and C2 is defined as the summation of weights to the concepts in the shortest path. After computing the I/O interface similarity between request and services, unsatisfactory services are removed. Contextual text of service is pre-processed using NLP techniques such as part of speech tagging and stemming. Different terms are identified in the service, and the similarity between all available service and request terms is computed.

QoS-BASED SEMANTIC WEB SERVICES SELECTION

Semantic web services have well defined semantics that offer automation in discovery, selection and composition of web services. Semantics of a web service description is represented using ontology. Ontology defines hierarchical relationship between concepts. Web services selection is performed after service discovery. A variety of techniques are proposed by the research community that assistances service requester to select the web services that most match the specific requirements of the individual tasks and the business process. In the web services selection procedure, the selection of web services that best suit the user requirements takes place from the set of functionally identical web services. QoS attributes like response time, availability, reliability, documentation and others can be used to differentiate among functionally identical web services.

Web Service Description Language WSDL does not facilitate specifying the QoS values. However, the Web Service Modeling Ontology (WSMO), Semantic Web Service Description (WSDL-S) and extension of OWL-S profile provide representation for the QoS properties. In **(Wang et al., 2006)**, the authors utilized Web Service Modeling Ontology (WSMO) model to present QoS model. Authors defined QoS ontology through a new class QoS which extends class nonFunctionalProperties. QoS selection model first filters web services based on functionality and returns a set contains web services similar in functionality. Then selection model considers all QoS attributes to select web services fulfilling the user's QoS requirements best. Simple additive method is used for web service selection based on QoS.

Baocai et al. (2006) design and construct QoS ontology to support the automatic web service discovery. Ontology provides understanding about a domain knowledge. It can resolve issues related to ambiguity and semantic descriptions. QoS attributes describes performance characteristics of web service. The QoS ontology presents the quality information of web services. QoS ontology includes the features such as QoS property, metric, and relationship. Property represents non-functional attributes of a web service in a specified domain. Some attributes are response time, reliability, security, throughput, availability etc. Value of a QoS property is measured using QoS metric. QoS relationship defines correlation between QoS attributes. The authors also proposed a QoS matching method that considers both the semantic and numerical aspects of the QoS attribute to determine the matching degree of QoS between service request and service advertisement. If there are many QoS parameters, weights are assigned to each QoS parameter. Values of QoS parameters are required to standardize for finding numerical matching.

In **(Chhun et al., 2016)**, the authors proposed automatic generation of web service ontology from tModel values of UDDI. Figure 5 shows an example of service ontology with QoS information. The QoS-based selection algorithm uses utility function computing the average weighted sum of QoS attributes to select the best service for a user request. It is defined as

$$U = \sum_{i=1}^{m1} wi * \frac{qi - \mu i}{\sigma i} + \sum_{j=1}^{m2} wj * \left(1 - \frac{qj - \mu j}{\sigma j}\right)$$

Where m1 is the number of QoS attributes whose higher values are desired, and m2 is the number of QoS attributes whose lower values are desired. The μ and σ represent the QoS attribute's mean and standard deviation values. Weights are assigned to QoS attributes such that sum of weights must be equal to 1.

Figure 5. Service Ontology

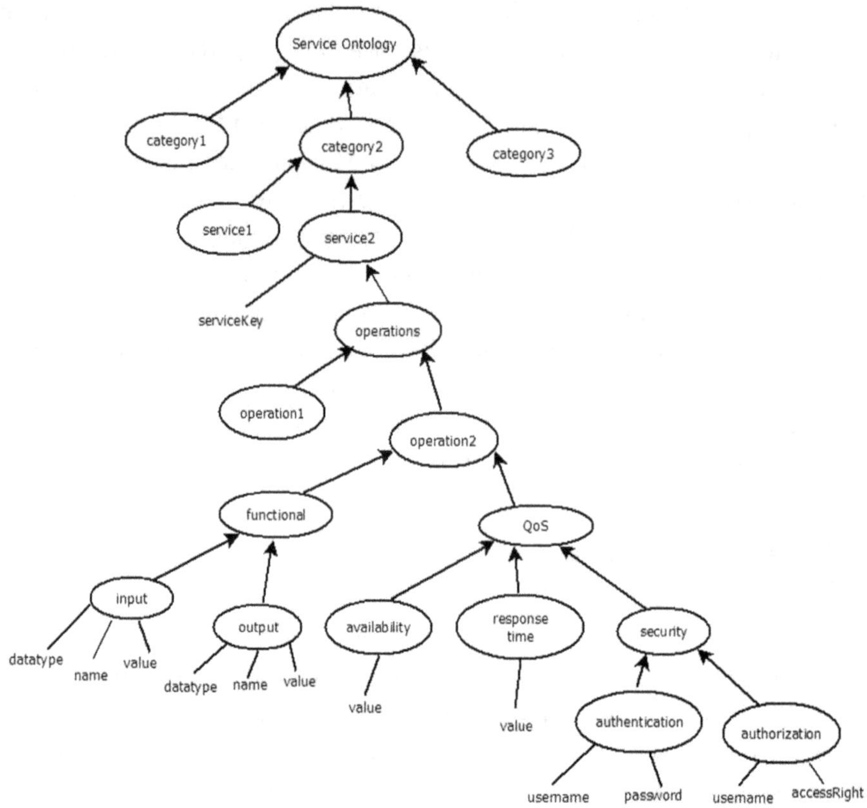

Feddaoui et al. (2021) described an approach for mining web services by incorporating opinions and selection history of users. Ontology-based querying for web services is used and collaborative filtering has applied to compute user's interest, and query similarity. Filtered web services are selected based on QoS criteria. The selection precision has improved by combining syntactic and semantic techniques.

Purohit and Kumar (2018) proposed a new approach for the selection of web service based on QoS parameters. Initially the candidate web services are prefiltered using the techniques of classification. The authors have modified the PROMETHEE method(called PROMETHEE Plus) and applied it to the most eligible of web services. Maximizing Deviation Method based hybrid weight evaluation mechanism is adopted. Top-k web services that match with the QoS requirements of the end user are identified. Experiments are conducted on the dataset of real world web services. The experiments performed show that this approach is better in terms of end user satisfaction and efficiency compared to existing similar approaches.

Purohit and Kumar (2019) apply clustering to candidate WS(web services) and make use of QoS information to find out similar web services. The research systematically analyses and evaluates the performance of three of the clustering techniques using Average Distance measure and Dunn Index. Based on the performance, the best clustering technique is used on the candidate web services. The Skyline based selection process is applied on the final set of web services. QoS parameters like documentation, availability and response time are prominent in the study. Using the QoS requirements of the end user and the QoS presented by the published web services, a QoS based matching is performed and similar web services are returned to the end user. Experiments are performed based on real world dataset for QoS of web services. It is found by comparing the time taken to select web services that the proposed method has performed better than the S-Sky approach of service selection.

BLOCKCHAIN TECHNOLOGY

Blockchain is an evolving technology that is used for various computer science applications such as cybersecurity, cryptocurrency, Internet of Things, distributed consensus mechanism etc. Blockchain was initially presented in 2008 and applied as the infrastructure of Bitcoin for secure transactions in 2009. Blockchain is mainly a distributed ledger or database that runs on a peer-to-peer environment. Each block in blockchain can be supposed a page in the ledger.

The traditional ledger technology requires a centralized trusted third party for verifying transactions. However, transactions in blockchain-based technology are created, verified, and stored in a secure and reliable network without involving a trusted third party. Figure 6 shows an example of blockchain technology.

Figure 6. Blockchain technology

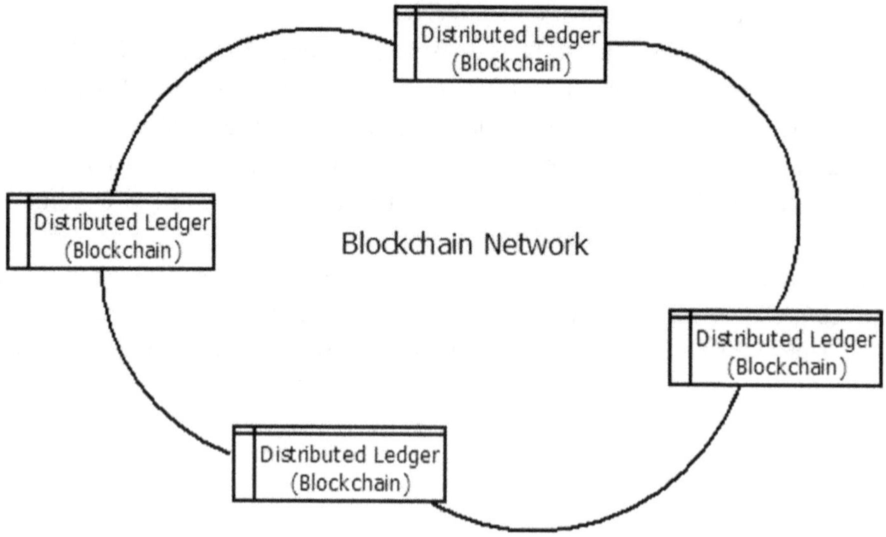

A blockchain can be realized either as permissioned or permissionless. In permissioned blockchain, participants' identities are maintained by an assigned authority. Quorum blockchain technology (Baliga et al.) creates permissioned blockchain network. Quorum uses QuorumChain consensus protocol and works fast as it is based upon majority voting. Permission-less blockchain network will be exposed to all participants. The proof-of-work is used as consensus protocol and it is slow. Ethereum is an example of public permissionless blockchain.

Every block in blockchain consists of a header and body parts. The block body comprises the information of transactions in detail. The block header includes Merkel Root, Nonce, Time Stamp, Difficulty and Version. A unique hash value is associated with every block which has a reference to the parent block's hash value. The blockchain logs all the changes in the block, hence the source of every transaction is traced by traversing the whole chain.

The digital signature is necessary to maintain secure communication between the users in the blockchain-based network. A pair of public and private keys is assigned to each user. User logins using private key and broadcast transaction and recipients verify the transaction using the user's public key. It maintains the integrity of the transaction in the blockchain-based network. The transaction is initiated by a peer node with a signed digital signature appended at the end of the transaction record. The completed transactions are put in the transaction pool maintained by the peer node. All the nodes then reach a consensus on a public ledger and synchronize the new block information, ensuring that all the nodes maintain same public ledger.

Iteratively, the miners package their own blocks following this particular block and consequently form a blockchain.

Smart Contracts

Smart Contracts **(Szabo, 1996)** were proposed by Szabo in 1996. Smart contracts are basically event driven promises written in languages understandable by computers. Different types of contractual clauses can be embedded in software and hardware. Smart contracts run automatically in a predefined manner once the preset conditions are met. This makes it possible to exchange electronic data in a safe and distributed environment. Every action on the smart record can be traced back as the blockchain records every operation. A smart contract possesses four key characteristics: observability, privity, verifiablity and enforcibility **(Li et al., 2021)**.

Ethereum uses Turing-complete byte language to write smart contract that executes by deploying on the Ethereum blockchain. Each participant in the blockchain knows about the smart contract details and is automatically applied after fulfilling the conditions. Miners validate the transactions associated to the smart contract. The "proof-of-work" consensus protocol is used to resolve any possible conflict. Hyperledger Fabric blockchain platform works for private network where smart contract is called chaincode. It is designed to build trustworthy, resilient, and scalable pluggable implementation for enterprise and business applications.

SEMANTIC WEB SERVICE AND BLOCKCHAIN

With the increasing number of web services, the corresponding challenges have increased significantly. Some of these challenges are discussed here.

- Risk of privacy leakage and security risk: The services providers might possess and collect the service consumer's confidential data without any explicit declaration. This privacy sensitive data can be abused. This data may be unintentionally or intentionally disclosed to others without the agreement of the service consumer. Also, the data centers themselves are vulnerable to security threats making the data insecure.
- Counterfeit values of QoS attributes: Some service providers may provide fake values of QoS parameters to attract service users. It lowers the satisfaction of service users and the performance of applications developed by these fake web services.

Blockchain-based selection and composition of web services promotes trustworthiness, resilience, and scalability in business applications across the organizations. Blockchain mechanism can be used to solve the challenges in web services as it provides the following features:

- Built-in digital signature and encryption mechanisms: Blockchain integrates with other security measures like access control and authentication to preserve data privacy and system security.
- Decentralization: Blockchain helps in the mitigation of vulnerabilities and security risks like SPFs and DDoS attacks. The auto execution of smart contracts helps in the updation of firmware for mitigating the security vulnerabilities.
- Intrinsic Incentive Mechanisms: These may be used to resort the service computing's incentive and pricing problems. The developers can be awarded a particular amount of digital currency for contributing codes and reporting bugs through automatic execution of Smart Contracts.

Ruta, et al. (2018) proposed a Semantic Web of Things (SWoT) blockchain to enable fundamental operations of service oriented architecture like registration, discovery and selection, and final execution. These tasks are implemented as smart contracts. The accomplishments of these smart contracts are distributed and the validation is reached through consensus. The proposed approach has a distinguished feature with respect to the basic blockchain infrastructures. This feature is the service discovery based on semantic matchmaking. The blockchain is used to improve trust and coordination management in SWoT. A resource discovery layer is incorporated in a blockchain infrastructure. Discovery results are grounded on logic and inferences web services for semantics-based matchmaking.

SWoT uses Description Logics based on First-Order Logic for knowledge representation and reasoning. Resource description and request description contain concept expressions belongs to a common ontology. Semantic matchmaking is used to retrieve the most relevant resources for a given user request. The subsume match checks whether all concepts in a user request are included in the resource description. Concept contraction identifies which part of user request is conflicting with resource description in case of incompatibility between user request and resource description. If a resource description is not fully matched with a user request, then concept abduction specifies what should be hypothesized in the resource description to satisfy the user request completely.

Each agent in the SWoT blockchain can execute a semantic-based resource discovery to use and transfer resources between accounts. Resources are annotated reagrding a domain ontology. Annotated resources are stored in the blockchain.

Smart contracts include matchmaking mechanism. Resources are registered in the blockchain and are described with following attributes: Uniform Resource Identifier (URI), the URI of the reference ontology, Semantic Annotations in OWL, and Resource Price. To search a set of resources requester send request to a number of peers. Request specifies URI of the reference ontology, semantic annotations, price constraint value, threshold value for semantic relevance, and number of services to be returned. The receiving node replies with semantic affinity score and concept contraction and concept abduction results. Semantic affinity score is in range 0 and 1 where one means full match and zero indicates no match between user request and resource description,

Viriyasitavat, et al. (2021) proposed to integrate blockchain with a multi-agent approach that warrantees the trust of real-time data for measuring the QoS in IoT environment. The proposed method assures trust for all entities and processes that collect, monitor, provide and store the QoS information. This approach is demonstrated and verified by demonstrative examples that address QoS specifications patterns found in service based applications. The qualitative analyses are performed for evaluating the patterns. Security assurance is essential to utilize a service-based application. Security concern is raised due to the dynamic and decentralized nature of service-based application. One of the key advantages of blockchain technology is its decentralization; it allows business partners to keep an immutable ledger of transactions. QoS blockchain is presented as a service. Blockchain technology describes receiving and storage of QoS values into a block. The used blockchain technology is classified as a permissioned blockchain where only authorized partners are allowed to construct new blocks and chain them into the main chain.

The QoS-based blockchain has the same structure as other blockchains. QoS blockchain contain two types of transactions. One category has a transaction type of smart contract. A smart contract prepares the service contract between the service provider and a service client. Authorized partners may be included in the service contract to give the QoS feedback as service clients. Another category contains the set of transactions of QoS values. The authorized partners or service clients provide QoS value of event, response time, and utilization time of service for the service based application. Event keyword can be used 1) as notifying events that connect virtual machine to the blockchain network and 2) as storing transactions logs to be used by the smart contract. Event value can be success, error, and incompleteness.

The QoS blockchain scheme is functioned as follows:1) creation of smart contract 2) invocation of web service and 3) providing feedback. Creation of smart contract function describes that service provider generates smart contract for QoS based on the agreement and includes the smart contract in a block. Public keys of service provider, service consumer, and other authorized partners are added to the block. The cryptographic hash is based on SHA-256 and ECDSA is accepted for digital signatures. Service provider invokes service and specifies the time timeRe of feedback. Service client verifies and approves the service instance. Service client or other authorized parties can provide feedback of QoS values after timeRe. Basic Practical byzantine fault tolerance consensus mechanism is used for voting purpose..

Viriyasitavat, et al. (2020) proposed a blockchain based approach to establish the trust of process execution fitting into the IoT environment. An IoT thing can be viewed as a service concept and encapsulated it with Service Oriented Architecture (SOA). IoT environment with SOA is beneficial for discovery and composition of IoT thing services, and data exchanges between services. IoT thing service acts as a gateway to provide a RESTful interface for accessing a smart object. Many similar functionality services by IoT things are available that are evaluated and selected based on QoS requirements of service consumer. The evaluation of IoT services is challenging due to unavailability of central authorization for examine the trustiness of data gathered in dynamic and distributed environment.

Blockchain technology is adopted for QoS management. A smart contract is used to establish the trust of process execution fitting into the IoT environment. The types of QoS measurement are defined in smart contract that is created by the cental blockchain hub. This avoids including malicious, unrelated, or unidentified data into a ledger. Smart contract is also used for the control of hub and agents. Central blockchain comprises federated nodes. The smart contract of federated nodes decides adding, eliminating, or replacing agents. After verification and agreeing upon initiation by federated nodes, the contract is broadcasted to nodes in a network. A consensus approach is introduced to resolve the prejudice and time challenges. This approach uses selected validators extended from Practical Byzantine Fault Tolerance (PBFT). The measurement of performance favours the use of this approach as it decreases the communication within limited number of validators.

Blockchain-based IoT services comprise lots of connected devices working in different scenarios. A blockchain-based platform for these applications requires managing a large number of smart contracts and distributed ledgers in diverse domains and environments. Formal semantics can be added to smart contracts for finding a smart contract with policies associated to a specific device.

Baqa, et al. (2019) introduced RESTful semantic web services allowing the indexing and invoking of Smart Contracts on Ethereum blockchain via a Universal Resource Index (URI). The researchers extended EthOn combining it with OWL-S ontology supporting Ethereum Smart Contracts. These smart contracts are deployed in Ethereum blockchain framework. This results in the execution of semantic queries over Smart Contracts using the existing semantic web platform.

A centralized platform is used to select services for service composition. There may be a chance that service providers provide counterfeit values for QoS and Quality of Experience (QoE) to include their services in the selection list. The centralized platform could not estimate the provider's non-functional values. The service user also is not able to verify the service quality.

Sridevi, et al. (2021) proposed extended Personalised and Generalised fuzzy TOPSIS for the Quality of Experience (QoE) prediction over the Ethereum blockchain for the removal of unreliable services. Semantic web services are created as OWL-S file and QoS information are created as OWL-O file. Ethereum smart contract is used to manage complete semantic web services selection process. The smart contract verifies the authenticity of non-functional attributes QoS and QoE values. When the smart contract detects the counterfeit of QoS and QoE values in a particular web service instance, it rejects the participation of unreliable web service in the composition process. Pi and Ui are the hash values obtained from the authorized service provider's prediction and user's prediction, respectively, by blockchain. These values are compared with the hash values stored on the blockchain server. After finding a match for hash values, a confirmation transaction is sent to the blockchain. The confirmed values are broadcasted to other trusted parties in the blockchain network. The smart contract performs secured and smooth web services interactions. Thus there is no chance that service providers or random users publish fake values of non-functional attributes of web services. The participating peers can find the web service source and its QoS and QoE values. It allows building a trusted web service selection system. The generalized and personalized criteria are used to predict non-functional attributes values. The generalized factor determines the significant QoS attributes such as response time, cost etc., for the domain. The personalized factor determines the user experienced QoE values such as likelihood, support, etc. Weights are assigned to the generalized and personalized parameters according to their importance. The discovered web services are ranked using the Fuzzy TOPSIS method based on the generalized and personalized criteria. The blockchain based extended fuzzy TOPSIS is used to evaluate the given approach on real world web services datasets. Experiments conducted on the proposed methodology show that it outperforms the existing approaches. This makes the proposed approach more effective as compared to other conventional techniques.

CONCLUSION

This chapter presents an overview of Semantic Web Services and use of Blockchain technology in service based computing. Semantic web services support the automatic discovery, selection and composition of web services in a business application. Various advancements in discovering and selecting semantic web services are also reviewed. Further, we discuss the utility of blockchain technology in web services applications. Permissioned Blockchains are used to establish trust among blockchain users. Blockchain technology helps identify unreliable web services, find smart contracts associated with a specific domain of web services, manage QoS parameters of web services, and implement smart contracts for discovery, selection, and execution of web services.

REFERENCES

Akkiraju, R., Farrell, J., Miller, J. A., Nagarajan, M., Sheth, A. P., & Verma, K. (2005). *Web service semantics-wsdl-s*. Academic Press.

Averbakh, A., Krause, D., & Skoutas, D. (2009, October). Exploiting user feedback to improve semantic web service discovery. In *International Semantic Web Conference* (pp. 33-48). Springer. 10.1007/978-3-642-04930-9_3

Baliga, A., Subhod, I., Kamat, P., & Chatterjee, S. (2018). *Performance evaluation of the quorum blockchain platform.* arXiv preprint arXiv:1809.03421.

Baocai, Y., Huirong, Y., Pengbin, F., Liheng, G., & Mingli, L. (2010, July). A framework and QoS based web services discovery. In *2010 IEEE International Conference on Software Engineering and Service Sciences* (pp. 755-758). IEEE. 10.1109/ICSESS.2010.5552261

Baqa, H., Truong, N. B., Crespi, N., Lee, G. M., & Le Gall, F. (2019, September). Semantic smart contracts for blockchain-based services in the Internet of Things. In *2019 IEEE 18th International Symposium on Network Computing and Applications (NCA)* (pp. 1-5). IEEE. 10.1109/NCA.2019.8935016

Chen, F., Lu, C., Wu, H., & Li, M. (2017). A semantic similarity measure integrating multiple conceptual relationships for web service discovery. *Expert Systems with Applications*, *67*, 19–31. doi:10.1016/j.eswa.2016.09.028

Chhun, S., Moalla, N., & Ouzrout, Y. (2016). QoS ontology for service selection and reuse. *Journal of Intelligent Manufacturing*, *27*(1), 187–199. doi:10.100710845-013-0855-6

De Bruijn, J., Bussler, C., Domingue, J., Fensel, D., Hepp, M., Kifer, M., & Stollberg, M. (2005). Web service modeling ontology (wsmo). *Interface: a Journal for and About Social Movements*, *5*(1), 50.

Feddaoui, I., Felhi, F., Algarni, F., & Akaichi, J. (2021). QoS-Based Collaborative Filtering for Web Service Mining. *International Journal of Web Portals*, *13*(1), 40–61. doi:10.4018/IJWP.2021010103

Jiao, H., Zhang, J., Li, J. H., & Shi, J. (2017). Research on cloud manufacturing service discovery based on latent semantic preference about OWL-S. *International Journal of Computer Integrated Manufacturing*, *30*(4-5), 433–441.

Kaouan, M., Bouchiha, D., Benslimane, S. M., & Boukli-Hacene, S. (2020, December). Towards Service Ontology for Web Services Storage and Discovery. In *2020 4th International Symposium on Informatics and its Applications (ISIA)* (pp. 1-6). IEEE. 10.1109/ISIA51297.2020.9416537

Klusch, M., Fries, B., & Sycara, K. (2006, May). Automated semantic web service discovery with OWLS-MX. In *Proceedings of the fifth international joint conference on Autonomous agents and multiagent systems* (pp. 915-922). 10.1145/1160633.1160796

Kopecký, J., Vitvar, T., Bournez, C., & Farrell, J. (2007). Sawsdl: Semantic annotations for wsdl and xml schema. *IEEE Internet Computing*, *11*(6), 60–67. doi:10.1109/MIC.2007.134

Li, X., Zheng, Z., & Dai, H. N. (2021). When services computing meets blockchain: Challenges and opportunities. *Journal of Parallel and Distributed Computing*, *150*, 1–14. doi:10.1016/j.jpdc.2020.12.003

Lu, C. (2021, March). A Novel Web Service Discovery Method Combining Semantic Interface Similarity and Context Similarity. *IOP Conference Series. Earth and Environmental Science*, *693*(1), 012006. doi:10.1088/1755-1315/693/1/012006

Martin, D., Paolucci, M., McIlraith, S., Burstein, M., McDermott, D., McGuinness, D., & Sycara, K. (2004, July). Bringing semantics to web services: The OWL-S approach. In *International Workshop on Semantic Web Services and Web Process Composition* (pp. 26-42). Springer.

Meditskos, G., & Bassiliades, N. (2009). Structural and role-oriented web service discovery with taxonomies in OWL-S. *IEEE Transactions on Knowledge and Data Engineering*, *22*(2), 278–290. doi:10.1109/TKDE.2009.89

Paolucci, M., Kawamura, T., Payne, T. R., & Sycara, K. (2002, June). Semantic matching of web services capabilities. In *International semantic web conference* (pp. 333-347). Springer.

Purohit, L., & Kumar, S. (2018). A classification based web service selection approach. *IEEE Transactions on Services Computing*, *14*(2), 315–328. doi:10.1109/TSC.2018.2805352

Purohit, L., & Kumar, S. (2019, July). Clustering based approach for web service selection using skyline computations. In *2019 IEEE international conference on web services (ICWS)* (pp. 260-264). IEEE.

Ruta, M., Scioscia, F., Ieva, S., Capurso, G., Pinto, A., & Di Sciascio, E. (2018, August). A Blockchain Infrastructure for the Semantic Web of Things. SEBD.

Sangers, J., Frasincar, F., Hogenboom, F., & Chepegin, V. (2013). Semantic web service discovery using natural language processing techniques. *Expert Systems with Applications*, *40*(11), 4660–4671. doi:10.1016/j.eswa.2013.02.011

Skoutas, D., Simitsis, A., & Sellis, T. (2007, July). A ranking mechanism for semanticweb service discovery. In *2007 IEEE Congress on Services (Services 2007)* (pp. 41-48). IEEE. 10.1109/SERVICES.2007.8

Sridevi, S., Karpagam, G. R., & Kumar, B. V. (2021). Incorporating blockchain for semantic web service selection (SWSS) method. *Sadhana*, *46*(2), 1–14. doi:10.100712046-021-01619-y

Szabo, N. (1996). Smart contracts: building blocks for digital markets. *EXTROPY: The Journal of Transhumanist Thought*, *18*(2).

Viriyasitavat, W., Da Xu, L., & Bi, Z. (2020). Specification Patterns of Service-Based Applications Using Blockchain Technology. *IEEE Transactions on Computational Social Systems*, *7*(4), 886–896. doi:10.1109/TCSS.2020.2999574

Viriyasitavat, W., Da Xu, L., Bi, Z., Hoonsopon, D., & Charoenruk, N. (2019). Managing qos of internet-of-things services using blockchain. *IEEE Transactions on Computational Social Systems*, *6*(6), 1357–1368. doi:10.1109/TCSS.2019.2919667

Wang, X., Vitvar, T., Kerrigan, M., & Toma, I. (2006, December). A qos-aware selection model for semantic web services. In *International Conference on Service-Oriented Computing* (pp. 390-401). Springer. 10.1007/11948148_32

Xue, Y., Zhang, C., & Ji, Y. (2015, September). RESTful web service matching based on WADL. In *2015 International Conference on Cyber-Enabled Distributed Computing and Knowledge Discovery* (pp. 364-371). IEEE. 10.1109/CyberC.2015.89

Chapter 13
Improved Hybrid Sampling Strategy for Software Defect Prediction of Imbalanced Data Distribution

K. Nitalaksheswara Rao
GITAM University, India

ABSTRACT

Software defect prediction using data mining techniques is one of the best practices for finding defective modules. The existing classification techniques can be used for efficient knowledge discovery on normal datasets. Most of the real-world data sources are biased towards any one of the classes. This type of data source is known as class imbalance or skewed data sources. The defect prediction rate for the class imbalance datasets reduces with the increases in the class imbalance nature. To handle such type of datasets, an approach with specific designing technique is required for improved performance. In this chapter, the authors propose an algorithm known as improved integrated sampling strategy (IISS) for improved performance using noisy removal strategy for software defect prediction. The experimental analysis conducted on skewed software defect prediction datasets provides the results that IISS algorithm have performed well when compared with C4.5, C4.5+Balance, RF, and RF+Balance algorithms with various class imbalance evaluation measures.

DOI: 10.4018/978-1-6684-4225-8.ch013

INTRODUCTION

Software engineering is the process of building software with the desired properties of the user. The complete process of software engineering consists of different phases such as requirement analysis, designing, coding and testing. The complete or exhaustive testing for finding all the errors in the software modules is a tedious job.

This research uses a unique strategy for replicating and generating instances in the minority subset and at the same time reducing the instances from majority subset. The proposed technique is known as improved integrated sampling strategy (IISS) as it integrates both sampling strategies in a single method. This rationale behind combining both the strategies is to address the issues of both majority and minority subsets. The task of combing these strategies in the single class is a challenging task as the counter effects need to be properly under taken for consideration of the learning process for class imbalance problem of software defect prediction.

Figure 1. A common software defect prediction process

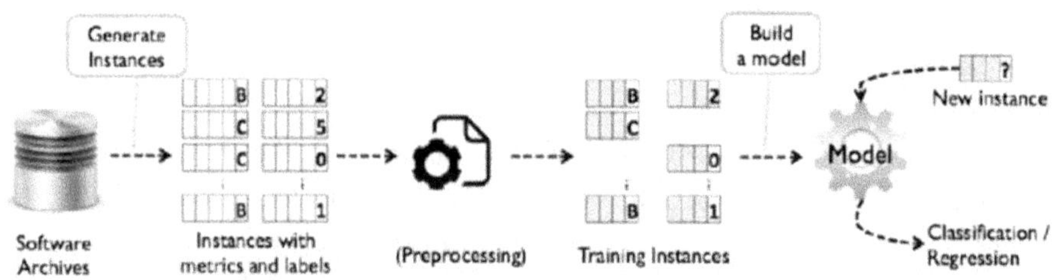

A common method for software defect prediction of class imbalance nature, need to be very accurate and precise, in spite of very less number of defective module instances. There by developing such a model is ineffective in the practical implementation due to a very high Imbalance ratio. In this study, we propose to use correlation based oversampling, instance ranges specific under sampling strategy and Improved integrated sampling techniques to help improve both majority and minority sub sets. The main rationale behind the approach is feature to feature correlation index and feature to class correlation index in the implementation of improved correlation based over sampling algorithm to learn range of instance. The proposals are supported with sound experimental setup for effective evaluation of class imbalance software defect datasets significantly improves classification over a decision tree as baseline.

The recent research in software defect prediction learning has not laid much stress to consider the software defect prediction as an efficient implementation in all the scenarios (Zi,Li, 2018) . The software defect prediction is also considered in the class balance framework where all the class are regard as equally. The main focus of our research is to overcome the issues with high imbalance ratio scenario in the knowledge discovery process of software defect prediction. The proposal, Improved Integrated Sampling Strategy (IISS) is well capable of handling effectively the process of knowledge discovery from the skewed software prediction datasets.

The remaining paper is presented as given: The recent literature on class imbalance learning in connection with software defect prediction is presented in section 2. The detailed problem statement with objectives is presented in section 3. The proposed approach of Integrated Sampling Strategy is presented in the section 4. The proposed approach is presented in detail in section 5. The dataset and evaluation criteria's are presented in section 6. The proposed algorithm is compared with benchmark algorithms and the details are presented in section 7. In section 8, the conclusion remarks and extension of work for future scope are resented. .

RELATED WORK

The special purpose databases execution traces analysis is conducted for software development process for quality improvement(Florian,2019). Different inherent properties, strategic learning tools and performance evaluation models are used to conduct a review of software quality analysis (Ayse Tousan 2009). A better paradigm of software testing strategy is presented recently for consumer support and maintaince in the field of durable electronic systems (Hasan,2012).

Florian (Wininger.,2009) have proposed a framework for execution traces analysis using custom build model from a collection of special purpose databases. (AyseTosun, 2008) has summarized a review for software quality analysis using application of inherit characteristics of software, parameters, and strategic learning tools with performance evaluation models. (HasanSo¨zer,2010) have presented a testing strategy for better quality improvement of software using model based paradigm. The unique applicability of testing is done in the wide area of durable electronic systems for better consumer support and maintaince.

Benjamin (Siegmund,2018) have prepared a unified opinion from different community of software testers for collection of recent approaches in the software defect analysis. (DuksanRyu,2007) have proposed a framework which considers both class imbalance nature and knowledge transfer techniques for boosting the target class. The main effect of the boosting is considered in the training phase by providing weightage to minority instances. Raphael Pham et al. [27] have conducted research to find the gap between recent university graduates and software testing professionals. They have also indicated formal methods to fill up the gap for benefiting the graduates in the hiring process.

(VahidGarouse,2017) have studied a set of software R&D projects for quality analysis to identify limitations of industry and academia collaborations. (Ahmed Gario,2019) have analysed unique models for fault analysis of safety critical applications. A rule set is prepared from the summary of the fault models identified and rectifying process is also built. (EmelieEngstro,2016) have presented the construction and evaluation of SERP-test which constructs an informal method for establishing communication between researcher and developer community for quality assurance improvement. (Arilo Claudio Dias-Neto,2017have analyzed different surveys indicating results penetrating to develop test case libraries for reusability, testing types, testing tools for bug reporting, need for isolated testing team and finally implementation of automation procedures for test case defining and building.

(Vincent Blondeau,2018)have worked in the direction to identify a group of errors which occur for specific events and also proposed justified reasons for rectification. The real time existences of specific case studies are also provided as examples for model presentation. (Feras A. Batarseh,2012) have introduced a method analytics-driven testing (ADT) for projection of mean time between failures, identification and rectification of errors. (FatmahYousefAssiri,2011) have worked on building a tool for fault detection in a specific location for performance improvement by identifying and repairing a potential failure.

The above discussed recent literature suggests that a limited amount of research is conducted on the class imbalance scenario of software defect prediction.

PROBLEM STATEMENT

Software development industry is one of the main technological applicability of the recent developments. Knowledge acquire from software developed can be used for different applications in real world scenario. The analyzed results of software defect analyses can be used for the below contexts:

- To identify the modules with errors.
- To identify total amount of defects.
- To analyze and identify the reasons and solutions for effective defective case prediction.
- To allocate error prone sub blocks further developing.

By using this algorithm minority classes can be oversampled and majority class is under sampled to populate the further required instances in the software defect prediction datasets for effective knowledge discovery.

The main objectives are:

- To apply novel under sampling technique to remove noisy and outlier instances.
- To apply novel over sampling strategies to over sample instances in the minority subset.
- To use advance filtering techniques to identify the correlated range of features from the data sources.
- To use efficient class imbalance validation measures for better results analysis.
- To decrease the overall time required for debugging the software.

In this research, we aim to answer several questions to analyze the performance of test models on software defect analysis datasets.

RQ1: What is the effect of class imbalance nature on the classification process?
RQ2: What is the level of class imbalance in software defect prediction datasets?
RQ3: What is the direct or indirect effect of imbalance nature in the software defect prediction?
RQ4: What is the effect of base learner on the class imbalance nature?
RQ5: What are the implications of class imbalance nature on the scalability of the dataset?

PROPOSED TECHNIQUES

Maximally Collapsing Metric Learning for Under Sampling

Maximally Collapsing Metric Learning (MCML) proposed is an innovative approach to learning metric(kims,2000). The instances in the same class are close, and those in different classes are far. The metric after learned would make instances in the same class close and those in different classes far. To begin with, the approach assumes that distance between instances in the same class is zero while distance between instances in different classes is infinite. In other words, the approach tries to map same class instances into a single point using a linear projection. The goal of the approach is to find matrix M such that set of instances is as close as possible to another set of instances. In order to match those distributions, KL divergence (Fukushima,2005), which measures the difference between two probability distributions, is minimized. The above said technique can be used for identifying noisy or borderline range of instances in majority subset for performing under sampling. The above said all techniques and other novel ideas are well utilized in the proposed model for improved performance on skewed software defect dataset predication.

Over Sampling for Imbalanced Learning

Learning from class-imbalanced data continues to be a common and challenging problem in supervised learning as standard classification algorithms are designed to handle balanced class distributions. While different strategies exist to tackle this problem, methods which generate artificial data to achieve a balanced class distribution are more versatile than modifications to the classification algorithm. Such techniques, called over samplers, modify the training data, allowing any classifier to be used with class-imbalanced datasets. Many algorithms have been proposed for this task, but most are complex and tend to generate unnecessary noise. The two classes existing in the software defect datasets are: defective modules and non-defective modules.

The proper learning of the defective modules is very crucial for the improved reliability of the quality assurance. The benchmark classification algorithms build the model from the software engineering datasets which are having very limited instances for defective modules. The class imbalance nature of the data source can be reduced either by increasing the instances in the minority subset, i.e. Oversampling or by decreasing the instances in the majority subset, i.e., under sampling. In the phase, it is proposed to oversample the minority subset, i.e., the number of instances in the minority subset will be increased. The oversampling will be performed using different techniques such as replication of some percentage of instances in the minority subset, synthetic instances generation, hybrid instances generation with properties of two or more instances.

PROPOSED IMPROVED INTEGRATED SAMPLING STRATEGY FOR SOFTWARE DEFECT PREDICTION (IISS)

The integrated hybrid sampling approaches are required as the user cannot predict where the noise situation existsi.e in the majority subset or minority subset. So, it is advised to apply both over sampling and under sampling for a finer results generation.

The main contribution of the research i.e the proposed framework of Integrated Sampling Strategy (IISS) is presented in Figure 2.

The IISS algorithm consists of three phases called Processing Phase, Adaptation phase and Building predictive Model Phase. The proposed framework is an extended implementation of our own previous proposal ICOS [37] and USS [38].

Processing Phase

In Processing Phase, the class imbalance software defect dataset which contains both majority and minority sub classes is divided into two separate classes. There by the majority class which has more percentage of instances is applied with under sampling and the minority class with less percentage of instances is applied with over sampling. In this proposed approach we have applied both majority and minority for better improvement of the data source. The under sampling technique used in the proposed approach uses correlation based feature selection technique to find the influential features of the data source. The weak range of instances for elimination can be identified using the irrelevant features identified in the filtering technique. Removal of these instances have the dual effect on the data source, one is the unnecessary instances ranges are removed helping for quality synthetic instances generation in the later on stages. The other is reducing the percentage of instances from the majority subset to limit the problem of class imbalance.

Adaptation Phase

The minority subset is further analyzed to form any existing outliers and noisy instances. The removal of noisy and outlier instances will improve the quality of the dataset and reduce the risk of magnifying the influence of outliers and noisy instances in the over sampling stage. In oversampling process different type of techniques are applied such as replicating the existing instances, generating synthetic instances, hybrid techniques for generating novel instances. The synthetic instances generation technique uses different level of generation, ranging from 10 to 100 percent instances generation depending on the unique properties of the datasets.

Building Predictive Model Phase

The improved majority and minority subsets are combined to form a almost balanced software defect prediction dataset and is applied to a base algorithm. In this case we have selected random forest as the base algorithm and evaluate on different evaluation metrics.

The algorithm of improved integrated sampling strategy can be elaborated as given in the below:

Algorithm: Improved Integrated Sampling Strategy (IISS)
Input: S: data stream of examples partitioned into chunks,
P: A set of minor class examples,
N: A set of major class examples,

jPj<jNj, and *Fj*,the feature set, j > 0.

where*jPj* are the number of minority instances and *jNj* are the number of majority instances.

Output: Average Measure {AUC, Precision, F-Measure, TP Rate, TN Rate}

Procedure:

Processing Phase:

Step 1. Take the datasets and find the important features in the dataset by apply Correlation based feature subset filter.

Step 2. Divide the dataset into majority and minority subsets.

Let the minority subset be P € pi (i = 1,2,..., *pnum) and majority subset be N € ni(i* = 1,2,..., *nnum*).

Let us consider

m' = the number of majority nearest neighbors

T= the whole training set

m= the number of nearest neighbors

Step 3. Find mostly misclassified instances pi

pi = m'; where m' (0 £ m'£ m)

if m/ 2 £ m'<m then pi is a mostly misclassified instance. Then remove the instances m' from the minority set.

Let us consider

m' = the number of minority nearest neighbors

Step 4. Find noisy instances pi'

pi' = m'; where m' (0 £ m'£ m)

If m'= m, i.e. all the m nearest neighbors of pi are majority examples, pi' is considered to be noise or outliers or missing values and are to be removed.

Step 5. Take the datasets and find the important features in the dataset by apply Correlation based feature subset filter.

begin

k ¬ 0,j¬1.

Apply CFS on subset N,

Find Fjfrom N, k= number of features extracted in classifier subset evaluator

repeat

k=k+1

Select the range for weak or noises instances of Fj.

Remove ranges of weak attributes and form a set of major class examples Nstrong

Until j = k

Form a new dataset using P and Nstrong

End

Adaptation Phase:

for all data *S* **do**

if input data is empty **then**
Generate model
else
Compute
MISSCLASS=Pm' using *m/ 2 £ Pm'<m*in the minority class P
*MISSCLASS=Nm'*using*m/ 2£ Nm'<m*in the majority class N
Remove
Pm'&Nm' from minority class P and majority class N respectively
Generate
PR= {p'1, p'2, ..., p'dnum}, 0 £ dnum£ pnum
s× dnum;
synthetic positive examples from the pr examples in minority set
Update
PR=s × dnum
endif
endfor
Building Predictive Model Phase:

```
Create a node N
  If samples in N are of same class, C then
   return N as a leaf node and mark class C;
      If A is empty then
return N as a leaf node and mark with majority
class;
else
            apply Random Forest
endif
endif
 Return N
```

In the Processing Phase, the dataset is passed on through a attribute evaluation filter and important features are selected, thereby removing the unnecessary features from the dataset.

Step 1. Take the datasets and find the important features in the dataset by apply Correlation based feature subset filter.

Then the dataset is split into majority and minority subsets for further processing. The minority subset is indicated with 'P' and the majority subset is indicated with 'N'.

Step 2. Divide the dataset into majority and minority subsets.

Let the minority subset be P € pi (i = 1,2,..., *pnum) and majority subset be N € ni(i =* 1,2,..., *nnum).*

The different terms used in the algorithm such as follows

Let us consider

m' = the number of majority nearest neighbors
T= the whole training set
m= the number of nearest neighbors

The noisy and outlier instances from majority and minority subsets are detected and removed.

Step 3. Find mostly misclassified instances pi
pi = m'; where m' (0 £ m'£ m)

if m/ 2 £ m'<m then pi is a mostly misclassified instance. Then remove the instances m' from the minority set.

Let us consider

m' = the number of minority nearest neighbors
Step 4. Find noisy instances pi'
pi' = m'; where m' (0 £ m'£ m)

If m'= m, i.e. all the m nearest neighbors of pi are majority examples, pi' is considered to be noise or outliers or missing values and are to be removed.

The weak range of instances from majority and minority subsets are detected and removed.

Find Fjfrom N, k= number of features extracted in classifier subset evaluator

repeat
k=k+1

Select the range for weak or noises instances of Fj.

Remove ranges of weak attributes and form a set of major class examples Nstrong

Until j = k

Form a new dataset using P and Nstrong

The misclassified instances from majority and minority subsets are detected and removed.

MISSCLASS=Pm' using *m*/ 2 £ *Pm'*<*m*in the minority class P
*MISSCLASS=Nm'*using*m*/ 2£ *Nm'*<*m*in the majority class N

The synthetic instances are generated in the minority subset depending upon the unique properties of the datasets

PR= {p'1, p'2, ..., p'dnum}, 0 £ dnum£ pnum
s× dnum;

synthetic positive examples from the pr examples in minority set

Update
PR=s × dnum

Then the decision tree is build using the Random forest base classifier.

There are many advantages of the proposed technique on the existing prediction methods. First, the proposed approach is an independent entity for implementation on any of the decision tree approaches as if it generates classification of classes. The only thing, which should be considered, is efficient model building techniques for defective modules prediction.

DATASETS AND EVALUATION CRITERIA'S

The experimental simulation is conducted on the 16 varied datasets of software defect prediction with class imbalance nature. The nature of the datasets is class imbalance, where one class has predominately more number of instances than the other class. The datasets are obtained from the PROMISE repository. The complete details of all the simulated datasets are presented in the Table 1.

Table 1. Summary of the defect prediction datasets used in the study

S.No	1	2	3	4	5	6	7	8	9	10	11	12	13	14	15	16	17
System	AR1	AR3	AR5	CM1	DR	DE	JM1	KC1	KC1D	KC1T	KC3	MC1	MC21	MW1	PC1	PC3	REUSE
Attributes	29	29	29	37	9	11	21	21	94	94	39	38	39	37	37	37	27
Modules	121	63	36	327	130	81	7782	2109	145	145	194	1988	125	253	705	1077	24
Defective	9	8	8	42	11	10	1672	326	60	8	36	46	44	27	61	134	9
IR	13.4	7.87	4.5	7.78	11.81	8.1	6.46	6.46	2.41	18.12	5.38	43.21	2.84	9.37	11.55	8.03	2.66

The performance evaluation measures used in this paper are presented below:

In class imbalance learning scenario, the accuracy measure is not opt way to evaluate the performance of an algorithm. Since the accuracy provides the percentage of correctly classified instances from both the majority and minority subsets in a combined way. In the case of class imbalance scenario we need to find the exact percentage of instances classified correctly in individual classes .ie subsets. This can be done by calculating the values such as True Positive (TP), True Negative (TN), False Positive (FP) and False Negative (FN) respectively.

The detail of these measures can be given as follows:

True Positive (TP): The instance is a positive instance and is classified as a positive instance.

True Negative (TN): The instance is a negative instance and is classified as a negative instance.

False Positive (FP): The instance is a negative instance and is classified as a positive instance.

False Negative (FN): The instance is a positive instance and is classified as a negative instance.

This above basic units of performance can be used to build the evaluation measures such as AUC, Precision, Recall and F-measure

EXPERMENTAL RESULTS

In this section, empirical comparisons are carried out for the proposed algorithm with the benchmarks. The results comparison of the proposed approach is one of the important parts of the manuscript, which is presented for highlighting the strengths and weakness of the proposed framework. The mean and standard deviation results are summarized for all the experimental conducted.

The results of AUC are presented in Table 3, the proposed algorithm results are better for all 16 software defect datasets. The C4.5 algorithm has also not performed well than the proposed IISS algorithm in terms of AUC. IISS verses REP has won on all 16 software defect prediction datasets in terms of AUC. The results of AUC metric for IISS when compared with CART are also improved on all the datasets.

Thus, compared to C4.5, REP and CART algorithms, the IISS algorithm lays more stress on identifying and improving the minority and majority subsets. In this situation, our proposed approach IISS have gained significant improvement in terms of precision (Table 4). The efficiency of the IISS can be shown in the form of

Table 2. Results of IISS model on SDP datasets in terms of AUC

Data set	C4.5	REP	CART	IISS
AR1	0.584 ±0.250●	0.503 ±0.045●	0.498 ±0.015●	0.936±0.130
AR3	0.738 ±0.259●	0.570 ±0.191●	0.604 ±0.218●	0.893±0.171
AR5	0.747 ±0.266●	0.610 ±0.216●	0.729 ±0.267●	0.926±0.156
DATA	0.481 ±0.068●	0.509 ±0.075●	0.500 ±0.048●	0.772±0.095
DESH	0.685 ±0.184●	0.694 ±0.177●	0.733 ±0.159●	0.844±0.132
KC1	0.696 ±0.067●	0.749 ±0.052●	0.683 ±0.064●	0.888±0.021
KC1DEF	0.704 ±0.126●	0.751 ±0.119●	0.704 ±0.110●	0.860±0.092
KC1TOP	0.654 ±0.230●	0.557 ±0.152●	0.500 ±0.000●	0.921±0.146
KC2	0.693±0.104●	0.767±0.087●	0.772±0.070●	0.871±0.046
KC3	0.604 ±0.175●	0.533 ±0.113●	0.540 ±0.117●	0.819±0.093
MC1	0.582 ±0.133●	0.585 ±0.135●	0.503 ±0.023●	0.918±0.057
MC21	0.596 ±0.166●	0.583 ±0.119●	0.604 ±0.129●	0.821±0.105
MW1	0.461 ±0.219●	0.586 ±0.149●	0.512 ±0.082●	0.865±0.080
PC1	0.675 ±0.141●	0.663 ±0.148●	0.515 ±0.062●	0.922±0.038
PC3	0.626 ±0.110●	0.657 ±0.133●	0.502 ±0.019●	0.874±0.039
REUSE	0.939 ±0.165●	0.500 ±0.000●	0.500 ±0.000●	0.995±0.050

●Bold dot indicate the win of IISS
○Empty dot indicate the loss of IISS

potential learning technique as it has performed well in the performance evaluation of recall measure (Table 5).

The results of IISS algorithms are efficient when compared with C4.5; out of the all 16 software defect prediction datasets, our proposed IISS algorithm have won on 10 datasets and losses on 6 datasets. When compared with REP Tree our proposed IISS algorithm have achieved 12 wins, 1 tie and 3 losses out of 16 software defect prediction datasets. The trends of AUC are shown in the Figure 3 for C4.5, REP, CART and IISS on software defect prediction datasets respectively.

Answer for RQ1: The RQ1 can be justified as follows that there is direct effect of class imbalance nature on the classification process. The claim can be observed in the results of the experimental simulation where there is less class imbalance nature the variation of performance is less and for high level of class imbalance the performance is high, directly proportional to the class imbalance nature of the data source.

Answer for RQ2: There is high level of class imbalance nature in the software defect prediction datasets as it is also clearly indicate that any real world data sources

Table 3 Results of IISS model on SDP datasets in terms of Precision

Data set	C4.5	REP	CART	IISS
AR1	0.932 ±0.036●	0.925 ±0.025●	0.925 ±0.025●	0.954±0.054
AR3	0.938 ±0.082○	0.894 ±0.078●	0.902 ±0.081●	0.917±0.116
AR5	0.881 ±0.200 ○	0.829 ±0.167●	0.884 ±0.174○	0.876±0.210
DATA	0.914 ±0.024○	0.915 ±0.024○	0.916 ±0.025○	0.838±0.027
DESH	0.751 ±0.165●	0.741 ±0.152●	0.752 ±0.142●	0.788±0.166
KC1	0.880 ±0.013○	0.869 ±0.011○	0.864 ±0.011	0.864±0.016
KC1DEF	0.628 ±0.153●	0.611 ±0.163●	0.589 ±0.139●	0.801±0.082
KC1TOP	0.049 ±0.144●	0.045 ±0.160●	0.000 ±0.000●	0.635±0.402
KC2	0.870±0.041○	0.861±0.038○	0.891±0.039○	0.842±0.040
KC3	0.374 ±0.298●	0.165 ±0.305●	0.184 ±0.324●	0.706±0.148
MC1	0.158 ±0.287●	0.066 ±0.207●	0.000 ±0.000●	0.780±0.172
MC21	0.503 ±0.271●	0.484 ±0.398●	0.567 ±0.352●	0.752±0.117
MW1	0.249 ±0.370●	0.140 ±0.305●	0.021 ±0.082●	0.685±0.188
PC1	0.359 ±0.246●	0.216 ±0.317●	0.077 ±0.218●	0.704±0.125
PC3	0.384 ±0.170●	0.176 ±0.293●	0.000 ±0.000●	0.626±0.079
REUSE	0.953 ±0.136○	0.617 ±0.151●	0.617 ±0.151●	0.935±0.183

●Bold dot indicate the win of IISS
○Empty dot indicate the loss of IISS

will be in class imbalance nature. The need of the proposed study on class imbalance software defect data source is necessary for proper extraction of the knowledge.

Answer for RQ3: The prediction accuracy of the modules with high probability of software defects can be easily identified for class balance data sources. Whereas the data sources with high level of class imbalance nature, are not well identified in the case of traditional approaches. The proposed approach has mitigated the problem by performing efficient level of sampling in both the subset of classes.

Answer for RQ4: The effect of base classifier can be clearly noticed for different approaches which use their unique ways for model building. The less percentage of instances in the minority subsets raises the problem of improper model building leading to inefficient performance, exclusively for minority subset.

Answer for RQ5: The scalability of the software prediction datasets is a virtue for minority subsets where there is scarcity of instances. The larger the data sources the more instance are available for better performance of the models which use training subset of instances for performing testing.

Table 4. Results of IISS model on SDP datasets in terms of recall

Data set	C4.5	REP	CART	IISS
AR1	0.964 ±0.057●	0.997 ±0.020○	0.995 ±0.034○	0.985±0.044
AR3	0.944 ±0.087●	0.973 ±0.078○	0.962 ±0.084○	0.925±0.133
AR5	0.867 ±0.226○	0.913 ±0.192○	0.903 ±0.203○	0.847±0.249
DATA	0.984 ±0.039●	0.992 ±0.034●	0.990 ±0.032●	1.000±0.000
DESH	0.799 ±0.198○	0.817 ±0.182○	0.855 ±0.172○	0.778±0.195
KC1	0.940 ±0.021○	0.964 ±0.022○	0.970 ±0.018○	0.932±0.019
KC1DEF	0.602 ±0.174●	0.745 ±0.255●	0.745 ±0.227●	0.910±0.083
KC1TOP	0.110 ±0.314●	0.080 ±0.273●	0.000 ±0.000●	0.635±0.401
KC2	0.901±0.057○	0.939±0.045○	0.913±0.054○	0.881±0.054
KC3	0.308 ±0.243●	0.115 ±0.216●	0.125 ±0.218●	0.689±0.163
MC1	0.077 ±0.135●	0.027 ±0.078●	0.000 ±0.000●	0.467±0.174
MC21	0.421 ±0.244●	0.263 ±0.242●	0.332 ±0.242●	0.809±0.129
MW1	0.170 ±0.239●	0.103 ±0.218●	0.027 ±0.100●	0.609±0.194
PC1	0.273 ±0.196●	0.099 ±0.150●	0.030 ±0.082●	0.649±0.140
PC3	0.284 ±0.138●	0.049 ±0.082●	0.000 ±0.000●	0.634±0.103
REUSE	0.995 ±0.050○	1.000 ±0.000○	1.000 ±0.000○	0.990±0.100

●Bold dot indicate the win of IISS
○Empty dot indicate the loss of IISS

The visualization results of AR1, KC1, MC1 and PC1 datasets generated using the IISS algorithm are shown in the Figure 2. The results show the improvement gained by the IISS algorithm on the software defect prediction datasets.

Above Figures. Class Representations of different class imbalance software defect datasets after applying to IISS algorithm.

In summary, the results are bias towards the better performance of the IISS, the reason lies in the capability of considering the software defect prediction in the class imbalance scenario.

Figure 3 presents the AUC results of all the datasets on the on the compared and proposed IISS algorithm. In the Figure 3, one can easily understand the overall improvement gained by the IISSalgorithm on all the datasets when compared with C4.5, REP and CART algorithms.

Table 7, summaries the results of wins, ties and losses of compared algorithms on IISS on all the datasets. The first row and first column of the table 7, presents the wins, ties and loses and evaluation measures of AUC, Precision, Recall and F-measure.

Table 5. Results of IISS model on SDP datasets in terms of F-measure

Data set	C4.5	REP	CART	IISS
AR1	0.947 ±0.039●	0.960 ±0.017●	0.958 ±0.024●	0.968±0.036
AR3	0.938 ±0.067○	0.929 ±0.058○	0.927 ±0.062○	0.913±0.099
AR5	0.861 ±0.195○	0.851 ±0.147○	0.879 ±0.169○	0.835±0.198
DATA	0.947 ±0.025○	0.952 ±0.023○	0.951 ±0.019○	0.912±0.016
DESH	0.759 ±0.152●	0.765 ±0.136	0.790 ±0.129○	0.765±0.145
KC1	0.909 ±0.011○	0.914 ±0.010○	0.914 ±0.007○	0.896±0.012
KC1DEF	0.602 ±0.134●	0.651 ±0.173●	0.646 ±0.155●	0.849±0.064
KC1TOP	0.067 ±0.195●	0.057 ±0.196●	0.000 ±0.000●	0.607±0.369
KC2	0.883±0.029○	0.897±0.024○	0.900±0.027○	0.860±0.033
KC3	0.322 ±0.236●	0.125 ±0.220●	0.140 ±0.237●	0.685±0.125
MC1	0.096 ±0.161●	0.037 ±0.109●	0.000 ±0.000●	0.563±0.156
MC21	0.433 ±0.213●	0.315 ±0.257●	0.388 ±0.242●	0.772±0.095
MW1	0.188 ±0.260●	0.107 ±0.216●	0.023 ±0.086●	0.623±0.150
PC1	0.293 ±0.190●	0.127 ±0.179●	0.041 ±0.108●	0.667±0.108
PC3	0.317 ±0.136●	0.069 ±0.108○	0.000 ±0.000●	0.626±0.077
REUSE	0.968 ±0.096○	0.753 ±0.104●	0.753 ±0.104●	0.953±0.142

●Bold dot indicate the win of IISS
○Empty dot indicate the loss of IISS

Figure 2. AR1 dataset

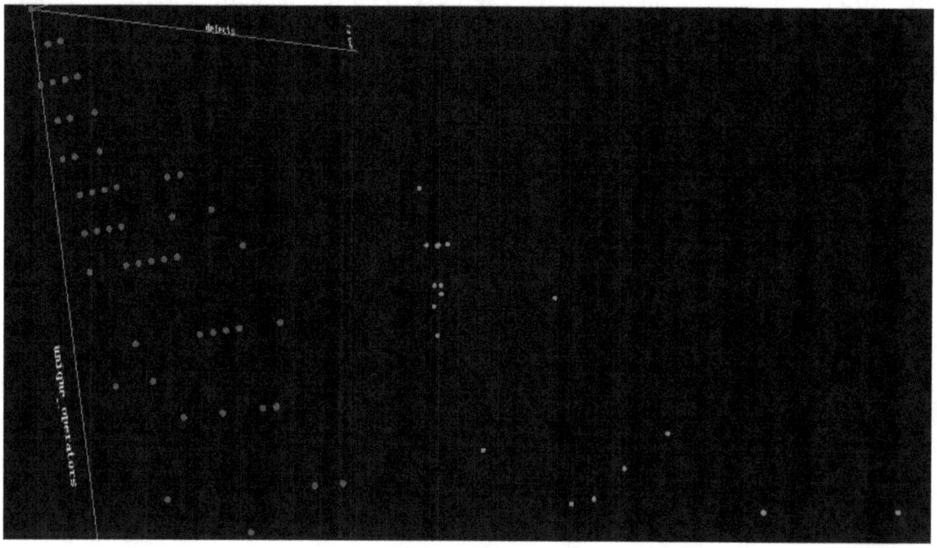

Figure 3. KC1 dataset

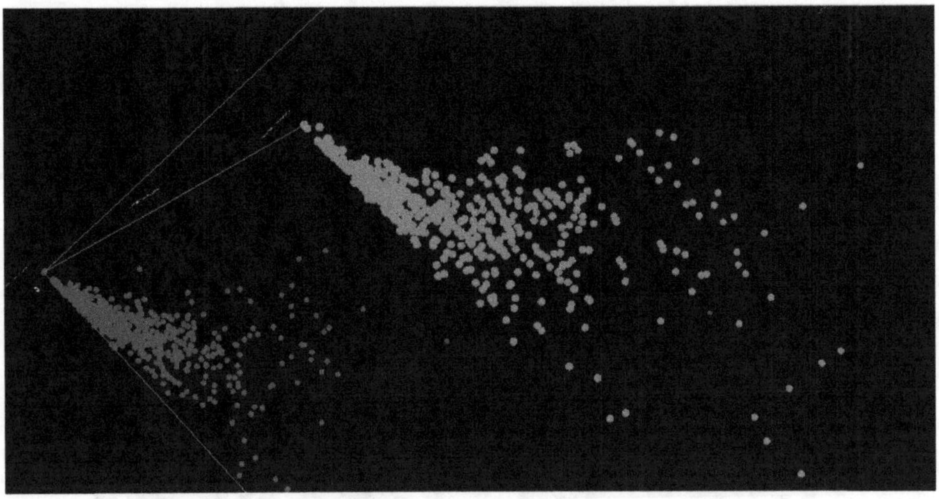

Figure 4. MC1 Dataset

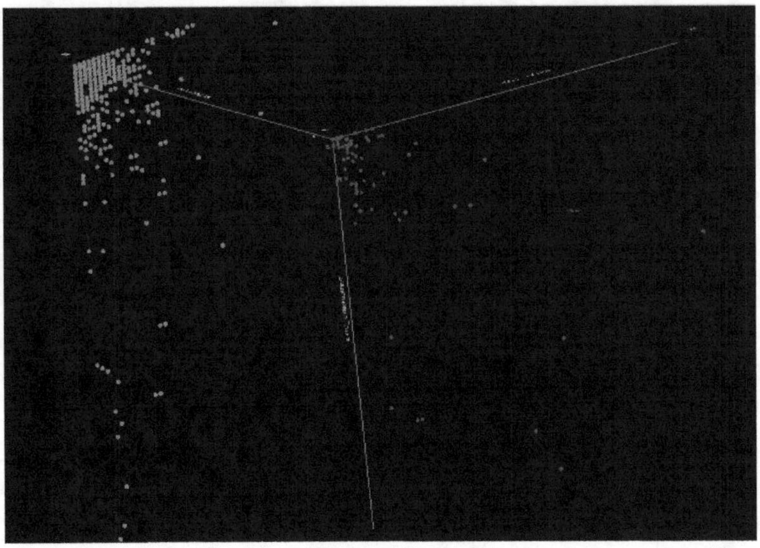

Figure 5. PC1 dataset

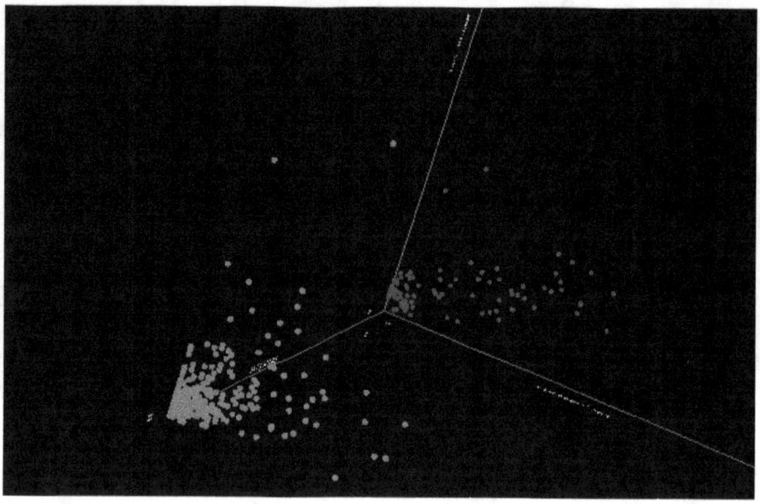

Table 6. Summary of experimental results for IISS

Results	Systems	Wins	Ties	Losses
AUC	IISS v/s C4.5	16	0	0
	IISS v/s REP	16	0	0
	IISS v/s CART	16	0	0
Precision	IISS v/s C4.5	10	0	6
	IISS v/s REP	13	0	3
	IISS v/s CART	12	1	3
Recall	IISS v/s C4.5	11	0	5
	IISS v/s REP	9	0	7
	IISS v/s CART	9	0	7
F-measure	IISS v/s C4.5	10	0	6
	IISS v/s REP	9	1	6
	IISS v/s CART	10	0	6

The IISS algorithm has performed well on all the measures. However, IISS is better in the aspect of class imbalance measures, which is the problem in hand for real world datasets. Finally, IISS have generated favourable results in terms of class

Figure 6. Trends in AUC for C4.5, REP, CART vs. IISS on SDP datasets

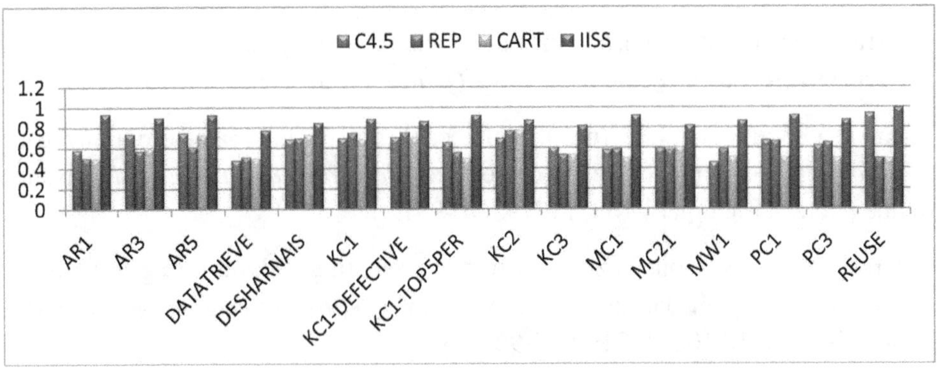

imbalance measures for software defect prediction. The total experimental simulation conducted on 16 class imbalance software defect datasets project that prominent recursive oversampling and intelligent under sampling approaches can improve the effectiveness when dealing with imbalanced data, as it has helped the IISS method to be the best performing algorithms when compared with benchmark algorithms.

CONCLUSION

In this session, a novel hybrid algorithm for imbalanced data with the application of software defect prediction has been proposed. This method uses unique oversampling and intelligent under sampling technique to almost balance dataset such that to minimize the imbalance effect in the software defect prediction model. The set of results run on 16 skewed software defect datasets show that the proposed approach IISS have performed better than the benchmark compared algorithms. In future extension of the work, the proposed approach can be implemented on more than 2 class skewed datasets in another domain of applicability.

REFERENCES

Assiri & Bieman. (n.d.). Fault localization for automated program repair: effectiveness, performance, repair correctness. *Software Qual J.* DOI 10.1007/ s11219-016-9312-z

Assiri, F. Y., & Bieman, J. M. (n.d.). Fault localization for automated program repair: Effectiveness, performance, repair correctness. *Software Quality Journal.* Advance online publication. doi:10.100711219-016-9312-z

Batarseh, F. A., & Gonzalez, A. J. (n.d.). Predicting failures in agile software development through data analytics. *Software Quality Journal.* Advance online publication. doi:10.100711219-015-9285-3

Batarseh, F. A., & Gonzalez, A. J. (n.d.). Predicting failures in agile software development through data analytics. *Software Quality Journal.* Advance online publication. doi:10.100711219-015-9285-3

Blondeau, V., Etien, A., Anquetil, N., Cresson, S., Croisy, P., & Ste'phane, D. (n.d.). Test case selection in industry: An analysis of issues related to static approaches. *Software Quality Journal.* Advance online publication. doi:10.100711219-016-9328-4

Blondeau, V., Etien, A., Anquetil, N., Cresson, S., Croisy, P., & Ste'phane, D. (n.d.). Test case selection in industry: An analysis of issues related to static approaches. *Software Quality Journal.* Advance online publication. doi:10.100711219-016-9328-4

Dias-Neto, A. C., Matalonga, S., Martı'n, S., Robiolo, G., & Travassos, G. H. (n.d.). Toward the characterization of software testing practices in South America: Looking at Brazil and Uruguay. *Software Quality Journal.* Advance online publication. doi:10.100711219-016-9329-3

Engstro, E., & Petersen, K. (n.d.). SERP-test: A taxonomy for supporting industry–academia communication. *Software Quality Journal.* Advance online publication. doi:10.100711219-016-9322-x

Fukushima, T., Kamei, Y., McIntosh, S., Yamashita, K., & Ubayashi, N. (2014). An empirical study of just-in-time defect prediction using cross-project models. *Proceedings of the 11th Working Conference on Mining Software Repositories.*

Gario, A., Andrews, A., & Hagerman, S. (n.d.). Fail-safe testing of safety-critical systems: A case study and efficiency analysis. *Software Quality Journal.* Advance online publication. doi:10.100711219-015-9283-5

Garousi, V., Eskandar, M. M., & Herkilog˘lu, K. (n.d.). Industry–academia collaborations in software testing: Experience and success stories from Canada and Turkey. *Software Quality Journal*. Advance online publication. doi:10.100711219-016-9319-5

Jing, X.-Y., Ying, S., Zhang, Z.-W., Wu, S.-S., & Liu, J. (2014). Dictionary learning based software defect prediction. *Proceedings of the 36th International Conference on Software Engineering*.

Kim, S., Whitehead, E. J., & Zhang, Y. (2008). Classifying software changes: Clean or buggy? *Software Engineering, IEEE Transactions on, 34*(2), 181-196.

Kim, S., Zhang, H., Wu, R., & Gong, L. (2011). *Dealing with noise in defect prediction*. Paper presented at the Software Engineering (ICSE), 2011 33rd International Conference on.

Li, Z., Jing, X., & Zhu, X. (2018). Progress on approaches to software defect prediction. IET Software, 12(3), 161-175. doi:10.1049/iet-sen.2017.0148

Menzies, T., Dekhtyar, A., Distefano, J., & Greenwald, J. (2007). Problems with precision: A response to "comments on 'data mining static code attributes to learn defect predictors'". *IEEE Transactions on Software Engineering, 33*(9), 637.

Mockus, A., & Votta, L. G. (2000). Identifying reasons for software changes using historic databases. *Software Maintenance, 2000. Proceedings. International Conference on*.

Nam, J., Pan, S. J., & Kim, S. (2013). Transfer defect learning. *Proceedings of the 2013 International Conference on Software Engineering*.

Nitalaksheswara Rao, K., & Satyananda Reddy, C. (2019). *A novel under sampling strategy for efficient software defect analysis of skewed distributed data*. doi:10.1007/s12530-018-

Pham, R., Kiesling, S., & Singer, L. (n.d.). On boarding inexperienced developers: Struggles and perceptions regarding automated testing. *Software Quality Journal*. Advance online publication. doi:10.100711219-016-9333-7

Rao & Reddy. (2018). An Efficient Software Defect Analysis Using Correlation-Based Oversampling. *Arabian Journal for Science and Engineering*.

Ryu, Jang, & Baik. (n.d.). A transfer cost-sensitive boosting approach for cross-project defect prediction. Software Quality Journal. Advance online publication. 10.1007/s11219-015-9287-1

Shatnawi, R. (2014). Empirical study of fault prediction for open-source systems using the Chidamberand Kemerer metrics. *IET Software, 8*(3), 113-119. Doi:10.1049/iet-sen.2013.0008

Siegmund, B., Michael, P. M. T., & Hirschfeld, R. (2014). Studying the Advancement in Debugging Practice of Professional Software Developers. *IEEE International Symposium on Software Reliability Engineering Workshops*. DOI 10.1109/ISSREW.2014.36

Smola, A., & Vishwanathan, S. V. N. (2008). *Introduction to machine learning*. Cambridge University.

Smola, A., & Vishwanathan, S. V. N. (2008). *Introduction to machine learning*. Cambridge University.

So¨zer, Ceren¸ & Gebizli. (n.d.). Model-based testing of digital TVs: An industry-as laboratory Approach. *Software Quality Journal*. Advance online publication. doi:10.100711219-016-9321-y

Tosun, A., Bener, A. B., & Akbarinasaji, S. (n.d.). A systematic literature review on the applications of Bayesian networks to predict software quality. *Software Quality Journal*. Advance online publication. doi:10.100711219-015-9297-z

Tosun, A., Bener, A. B., & Akbarinasaji, S. (n.d.). A systematic literature review on the applications of Bayesian networks to predict software quality. *Software Quality Journal*. Advance online publication. doi:10.100711219-015-9297-z

Wininger, F., Ezzati-Jivan, N., & Dagenais, M. R. (n.d.). A declarative framework for stateful analysis of execution traces. *Software Quality Journal*. Advance online publication. doi:10.100711219-016-9311-0

Wininger, F., Ezzati-Jivan, N., & Dagenais, M. R. (n.d.). A declarative framework for stateful analysis of execution traces. *Software Quality Journal*. Advance online publication. doi:10.100711219-016-9311-0

Zheng, J. (2010). Cost-sensitive boosting neural networks for software defect prediction. *Expert Systems with Applications, 37*(6), 4537-4543.

Chapter 14
Source Identification of Documents Based on LOOP Features

Pushpalata Gonasagi
Department of P.G. Studies and Research in Computer Science, Gulbarga University, Kalaburagi, India

Mallikarjun Hangarge
Karnatak Arts, Science, and Commerce College, Bidar, India

ABSTRACT

Pinpointing the ownership of documents based on printers is a challenging task. However, many methods have been proposed to identify the printers through printed documents. In this task, the chapter explores a method in simple way to identify the laser printer models. The LOOP (local optimal oriented pattern) method is applied to discriminate 10 laser printer models based on the character images. LOOP is an efficient descriptor to differentiate the images. A 10-fold cross-validation technique is used to classify the printers. The classifiers, namely Linear SVM (support vector machine) and Quadratic SVM, are applied to classify the laser printer models based on the documents at a character level. The experimental result shows that the proposed method is robust and outperforms comparable counterparts in the literature survey. The authors have achieved an average accuracy of Linear and Quadratic SVM classifiers of 99.2% and 99.8%, respectively.

DOI: 10.4018/978-1-6684-4225-8.ch014

INTRODUCTION

However, it is a difficult challenge since most present solutions are restricted to printing artifacts on printed documents produced by the printers and also need a high configuration computer system. Printing technologies are evolving at a rapid pace, resulting in the introduction of new printer models to the market. Laser printer models are the most often used printer models due to their high print quality and inexpensive cost. On the other hand, utilizing software to falsify printed papers is easy. The texture of the papers printed by the various printer models varies. As a result, the printer is a good source for determining the authenticity of a document. Hence, based on printed documents, this research provides a methodology for recognizing laser printer models.

The question document usually contains information like ownership of house, contracts, suicide note etc. (Hilton, 1992). During the examination of forgery documents, usually asking the questions as i) whether the documents are printed or photocopied, ii) whether the documents are printed by one printer or more than one printer, iii) which printer device is used to print the documents etc. Therefore, a document examiner needs the tool to determine the genuineness of the documents. The counterparts of the literature survey, many approaches have been developed in this area. In forensic document science, experts are inspecting question documents manually using chemical components (Braz, 2013). However, these approaches are expensive and also may be a chance to destroy the documents while inspecting. Later on, machine learning techniques are starting to be used to detect printed documents' genuineness using printer devices (Khanna, 2008). The printer has a specific intrinsic signature. It means that the printer has unique printing quality, and it can be observed at the margin of the characters of printed documents (Ali, 2003; Schreyer, 2009). This type of method is known as the intrinsic signature method. In this paper, we are finding the solution to find the printer models based on the texture of the whole documents at a character level instead of considering a particular region of the documents.

LBP (Local Binary pattern), LDP (Local Directional Method) and LOOP (Local Optimal Oriented Pattern) are the local binary methods. LDP and LBP are rotation variants, respectively, whereas LOOP is a novel and effective descriptor for image differentiation. The rotation invariant LOOP is illustrated using a numerical example in this paper. This chapter per the authors implemented the proposed solution using the LOOP algorithm.

The rest of the article is structured as follows: In section of RELATED WORK, counterparts of related work from the literature survey have been discussed. Section of PROPOSED METHOD explained the proposed method, including data collection, preprocessing procedure, and feature extraction method. In section of

EXPERIMENTAL RESULTS AND DISCUSSION discusses the experimentation and its results. Finally, the section of CONCLUSION concludes the concluding part of the implementation results.

RELATED WORK

(Elkasrawi, 2014) described a method to identify printers according to the noise produced by the printer. They have classified 20 different printer models and achieved 76.75% average accuracy. (Tsai, 2013) discussed a method to analyze the original digital device used to print the documents of the Chinese language. The features are extracted by applying GLCM(Gray-Level Co-occurrence Matrix) and DWT(Discrete Wavelet Transform). They achieved an accuracy of 98.64% using SVM classification. (Lampert, 2006) developed a system to identify the forged documents generated by different Laserjet printer devices. The features are based on the text line, edge roughness, area difference and correlation coefficient. They have used the SVM classifier to carry out the classification of the documents. To identify the printer, (Mikkilineni, 2009) examined the font type, font size, document's age, and paper type. The SVM classifier is used to classify the printers used in printing the documents. (Wu, 2009) expressed a method to detect the type of printer used in printing according to the geometric distortion of documents pages. The SVM classifier is applied to classify the ten printers and achieved an accuracy of 100%. (Bibi, 2019) described a method for identifying the printers according to GLCM texture features. The KNN classifier is used to classify ten Electrophotographic (EP) printers. (Devi, 2014) depicted an algorithm to distinguished inkjet printers and photocopiers. The skew and kurtosis of the histogram of document images are calculated and based on the decision, inkjet printer and photocopier are classified. (Tsai, 2016) presented the technique that analyses the printed character's microscopic images for source identification. This system used the SVM classifier to classify the printers using different descriptors such as GLCM and DWT. (Ferreira, 2017) developed a method to identify the printers using character images. The features are extracted using median and average filter techniques. CNN(Convolutional Neural Networks) are utilized to classify the printers. The classification accuracy of ten printers is 97.33%. (Jain, 2020) explored a method for the classification of fourteen printers. They used the geometric distortion method at the text-line level. They achieved 98.85% average accuracy to classify the printer models using the SVM classifier. (Shang, 2014) proposed a method for distinguishing the documents produced by the Laser printer, Inkjet printer and Electrostatic copier. They extracted features such as contour roughness, noise energy and average gradient from the characters in the documents. They have achieved an accuracy of 90% using the SVM classifier.

PROPOSED METHOD

We propose a method in this section to achieve the identification of ten different laser printer models based on the printed documents as the following process:

The first step is to remove the noise from the scanned document images at the character level using a filtering technique and then stored them into the computer system. In the second step, texture features are extracted using LOOP technique. Finally, all features of document images at the character level are combined into a single feature vector to classify ten printers separately using classifiers. The standard dataset is utilized to implement this new approach. The dataset contains the document images at character level 'e', which is segmented from the scanned printed documents images. The character images are segmented from the document's images, and these are Wikipedia web pages. The block diagram for the process of identification of the printer is shown in Figure 1.

Figure 1. Process of Laser Printer Identification

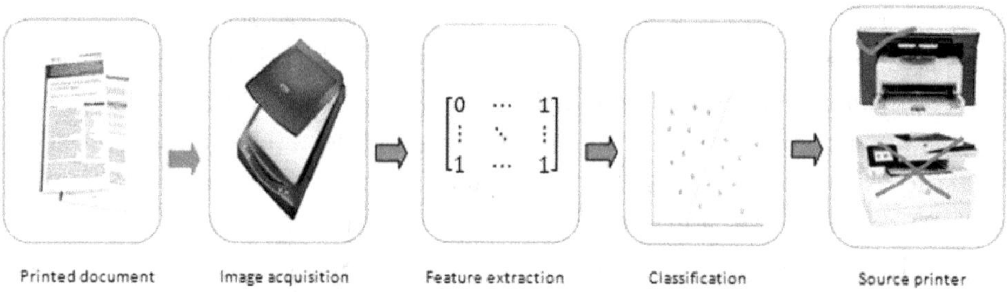

Data Collection

The standard dataset is applied to meet our goal as prescribed in the proposed method. The dataset contains the scientist biographies collected from the Wikipedia website, and these are printable versions. The documents contained text with different font sizes and pictures. Ten laser printers with standard resolutions print the documents, and the dataset is available at Figshare (Ferreira, 2017) website.

The names of ten printers are as follows: Brother.-HL.-4070CDW, Canon.-D1150, Canon.-MF3240, Canon.-MF4370DN. Hewlett Packard-CP1518, Hewlett Packard-CP2025A, Hewlett Packard-CP2025B. Lexmark-E260DN, OKI Data-C330DN, Samsung- CLP315. The document had been printed on white paper. Later these documents are scanned by a scanner at 600 dpi resolution. The character of 'e' is the common character in all English documents. Then character 'e' is segmented from each scanned document image. The dataset contains 122500 character images 'e' and is extracted from 600 scanned document images. We have selected 10,000 characters of 'e' images randomly, which are printed from ten laser printer models to implement the proposed model. Figures 2 and 3 represent the samples document pages printed from laser printers and their segmented characters in Figure 3.

Preprocessing

The character image 'e' contains the different scales and directions. Randomly we have selected 1000 character images from each ten laser printer models as discussed in section of Dataset. Hence totally of 10000 character images, 'e' are preprocessed. The size of the segmented character images 'e' is 38X47 pixels. We have resized the images by 28X28 and de-noised the images by applying the median filter technique with 3X3 neighboring pixels. The median filtering technique is a non-linear method used to preserve the neatness of the edge and remove the noise. Median filter enhanced the character images, and these character images are in PNG (Portable Network Graphics) format. Further, these de-noised images are processing for feature extraction.

Figure 2. Sample document pages printed from laser printers Brother.-HL.-4070CDW, Canon.-D1150, Canon.-MF3240, Canon.-MF4370DN

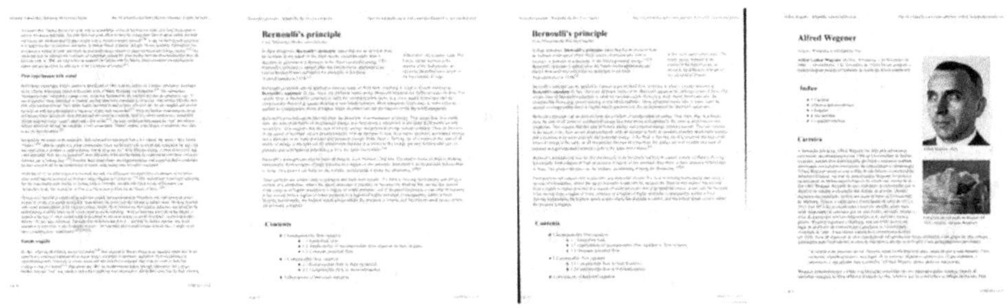

(a) (b) (c) (d)

Figure 3. Sample character 'e' images from laser printer model.

Dataset/ Printers	Brother -HL-4070CDW	Canon -D1150	Canon- MF3240	Canon-MF4370DN
Character images				

Feature Extraction

LOOP: A local binary descriptor is used to encode the repeated local patterns. LOOP is a non-linear descriptor and it is derived from the local binary descriptors, namely, LBP (Ojala. 1994) and LDP (Jabid, 2010). This descriptor is efficient descriptor for the discrimination of images. LOOP (Chakraborti, 2018) is overcoming the drawback of LDP and LBP methods. The drawback of LBP and LDP is that the direct local image patch and rotated local image patch do not provide the invariance property. It is elaborated with example as shown in Figure 4.

LBP descriptor computes a binary code for each pixel in an image by thresholding circularly symmetric neighborhood*Mn* with the value of central pixel *Ic*. The equation of the LBP is given below.

$$LBP(Xc,Yc)=\sum_{n=0}^{7}s(Mn-Ic). \tag{1}$$

Where

$$s(x)=\begin{cases}1, & x\geq 0\\ 0, & x<0\end{cases}$$

The equation of LOOP: Let *(Xc, Yc)* be the center pixel of the image *I. I* is the 3x3 neighboring pixel image. *Ic* is the intensity of images that occur as the most significant bit value. *Mn* is the corresponding magnitude to *In(n=0,1,...7). Mn(n=0,1....7)* is the 8 responses of kirsch mask as described in Figure 4. *s(x)* represents a function of x, where x indicates (Mn-Ic). If $(Mn\text{-}Ic)^3\ 0$, assign *s(x)=1*, otherwise s*(x)=0*. *Wn* is exponential and assigns digits from *0 to 7* as following to rank of magnitude among the eight Kirsch Mask outcomes.

$$LOOP(Xc,Yc)=\sum_{n=0}^{7}s(Mn-Ic).2^{Wn} \tag{2}$$

The eight Kirsch Mask directed in the direction of *(n=0, 1,...7)* measures the strength of intensity variations in various directions for eight adjacent pixels.

Here Pattern 1 and 2 from (Chakraborti, 2018), apply the LBP rule from equation (1) to create the binary words for LDP mask activation. Then binary weights are assigned from ascending order from equation (2). Finally, LOOP value is given by $2^7+2^6+2^5$= 224 for both the pattern. It shows that LOOP is an invariance rotation attribute, whereas the LBP and LDP have different Pattern 1 and Pattern 2. The values of LBP and LDP for Pattern 1 and Pattern 2 are 26 and 7, respectively, as in Figure 4.

EXPERIMENTAL RESULTS AND DISCUSSION

It is a ten-class problem to identify and classify the laser printer models based on the printed documents at a character level. We cannot observe the differences between the different printed documents images produced by the different printers by the eyes because there are minute differences in textures of printed characters. To evaluate the proposed system, we have carried out the experiment using Matlab tool and standard dataset. The dataset contains the printed character images of ‘e’ from ten laser printers as discussed in section of Dataset. The 60 article pages are printed from ten laser printer models. Total 600 pages are printed. Then these 600 printed pages are scanned by the scanner machine and stored into a computer according to ten laser printer models. The same scanner machine is used to scan all printed pages. Further, character ‘e’ is segmented from all the pages, and these segmented images are used to validate the performance of the LOOP features in discriminating the printer models. Total 10000 character images ‘e’ is used to identify the ten laser printer models.

Figure 4. The numerical calculation of LOOP, for instance

Pattern 1

250	245	15
240	100	5
10	20	0

Pattern 2(Rotated)

10	240	250
20	100	245
0	5	15

Kirsch Mask:

5	5	5
-3	0	-3
-3	-3	-3

North

5	5	-3
5	0	-3
-3	-3	-3

North West

5	-3	-3
5	0	-3
5	-3	-3

West

-3	-3	-3
5	0	-3
5	5	-3

South West

-3	-3	-3
-3	0	-3
5	5	5

South

-3	-3	-3
-3	0	5
-3	5	5

South East

-3	-3	5
-3	0	5
-3	-3	5

East

-3	5	5
-3	0	5
-3	-3	-3

North East

3525	1725	-235
1645	0	-2195
-195	-2115	-2155

Pattern1
LDP Mask Activation

1	1	0
1		0
0	0	0

LBP Binary Word

-195	1645	3525
-2115		1725
-2155	-2195	-235

Pattern2
LDP Mask Activation

0	1	1
0		1
0	0	0

LBP Binary Word

Table 1. The classification accuracy using linear and quadratic SVM classifiers

Ten Laser Printer models/classes	Linear-SVM classifier		Quadratic-SVM Classifier	
	Document identification in-terms of classification rate (%)	Error Rate (%)	Document identification in-terms of classification rate (%)	Error Rate (%)
Brother -HL-4070CDW	99.0	1.0	99.8	0.2
Canon -D1150	99.4	0.6	99.9	0.1
Canon- MF3240	97.4	2.6	99.9	0.1
Canon-MF4370DN	99.1	0.9	99.9	0.1
Hewlett Packard-CP1518	100.0	0.0	100.0	0.0
Hewlett Packard-CP2025A	100.0	0.0	100.0	0.0
Hewlett Packard-CP2025B	99.4	0.6	99.4	0.6
Lexmark-E260DN	99.2	0.8	99.9	0.1
OKI Data-C330DN	98.7	1.3	99.6	0.4
Samsung- CLP315	100.0	0.0	100.0	0.0
Average Accuracy	99.2	0.8	99.8	0.2

Table 2. Confusion matrix using Linear SVM classifier

Ten Laser Printer models/Classes	Brother .-HL.-4070CDW	Canon.-D1150.	Canon-. MF3240.	Canon.-MF4370DN.	Hewlett Packard-CP1518	Hewlett Packard-CP2025A	Hewlett Packard-CP2025B	Lexmark-E260DN	OKI Data-C330DN	Samsung-CLP315
Brother.-HL.-4070CDW	900	--	2	2	--	--	--	--	3	3
Canon.-D1150	--	994	--	1	--	--	5	--	--	--
Cano.-MF3240	--	--	974	22	--	--	--	--	--	4
Canon.-MF4370DN	--	--	8	991	--	--	--	1	--	--
Hewlett Packard-CP1518	--	--	--	--	1000	--	--	--	--	--
Hewlett Packard-CP2025A	--	--	--	--	--	1000	--	--	--	--
Hewlett Packard-CP2025B	--	--	--	4	--	--	994	1	1	--
Lexmark-E260DN	1	--	--	3	--	--	--	992	4	--
OKI Data-C330DN	--	4	--	6	--	--	--	3	987	--
Samsung- CLP315	--	--	--	--	--	--	--	--	--	1000

Table 3. Confusion matrix using Quadratic SVM classifier

Ten Laser Printer models/Classes	Brother --HL-4070CDW	Canon --D1150	Canon-- MF3240	Canon-- MF4370DN	Hewlett. PackardCP1518	Hewlett. packard-CP2025A	Hewlett ackard-CP2025B	Lexmark-E260DN	OKI Data-C330DN	Samsung-CLP315
Brother- -HL-4070CDW	998	--	--	2	--	--	--	--	--	--
Canon --D1150	--	999		1	--	--	--	--	--	--
Canon-- MF3240	--	--	999	1	--	--	--	--	--	
Canon-MF4370DN	--	--	--	999	--	--	--	1	--	--
Hewlett Packard-CP1518	--	--	--	--	1000	--	--	--	--	--
Hewlett Packard-CP2025A	--	--	--	--	--	1000	--	--	--	--
Hewlett Packard-CP2025B	--	--	--	4	--	--	994	1	1	--
Lexmark-E260DN	1	--	--	--	--	--	--	999	--	--
OKI Data-C330DN		4	--	--	--	--	--	--	996	--
Samsung- CLP315	--	--	--	--	--	--	--	--	--	1000

Table 4. Comparison performance of the proposed method.

Authors	Classifier	Accuracy (%)
(Ferreira, 2017)	CNN	97.33%
Proposed Approach	Linear SVM	99.2%
	Quadratic SVM	99.8%

CONCLUSION

The proposed system has shown outperforming results in identifying ten laser printer models using printed character images. The overall classification accuracy of printers is 99.8% using a quadratic SVM classifier. The developed method is the best explorer to extract the texture properties of the images of printed documents with multi-scale and multi-directions using the local binary descriptor, namely LOOP. This LOOP efficiently discriminated printed character images. It shows that the same document is printed by the different printers differently. In other words, each laser printer model has specific characters and printing technologies to print the documents. The benefit of this system is to automatically link to the document's ownership and detect the originality of the documents. In the future, we can enhance the experimental results by taking different printers such as dot-matrix printers and inkjet printers. Also, we can implement the system at the word, line and page level of documents to reduce the time complexity of the proposed system.

REFERENCES

Ali, G. N., Mikkilineni, A. K., Allebach, J. P., Delp, E. J., Chiang, P. J., & Chiu, G. T. (2003, January). Intrinsic and extrinsic signatures for information hiding and secure printing with electrophotographic devices. In NIP & Digital Fabrication Conference (Vol. 2003, No. 2, pp. 511-515). Society for Imaging Science and Technology.

Bibi, M., Hamid, A., Moetesum, M., & Siddiqi, I. (2019, September). Document Forgery Detection using Printer Source Identification—A Text-Independent Approach. In *2019 International Conference on Document Analysis and Recognition Workshops (ICDARW)* (Vol. 8, pp. 7-12). IEEE.

Braz, A., López-López, M., & García-Ruiz, C. (2013). Raman spectroscopy for forensic analysis of inks in questioned documents. *Forensic Science International, 232*(1-3), 206–212. doi:10.1016/j.forsciint.2013.07.017 PMID:24053882

Chakraborti, T., McCane, B., Mills, S., & Pal, U. (2018). Loop descriptor: Local optimal-oriented pattern. *IEEE Signal Processing Letters, 25*(5), 635–639.

Devi, M. U., Rao, C. R., & Jayaram, M. (2014). Statistical measures for differentiation of photocopy from print technology forensic perspective. *International Journal of Computers and Applications, 105*(15).

Elkasrawi, S., & Shafait, F. (2014, April). Printer identification using supervised learning for document forgery detection. In *2014 11th IAPR International Workshop on Document Analysis Systems* (pp. 146-150). IEEE.

Ferreira, A., Bondi, L., Baroffio, L., Bestagini, P., Huang, J., Dos Santos, J. A., ... Rocha, A. (2017). Data-driven feature characterization techniques for laser printer attribution. *IEEE Transactions on Information Forensics and Security, 12*(8), 1860–1873.

Hilton, O. (1992). *Scientific examination of questioned documents*. CRC Press.

Jabid, T., Kabir, M. H., & Chae, O. (2010, August). Gender classification using local directional pattern (LDP). In *2010 20th International conference on pattern recognition* (pp. 2162-2165). IEEE.

Jain, H., Joshi, S., Gupta, G., & Khanna, N. (2020). Passive classification of source printer using text-line-level geometric distortion signatures from scanned images of printed documents. *Multimedia Tools and Applications*, *79*(11), 7377–7400.

Khanna, N., Mikkilineni, A. K., Chiu, G. T. C., Allebach, J. P., & Delp, E. J. (2008, August). Survey of scanner and printer forensics at purdue university. In *International Workshop on Computational Forensics* (pp. 22-34). Springer. 10.1007/978-3-540-85303-9_3

Lampert, C. H., Mei, L., & Breuel, T. M. (2006, November). Printing technique classification for document counterfeit detection. In *2006 International Conference on Computational Intelligence and Security* (Vol. 1, pp. 639-644). IEEE.

Mikkilineni, A. K., Arslan, O., Chiang, P. J., Kumontoy, R. M., Allebach, J. P., Chiu, G. T. C., & Delp, E. J. (2005, January). Printer forensics using svm techniques. In NIP & Digital Fabrication Conference (Vol. 2005, No. 1, pp. 223-226). Society for Imaging Science and Technology.

Ojala, T., Pietikainen, M., & Harwood, D. (1994, October). Performance evaluation of texture measures with classification based on Kullback discrimination of distributions. In *Proceedings of 12th international conference on pattern recognition* (Vol. 1, pp. 582-585). IEEE.

Schreyer, M., Schulze, C., Stahl, A., & Effelsberg, W. (2009, March). Intelligent Printing Technique Recognition and Photocopy Detection for Forensic Document Examination. In Informatiktage (Vol. 8, pp. 39-42). Academic Press.

Shang, S., Memon, N., & Kong, X. (2014). Detecting documents forged by printing and copying. *EURASIP Journal on Advances in Signal Processing*, *2014*(1), 1–13.

Tsai, M. J., & Liu, J. (2013, May). Digital forensics for printed source identification. In *2013 IEEE International Symposium on Circuits and Systems (ISCAS)* (pp. 2347-2350). IEEE.

Tsai, M. J., & Yuadi, I. (2016, September). Printed source identification by microscopic images. In *2016 IEEE International Conference on Image Processing (ICIP)* (pp. 3927-3931). IEEE.

Wu, Y., Kong, X., & Guo, Y. (2009, November). Printer forensics based on page document's geometric distortion. In *2009 16th IEEE International Conference on Image Processing (ICIP)* (pp. 2909-2912). IEEE.

Chapter 15
Time Series Prediction for Stock Market Analysis

Akshita Kaushik
Bhagwan Parshuram Institute of Technology, India

Akanksha Dhamija
Bhagwan Parshuram Institute of Technology, India

Shriya Sharma
Bhagwan Parshuram Institute of Technology, India

Raghav Jindal
Bhagwan Parshuram Institute of Technology, India

ABSTRACT

The stock market is an ideal way to invest money and earn potential returns. But, even with advanced technology and computational power, this chapter still cannot predict the patterns of rise and fall in the stock market, and still, it is considered a risky business. To ease the process of investment and to provide better consciousness, this chapter proposes the prediction of stock market deviation using the ARIMA (auto-regressive integrated moving average) algorithm and long short-term memory (LSTM) algorithm. This chapter is using an algorithm which is trying to predict the future pattern for any stock based on the real-time analysis and data provided from Yahoo Finance data. The software provides pictorial and graphical representations. The objective is to provide short-term and long-term prediction competence to prepare for future potential investments.

DOI: 10.4018/978-1-6684-4225-8.ch015

INTRODUCTION

The stock market is the aggregation of buyers and sellers of stocks and it represents the ownership of a business. Stocks can be categorized by the company they belong to and where they are domiciled. Stocks are liquid commodities and indicate an interest in the ownership of a company. Transactions in a stock market mean the transfer of money from a seller to a buyer for equities when they agree on a price and it is facilitated by market exchange. There is a total of 60 stock exchanges in the world and amount to the US $76.3 trillion. Efficient Market Hypothesis (EMH) is used to explain the behavior of the stock market more effectively and this states that only changes in the fundamental factors like margins and profits affect the market beyond a short-term period. Ultimately, the values of stocks growth but also affects the economy of a country.

Market equilibrium and public reaction are the two factors that influence the stock markets around the globe. Investments tend to vary exponentially based on these two and thus drive prices unduly high or low. Web media has become a critical destabilizing factor and its results are quite hard to determine. But, with extensive research, it has become certain that social media plays a hand in the fluctuations of the market. Thus, the need to consider both historical data and the current market stand arises.

Many techniques in the past have been proposed to map the patterns of the stock market using statistical, regressive, and machine learning algorithms. Statistical models take one data source into account and cannot combine multiple bases and analyze the joint effect of them. With all the underlying assumptions being satisfied, these methods work very well. Regression models only take scalars as independent variables and thus information sources with high dimensions are reduced causing large amounts of information loss. Machine learning methods integrate all information sources into one super feature vector which causes weakening in the linearization of information or important associations to disappear between them. Though, the relative popularity of all of the above methods, they face considerable drawbacks. To circumvent a few of these disadvantages, the ARIMA algorithm was developed. It combined both statistical and regression methods and integrated the results to achieve better performance. It is a more popular and widely used statistical method for time series forecasting. ARIMA models can be used for the determination of autocorrelations in the data. Primarily ARIMA model is used to better understand the historical data or to predict future data points in time series.

Stationarity

A stationary time series data does not rely on the time, hence time series having trends or seasonality is not stationary. The trend and seasonality impact the time series at discrete intervals of time. On the contrary, for stationarity, it does not matter when you notice it, it should appear much identical at any point in time. ARIMA in its essence is the type of model that chronicles a set of distinct standard temporal structures in time series data. An ARIMA model is a statistical model employed for analyzing and prognosticating time series data. It is really simplified in terms of using it, still, it is a model that is really powerful. The algorithm is used to study time series sequences with a focus on statistics and econometrics data. It is a abstraction of the ARMA (Auto Regressive Moving Average) algorithm. It can be used for nonstationary data where a differencing step is applied repeatedly to eliminate it.

The algorithm includes three parts: AutoRegressive, Moving Average, and Integration. Autoregression is used to indicate that the evolving variable is dependent on its own lag. Moving Average indicates that the regression error is a combination of all previous error values and integration is the difference between the current and past values of data.

RELATED WORK

Various studies have shown that the implementation of some powerful Machine learning algorithms has remarkably improved the accuracy of stock price classification and prediction, thereby drawing the attention of researchers in the aforementioned field. Historical data of the share market plays a pivotal role in aiding the investors to get a gist of market behavior and trends over a certain period of time. This study shows historical data of the stock market being collected and trained using the ARIMA model fed with different parameters to understand the data articulately and thereby predicting future values.

Time-series forecasting, and analysis are of indispensable implication, due to the extensive use in different practical domains. A time series is a series of data points cataloged in the order of time. Thus, it is a sequence of data at discrete intervals of time. As we can see that the Stock market is known to be the one which is the most vastly intricate financial system which has numerous components/stocks, therefore the price differs greatly with reference to time.(Chen et al., 2001; Liu & Song, 2017; Tektas, 2010) The most efficacious way for the prediction of the future is mainly to acknowledge today's scenarios. To achieve the objective of forecasting stock returns, technical and fundamental analysis techniques have been used by several researchers.(Liu & Song, 2017) The author tried to come out with a pertinent model that surely helps to forecast the unobserved values of the stock market. They foretell the indices of future stock which entertain the efficient performance of the Indian economy through the ARIMA Model. It is essential to understand the current scenario of the stock market because for many economists/investors/researchers the Indian Stock market is considered to be the center of interest. Predictions are made that the overall performance of the Indian stock market shown an appropriate time series model (ARIMA Model) which will be helpful to create suitable values of upcoming indices. (Beveridge & Nelson, 1981; Pouzols & Lendasse, 2010)

Autoregressive integrated moving average (ARIMA) model has been one of the most eminent linear models in Time Series Forecasting over three decades. Researchers come up with their recent research activity in forecasting with ANNs suggesting that the Artificial Neural Network can be an optimistic alternative to other old linear methods. Both (ARIMA model and ANNs) are often compared with each other having mixed conclusions with reference to the excellence in forecasting performance. (Ding et al., 2019; Liu et al., 2016; Siami-Namini et al., 2018; Wu et al., 2016)

The performance of the ANN predictive model was compared with the traditional ARIMA model in this paper (Franses, 1991; Granger & Joyeux, 1980; Nelson & Plosser, 1982; Witt et al., 1998; Yang et al., 2018), which is extensively used for time series forecasting. It is observed that both of these models perform well in forecasting with respect to real-world problems and can hence be employed for the prediction of stock price. Arima model was perceived to be directional and the model developed with the Artificial Neural Network (ANN)-approach demonstrated better performance than the latter.

ALGORITHM USED

ARIMA Model

ARIMA model has following parameters:

Lag Order (p): The no. of lag observations integrated in the model.

Degree of Difference (d): It is the number of non-seasonal differencing required by the time series for transforming into a stationary time series without seasonality and trends.

Order of Moving Average (q): It is the number of lagged errors or lags in the prediction equation. The general form of representation is ARIMA (p,d,q) and when two of the three parameters become zero, it can be attributed to the non-zero parameter. ARIMA model can be generalized using this formula: (for drift value $\delta/(1 - \Sigma\varphi i)$). Differentiating is performed by computing the differences between consecutive observations. It can be represented using the order of differentiation done such as First-order differentiation:

$$y_t^{'} = y_t - y_{t-1}$$

Second-order differentiation:

$$\begin{aligned} y_t^{*} &= y_t^{'} - y_{t-1}^{'} \\ &= (y_t - y_{t-1}) - (y_{t-1} - y_{t-2}) \\ &= y_t - 2y_{t-1} + y_{t-1} \end{aligned}$$

LSTM Model

With the recent breakthroughs that have been happening in data science, it is found that for almost all of these sequence prediction problems, Long ShortTerm Memory networks, also known as LSTMs have been observed as the most effective solution. LSTM has an upper hand over traditional feedforward neutral network and RNN in numerous ways. LSTMs, special kind of RNNs, having additional features of retaining the order of data.

LSTM basically tackle the problem related to long term dependencies of Recurrent Neural Network where RNN is not able to predict the word that is stored in long term memory and at the same time can provide the more precise predictions from recent information.

LSTMs networks, special kind of Recurrent Neural Network, capable of retaining order dependance in different prediction problems. LSTM has the ability to remember the values from the previous stages that can be used in future.

Each LSTM can be seen as a set of system modules and here data streams are retained and stored. An LSTM module contains a cell state and three gates which attributes them the faculty of selectively learning, unlearning or retention of information from each unit. The cell state in LSTM aids the flow of information through the units without changing it. Each unit has an input, output and a forget gate which can be used for addition or removal of the information from the cell state.

Forget gate: It decides which information should be forgotten from the previous cell state for which a sigmoid function is used.

Input gate: It controls the flow of information using a pointwise multiplication operation of 'sigmoid' and 'tanh' high respectively to the current cell state.

Output gate: It decides which information should be relegated to the next hidden state.

Figure 1. It shows the workflow of this proposed technique.

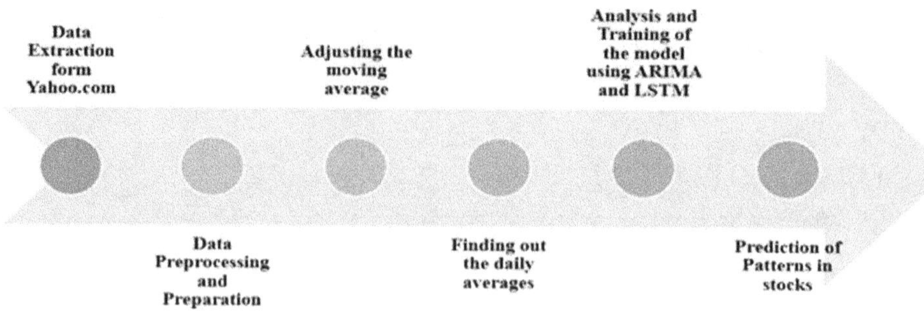

Figure 2. The data snippet

Date	High	Low	Open	Close	Volume	Adj Close	company_name
2020-05-20	1410.420044	1387.250000	1389.579956	1406.719971	1655400.0	1406.719971	GOOGLE
2020-01-13	79.267502	77.787498	77.910004	79.239998	121532000.0	77.923538	APPLE
2020-04-15	2333.370117	2245.000000	2257.679932	2307.679932	6866600.0	2307.679932	AMAZON
2020-11-25	215.289993	212.460007	215.110001	213.869995	21012900.0	213.869995	MICROSOFT
2020-11-30	120.970001	116.809998	116.970001	119.050003	169410200.0	119.050003	APPLE
2020-05-06	2357.449951	2320.000000	2329.439941	2351.260010	3117800.0	2351.260010	AMAZON
2020-05-12	1415.000000	1374.770020	1407.119995	1375.739990	1390600.0	1375.739990	GOOGLE
2020-03-12	153.470001	138.580002	145.300003	139.059998	93226400.0	137.978134	MICROSOFT
2020-02-06	1482.000000	1449.569946	1450.329956	1476.229980	1679400.0	1476.229980	GOOGLE
2020-03-25	1950.260010	1885.780029	1920.689941	1885.839966	6479100.0	1885.839966	AMAZON

IMPLEMENTATION AND RESULT

The proposed technique make use of the ARIMA model and LSTM model for the prediction of the future patterns in the stock prices of any company, we have consider APPL (Apple), GOOG (Google), MSFT (Microsoft), AMZN (Amazon) for our study. The technique starts from the live data collection from the Yahoo.com, the data consist of the various stock prices and parameters like high, low, open, close, volume like attributes about the stocks.

The biggest challenge with the time series data is moving average, we try to adjust the moving average of the various stocks by considering the number of days as 10,

20 and 50. Once done, we tried to find out what are the daily average returns of any stock under consideration, thus to find out the general trend about the stock. The system architecture is shown in figure 1. The brief explanation about each step is as:

Data Collection

The prediction works on obtaining the data, processes it and then showcases the result after prediction via a table and graph format. Data Collection module acquires real time data using pandas_datareader. Data is collected using the time and company's opening value and closing value (AAPLApple Inc, GOOGGoogle /Alphabet Corporation etc.). It then plots a graph with volume on y axis and date on x axis. The snippet of the data is shown in the figure 2. By looking at the pattern, one can easily identify the presence of the moving average in the data, as shown in figure 3.

Figure 3. The graph on the x and y axis for the data

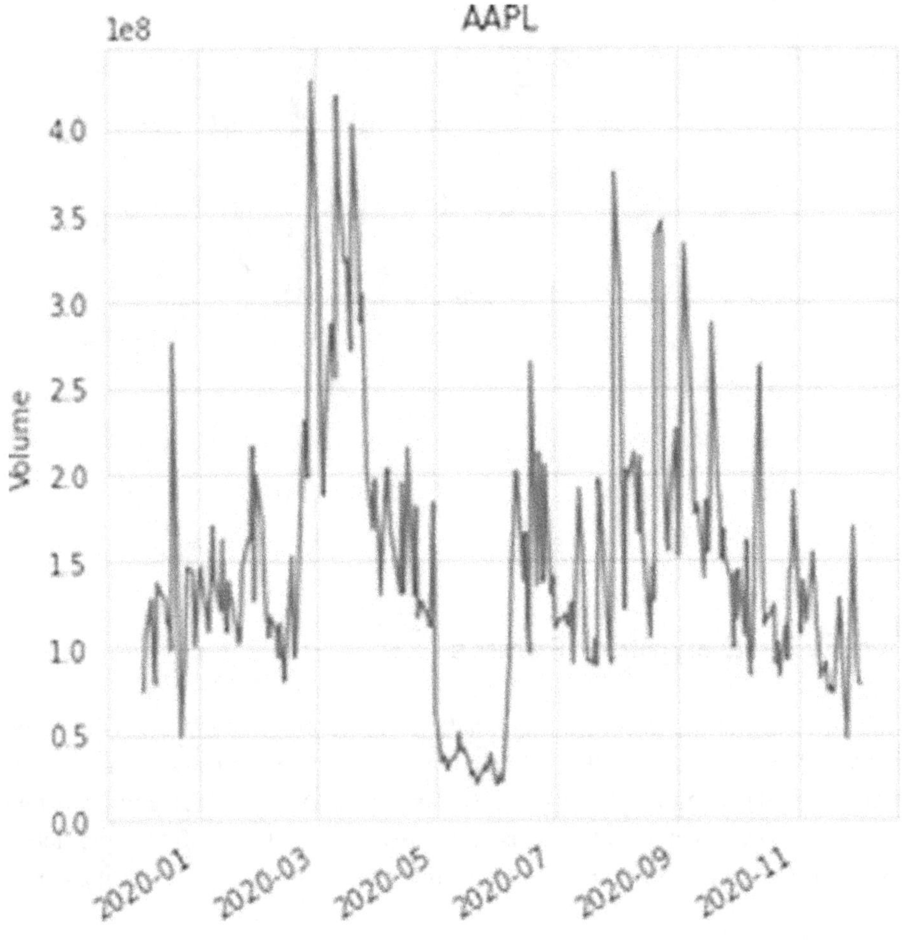

Removing the Moving Average

The moving average (MA) is a simple technical analysis tool that computes price data by creating a continually updated average price. The average is taken over a certain period of time, like 10 days, 20 minutes, 30 weeks or any time period the trader chooses. In our case we have taken it over 10, 20 and 50 days. The same can be seen in figure 4.

Figure 4. The moving average curves

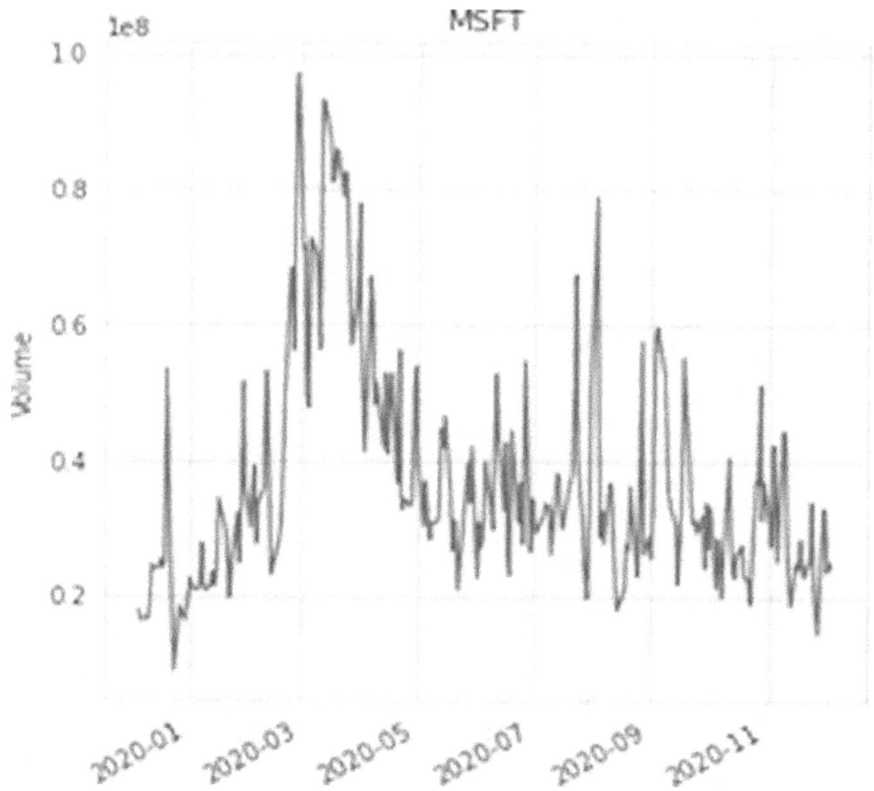

Finding the Daily Stock Returns

After some initial pre-processing and data exploration we found that Google will give fewer returns but it's riskier. Amazon will give more returns and is less risky. Microsoft will give more returns but its risk is greater. Apple will give the maximum returns compared to others but it's also the riskiest stock. The following predictions have been generated from the heatmap of the various stocks as shown in figure 5.

Figure 5. The heat map of the various stocks

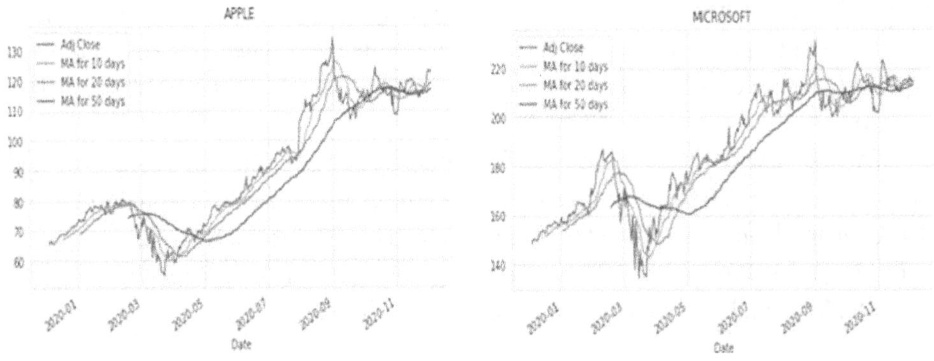

Figure 6. The results for prediction using LSTM model

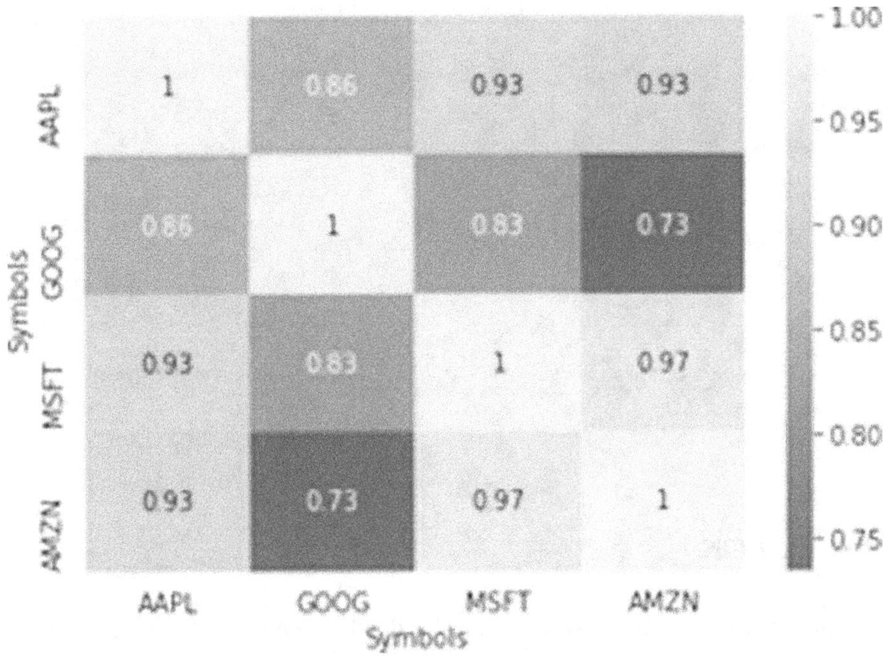

Model Training

The model is trained on the by taking a test train split of the data. The results have been evaluated for the AAPL stock prices for the next year. Firstly, we have used the

LSTM model for the analysis as it remembers the various pattern over a period of time and can be a useful technique when applied to stock market data. The results are shown in figure 6. The initial results are promising and we have reached up to the RMSE value of 3.28 only, which is very good, considering the amount of data we have.

Next up, we tried to apply the ARIMA model and put it to the test and as expected the results obtained from the models are very promising and can be improved further also with amalgamation of the techniques, which can we be the future scope of this study. The results from the LSTM model are shown in figure 7. The results are better than LSTM model and we have an average mean square error of 2.15.

Figure 7. The results for prediction using ARIMA model

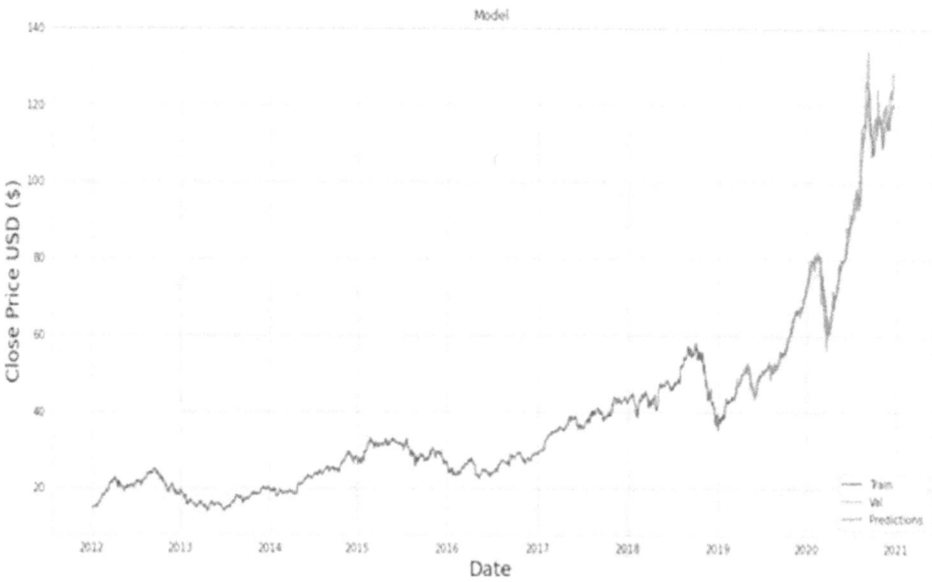

CONCLUSION

Prediction of stock market values is a risky deed as the market is highly prone to changes based on information portrayal and information perception. This system tries to display the range of values that a company can hit in the future based on previous values and the current position of the company in the market. But, despite upcoming resources attempting to make the stock market investment safer, there is a constant stigma that it is complex and incomprehensible. Awareness needs to be provided to reduce this idea and better education practices need to be cultivated to teach young minds on the advantages and disadvantages of the stock market investment. This system is an attempt in increasing the accuracy of prediction by combining both the ARIMA algorithm and LSTM model.

FUTURE SCOPE

Since the system dabbles with a lot of ongoing research work, further work can be done in the following areas: Improved accuracy can be achieved by classifying sentiment analysis data into classes other than just positive, negative and neutral. Graphical and tabular representation of data for more than 1 company. This can be used to compare the data from two or more companies (based on user requirements). The system can be built to provide alert messages for critical alerts in the prediction result. The system can be modified to add user favorites for stock analysis. Further work can be done in the system to add more companies as the current software takes into consideration only a few select companies. Implementation of the system with other Machine learning or deep learning techniques with better accuracies can also be done. Launching this system into a website for public use is another future prospect.

REFERENCES

Beveridge, S., & Nelson, C. R. (1981). A new approach to decomposition of economic time series into permanent and transitory components with particular attention to measurement of the 'business cycle'. *Journal of Monetary Economics*, *7*(2), 151–174. doi:10.1016/0304-3932(81)90040-4

Chen, J., Hong, H., & Stein, J. C. (2001). Forecasting crashes: Trading volume, past returns, and conditional skewness in stock prices. *Journal of Financial Economics*, *61*(3), 345–381. doi:10.1016/S0304-405X(01)00066-6

Ding, D., Zhang, M., Pan, X., Yang, M., & He, X. (2019, July). Modeling extreme events in time series prediction. In *Proceedings of the 25th ACM SIGKDD International Conference on Knowledge Discovery & Data Mining* (pp. 1114-1122). ACM.

Franses, P. H. (1991). Seasonality, nonstationarity and the forecasting of monthly time series. *International Journal of Forecasting*, *7*(2), 199–208. doi:10.1016/0169-2070(91)90054-Y

Granger, C. W., & Joyeux, R. (1980). An introduction to longmemory time series models and fractional differencing. *Journal of Time Series Analysis*, *1*(1), 15–29. doi:10.1111/j.1467-9892.1980.tb00297.x

Harvey, A. C., & Todd, P. H. J. (1983). Forecasting economic time series with structural and BoxJenkins models: A case study. *Journal of Business & Economic Statistics, 1*(4), 299–307.

Liu, C., Hoi, S. C., Zhao, P., & Sun, J. (2016, February). Online arima algorithms for time series prediction. *Thirtieth AAAI conference on artificial intelligence.*

Liu, H., & Song, B. (2017). Stock trends forecasting by multi-layer stochastic ANN bagging. *2017 IEEE 29th International Conference on Tools with Artificial Intelligence (ICTAI),* 322–329.

Nelson, C. R., & Plosser, C. R. (1982). Trends and random walks in macroeconomic time series: Some evidence and implications. *Journal of Monetary Economics, 10*(2), 139–162. doi:10.1016/0304-3932(82)90012-5

Pouzols, F. M., & Lendasse, A. (2010). Effect of different detrending approaches on computational intelligence models of time series. *The 2010 International Joint Conference on Neural Networks (IJCNN),* 1–8.

Siami-Namini, S., Tavakoli, N., & Namin, A. S. (2018, December). A comparison of ARIMA and LSTM in forecasting time series. In *2018 17th IEEE international conference on machine learning and applications (ICMLA)* (pp. 1394-1401). IEEE.

Tektas, M. (2010). Weather forecasting using ANFIS and ARIMA mod els. *Environmental Research, Engineering and Management, 51*(1), 5–10.

Witt, A., Kurths, J., & Pikovsky, A. (1998). Testing stationarity in time series. *Physical Review E: Statistical Physics, Plasmas, Fluids, and Related Interdisciplinary Topics, 58*(2), 1800–1810. doi:10.1103/PhysRevE.58.1800

Wu, Y., Pan, Z., Luo, X., Gao, J., & Zhang, Y. (2016). A hybrid forecasting method of electricity consumption based on trend extrapolation theory and LSSVM. *2016 IEEE PES AsiaPacific Power and Energy Engineering Conference (APPEEC),* 2333–2337.

Yang, J., Li, L., Shi, Y., & Xie, X. (2018). *An ARIMA model with adaptive orders for predicting blood glucose concentrations and hypoglycemia. IEEE J. Biomed. Health Inf.*

Chapter 16
EGODO Ontology:
Sharing, Retrieving, and Exchanging Legal Documentation Across E-Government

Fernando Ortiz-Rodriguez
https://orcid.org/0000-0003-2084-3462
Tamaulipas Autonomous University, Mexico

Jose Melchor Medina-Quintero
Tamaulipas Autonomous University, Mexico

Sanju Tiwari
Tamaulipas Autonomous University, Mexico

Vicente Villanueva
Tamaulipas Autonomous University, Mexico

ABSTRACT

The electronic government is a new application field for the Semantic Web, and the ontologies play a key role in the development of the Semantic Web. Consequently, if we focus on the law and the public policy that base the governments and build a documentation ontology that encompasses the electronic governments' needs of services at the front-office and mainly at the back office, we will provide a starting point to enable a broad class of semantic electronic government application. In this chapter, the authors show the e-Government Documentation Ontology as a part of the EGO model in order to describe how to access and manage official and non-official documentation across the governments. They also show a couple of applications using the documentation ontology on a semantic information retrieval system and on a semantic peer-to-peer.

DOI: 10.4018/978-1-6684-4225-8.ch016

INTRODUCTION

The electronic Government (eGov) is an important application field (Curtain et al., 2003) for the transformations that governments and public administrations will have to undergo in the next decades. Therefore, to transform the eGov into the eGovernance, the eGov research needs to be based on a robust theory, on modelling approaches, and on planning. In this scenario, a crucial issue is to manage in different ways the legal knowledge to improve and create semantic systems applications. The advancing technologies are being able to deal with real world legal knowledge contained by governments. In concrete, the Semantic Web has the capabilities to face this challenge. The Semantic Web was proposed by Tim Berners-Lee (Hendler et al., 2002; Villazón-Terrazas et al., 2020) as a new field of research, and according to the World Wide Web Consortium[1] (W3C) the Semantic Web is defined as "an extension of the current Web in which information is given well-defined meaning, better enabling computers and people to work in cooperation. It is based on the idea of having data on the Web defined and linked such that it can be used for more effective discovery, automation, integration, and reuse across various applications".

The application of the Semantic Web to the eGov domain is completely new; it features knowledge representation, knowledge engineering, database design, information systems, database integration, natural language understanding, information retrieval and semantic portals, among others. The Semantic Web is considered to be the infrastructure upon which all intelligent eGov applications will be built in the near future. Within the objectives of the Semantic Web the ontologies play an important role. This paper is organized as follows: section 2 shows the reference model used; section 3 describes the roles, conceptualization and modelling components of EGODO ontology; section 4 enclose a couple of applications that use EGODO; section 5 deals with the related work done. And finally, section 6 is devoted to the conclusions.

REFERENCE MODEL

A reference model was used to focus on and build a common understanding of the problem stated; figure 1 shows the different actors within the eGov. At the Back-office, the main actor is the Public Administration; it has many processes inside which should work properly to provide efficient services. The dynamics of the Public Administration provides a huge amount of information to be processed and these data should be managed in a transparent and efficient way.

Figure 1. Egovernment reference model

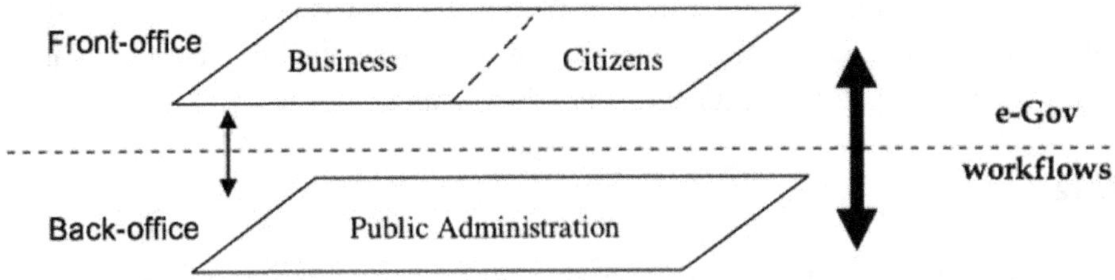

Within the Public Administration many processes take place and these must be carried out properly to provide efficient services; since the Public Administration functions in a decentralized way and the dynamics of this field generates a huge amount of information to be processed, it is necessary to manage this vast amount of information in a transparent and efficient way. Therefore, the implementation of eGov ontologies and applications is crucial. The main objective of designing this initial model in the eGov domain is aims to represent the legal issues behind the governments. This model should work as a framework to deploy semantic eGov systems, under the law and public policy approach.

The main objective of designing this initial model in the eGov domain is aims to represent the legal issues behind the governments. This model should work as a framework to deploy semantic eGov systems, under the law and public policy approach

Ego Model

The EGO Ontology Model reuses parts of the first two layers of LRI-Core model (Breuker, 2004) and is being adapted to the legal system of the Spanish and Mexican government. The EGO Ontology Model (Ortiz-Rodríguez & Villazón-Terrazas, 2006; Panchal et al., 2021) is one of the first efforts not intended for legal domain but for eGov domain instead, which is a domain that needs to consider the law, public policy, citizen services, administrative processes, best-practices, and also the different languages spoken within the nation.

EGODO ONTOLOGY

There are many things to manage at governments, this work focusses just in one important issue, the documentation. All the existing documentation and the growing one created and stored could increase the inaccessibility of it across government. The documentation ontology EGODO aims to cover a difficulty on document management at government. One main objective of this Documentation Ontology is to make all documents, regardless of the format, accessible and secure through any point of access across governments; as a secondary objective is to regulate the access of the role actors within eGov and the documentation

Roles

In Valente (2005), the five main roles of ontologies are identified: organizing and structuring information; reasoning and problem solving; semantic indexing and searching; semantics integrating and interoperating; and understanding the domain. Before building the EGODO ontology, we think it should be useful to settle the proper roles that the ontology will play. The EGODO will perform three of the five roles mentioned above: the first role is organizing and structuring information in the eGov domain, mainly by defining the terms used. The second role is reasoning and problem solving; this role basically represents the knowledge of the domain subsequently an automated reasoner can represent problems and generate solutions for these problems, what implies the use of an inference engine to achieve specific goals. The third role is semantic indexing and searching (where the ontology will represent the contents of documents) that will enable semantic search for content.

Main Ontology Modelling Components

In this section we show the EGODO ontology; this was built with the methodology METHONTOLOGY and the workbench WebODE (Gómez-Pérez et al., 2003), created to give technological support to METHONTOLOGY.

METHONTOLOGY proposes to conceptualize ontologies with a set of tabular and graphical intermediate representations. These allow modelling the components described in this section.

Concepts are taken in a broad sense. They are identified by their name, though they can also have synonyms and abbreviations. A natural language (NL) description can be also included (e.g., Official_Documentation, NonOfficial_Documentation, Legislation_Documents, etc.).

Relations represent a type of association between concepts of the domain. If the relation links two concepts, it is called a binary relation (e.g., Official_Documentation has a Location).

Groups are sets of disjoint concepts that are also known as partitions. They are used to create disjoint and exhaustive class partitions (e.g., Government_Documentation_Group).

Constants are numeric values that do not change in long periods of time (e.g., quantity of legal birth record).

Attributes describe properties of instances and concepts. We can distinguish two types of attributes: instance and class attributes. Instance attributes describe concept instances, from where they take their values. Class attributes describe concepts and take their values from the concepts where they are defined. Class attributes are neither inherited by subclasses nor by instances (e.g., chapter of Civil Code as a part of legislation).

Formal axioms are logical expressions that are always true and normally used to specify constraints in the ontology (e.g., a non official document could be retrieved by any role of government).

Rules are generally used to infer knowledge from the ontology, such as attribute values, relation instances, etc. (e.g., a birth certificate as an instance of Official_Documentation just can be retrieved by the citizen who belongs).

Ontology Conceptualization

We provide the conceptualization of the ontology, which is based on the set of intermediate representations proposed in METHONTOLOGY. The intermediate representations used in this document are: the concept classification tree and the ad-hoc relationships between different concepts

The main concept of this ontology is 'Government Documentation', which is organized according to the nature of document (Official Documentation or Non-official Documentation) within taxonomy.

This taxonomy is shown in figure 2. As it can be seen in the figure, the Official Documentation related are:

Administrative Document, Jurisprudence Document, Legislation Document, Notary Document, Registration Document. For example, Civil Code could be an instance under Legislation.

Figure 2. EGODO's taxonomy

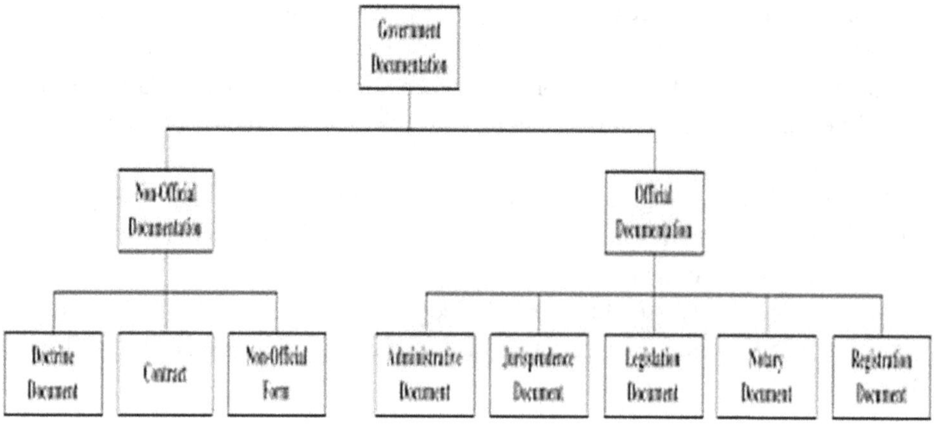

EGODO imports concepts from the Location Ontology, the Government Roles Ontology, and the Civil Personality Ontology, and these imported concepts (Gómez-Pérez et al., 2005; Ortiz-Rodríguez et al., 2009)] are used to connect the Documentation ontology with different ontologies. Examples of these relations can be the following: 'Civil Personality is author of Government Documentation', 'Government Documentation is associated with Civil; Personality', 'Government Roles Grant Access to Government documentation', 'Civil Personality has Government Roles', etc. Figure 3 shows all the ad-hoc relations whose domain is one concept that belongs to the Documentation Regulatory Ontology.

Figure 3. The ad-hoc relationships of EGODO

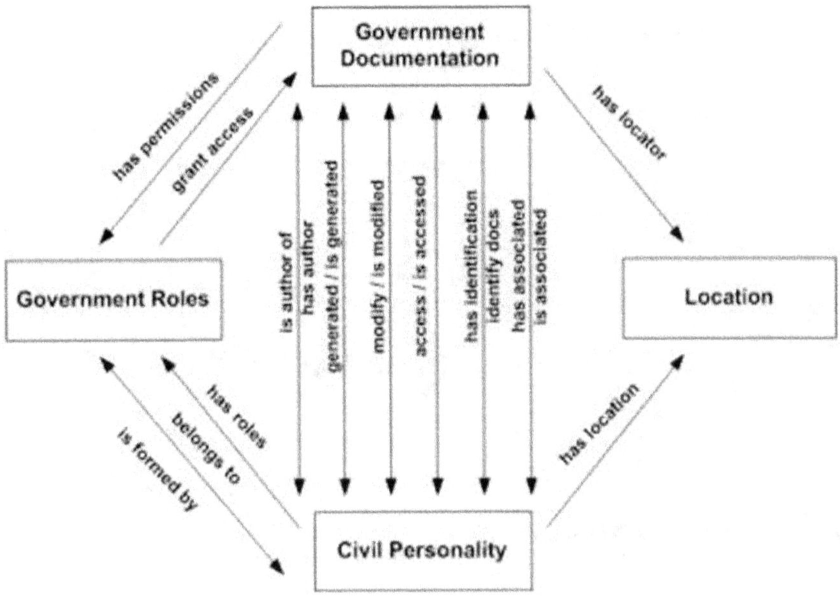

APPLICATIONS

We now share two applications that are being developed that employ the proposed EGODO Ontology. In detail, we present two complementary applications, the peer-to-peer (P2P) system Egoster (Ortiz-Rodríguez et al., 2006) and the Information Retrieval EgoIR (Gómez-Pérez et al., 2006). In general, the two tools differ in their usage perspective and are appropriate for different tasks. However, only the combined application of both tools will offer users the full potential of document management across government.

EgoIR

EgoIR is a java-based system that offers an ontology-based approach to Information Retrieval and its main goal is to retrieve eGov documentation. The system deals with government documentation, and gives citizens, business and governments the opportunity to integrate and to recover documents. For this purpose, EgoIR provides facilities that manage, search, and share eGov documentation. EgoIR also offers an ontology browsing capabilities (see figure 4) using the ontology described in section 3.

Figure 4. The ad-hoc relationships of EGODO

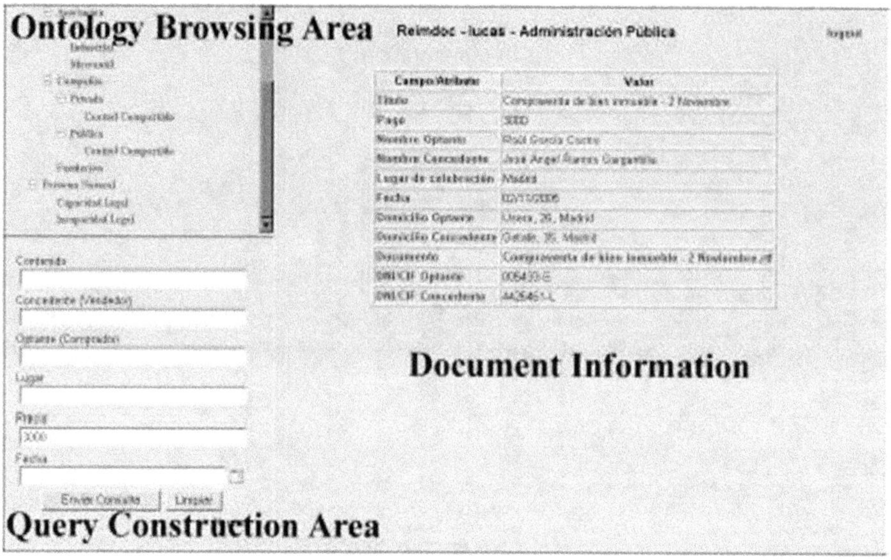

EGODO is stored in WebODE (workbench for ontological engineering). Besides, EgoIR allows the construction of a query from the ontology concepts; the query obtained is composed of a set of concepts extracted from the ontologies. EgoIR (see figure 5) connects to WebODE throughout WebODE's ODE service to obtain ontology concepts and it employs Lucene (search engine library) to retrieve the documents that match the given query.
The possibly main users of EgoIR are:

1. end users, who require consulting government documentation;
2. agencies, which need to know the current legislation; and
3. lawyers, who have to consult concrete aspects.

Figure 5. EgoIR system architecture

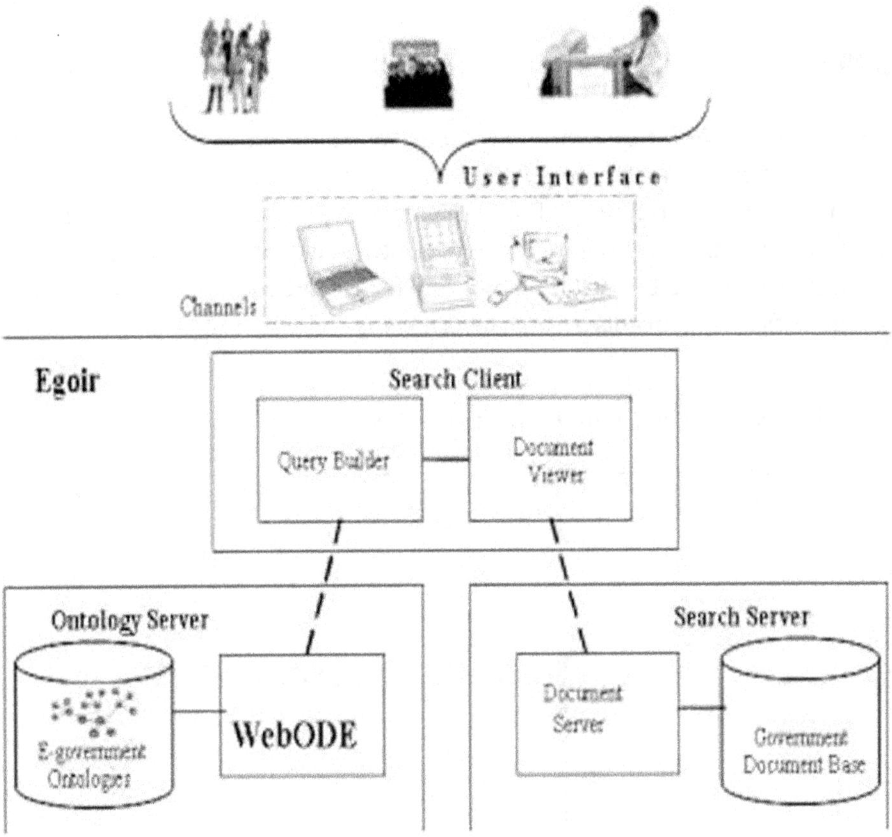

EgoIR provides the following main functionalities: query construction, document retrieval and type of documents inside a personal computer within a public administration. Query construction. In this option the user, through a guided interface, navigates among the ontologies, exploring concepts and relationships; and with the selected concepts, the user builds the query.

A query consists of a set of concepts extracted from the Legal Ontologies. Once the user has finished adding concepts to the query, the user sends the query to the document server. Document retrieval. In this option the system retrieves the legal documents that match the user query constructed.

The documents coming from the document server are retrieved and showed in a list form. If the user wants to choose a given document, he only has to click on the document. Then, the system launches a window with the selected document. Types of document. The EgoIR can deal with different types of documents for instance it can retrieve rtf, html, txt, pdf.

Egoster

Egoster is a java-based system that exploits semantic web techniques in order to provide an innovative and useful solution for exchanging and retrieving Government Documentation (see figure 6). For this purpose, Egoster provides facilities for managing, searching and sharing Government Documentation in a P2P network, thereby implementing EGODO as a proposal for a standard base to manage Official and Non-official documents across Governments.

Figure 6. A snapshot of Egoster retrieving a document

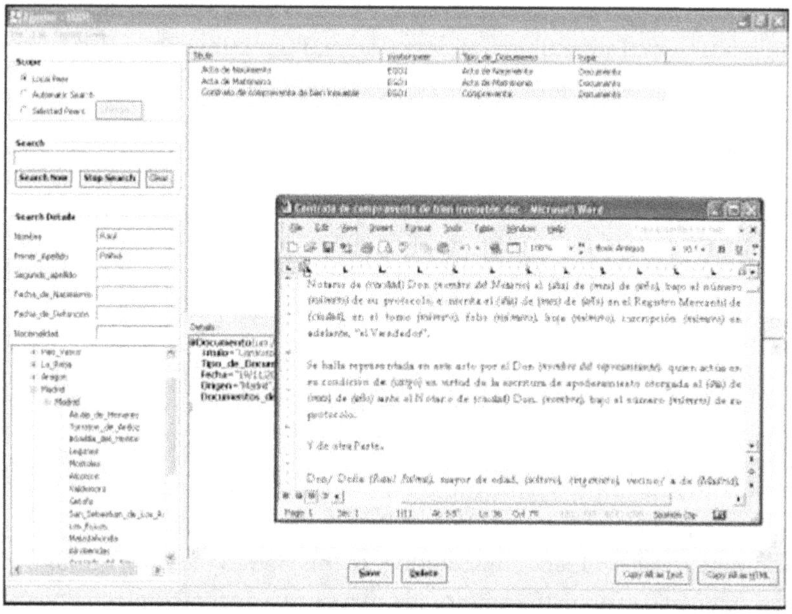

Egoster offers a user driven approach where each peer has its own local repository of documents and also has access to the information of others repositories, thus creating a virtual decentralized document repository. The Egoster client on its own (e.g. disconnected from the P2P network) will already provide added value to its users as it will give developers an overview and search facilities of his/her own government documentation stored in its local repository.

Figure 7. Egoster system architecture

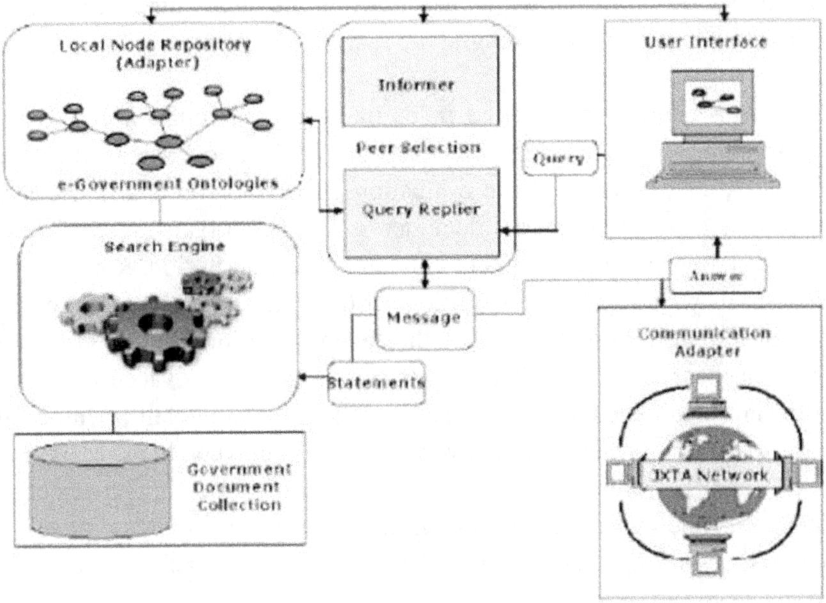

The Egoster is an instance of the Swapster system architecture (figure 7). It uses ontologies extensively in order to provide some of its main functions importing government documentation, formulating queries, routing queries and processing answers. This system in further development will consider electronic signature and security issues in order to function properly in real environments. The government document collection retrieved and exchanged by Egoster is located at each public administration office and is stored locally in the file system of each peer (e.g., birthday certificate, real-estate transaction documents, etc.). Each peer also implements a search engine for the information retrieval capabilities. The search engine processes the document collection to create internally access structures that allow fast document location. These access structures, called indices, are stored locally in the file system. The search engine connects to the ontologies in order to obtain synonyms of the query concepts. In Egoster system the search engine is based on EgoIR.

Egoster provides the following main functionalities: formulating queries, routing queries and processing queries. Formulating queries: The user can search for documents using simple keyword searches, or using more advanced, semantic searches. Here, queries are formulated in terms of two ontologies:

1. the EGODO ontology which describes the properties of the government documents,

2. a geographic hierarchy of local administration offices (e.g., the Spanish local administration offices), which describes all the local administration offices classified by the city and geographic region.

This means queries can refer to fields like name, nationality, etc. (using the EGODO ontology) or queries may refer to geographic location (using the geographic hierarchy). After we conclude the evaluation of the first version of Egoster, we expect to have a bigger number of documents of different types, so we foresee the inclusion of the type of the document as another search criterion. Routing queries:

The user may query a single specific peer (e.g. their own computer, because it is the local administration office where the document resides, or users may want to query another peer in particular because this peer is known in advance as the location where the information resides), or a specific set of peers (e.g. all the members under the same manager), or the entire network of peers (e.g. when the user has no idea where to search), in which case queries are routed automatically in the network. In the latter case, queries are routed through the network depending on the expertise of the peers, describing which geographic space of the geographic hierarchy a peer is knowledgeable about.

In order to achieve this expertise-based routing, a matching function determines how closely the semantic content of a query matches the expertise of a peer. In the future, as we mention above, the type of the documents (e.g., birth certificate) that each peer has is going to be considered for the definition of the expertise of the peer as well. Processing queries: In order to process a query, whether the query comes from the user interface or from other peers, the peer tries to answer the query or distribute it further according to the content of the query.

The decision about which peers a query should be sent is based on the knowledge about the expertise of other peers. If the peer can answer the query, it uses its local search engine (e.g., the information retrieval system). Egoster aims to exchange and retrieve the government documentation from eGov back office to allow final users (citizen and business) to get what they need in any time and any place where a peer exists. In particular, Egoster goals are:

- To aid Public Administration to find and share government documents within the back-office.
- To provide interoperability between eGov offices.
- To improve citizens-government interaction.
- To offer a scalable technology that could be adapted to future requirements of the Public Administration.

RELATED WORK

Nowadays the joint efforts put in by different research communities have made possible the birth of the semantic eGov. Since eGov ontologies are still in their initial state, only a few works carried out in this field are known; thus, in this section we provide a brief state of the art of those works performed in Artificial Intelligence (AI) field, in the Law field and in the Semantic Web. The sum up of all these efforts will produce robust ontologies for the eGov domain in the near future.

Law and eGov within the Semantic Web

Currently, the Semantic Web is a new area of research and applications within the legal system and eGov domains and is a promise for the Web of the next generation; this new area, which is now used mainly to communicate with people but not with machines, will transform the current web since the capability of communication with machines is one of the main objectives of the Semantic Web.

If the Web were equipped with more meaning, every citizen would extract answers in a new, easy and simple way and this action could be carried out by web powered semantics, which would enable citizens and businesses to obtain better information from the government. Web powered semantics could help the eGov in two ways: first, by allowing the government to delegate more intelligent tasks to computers and second, by solving daily problems with logic deductions and reasoning.

At present time, the web is merely a common framework that allows data to be shared and reused. Currently the legal and eGov Semantic Web applications are still in an experimental phase, but their potential impact on social, economic and political issues is extremely significant.

The main goals of eGov are to develop user-friendly and efficient services for the public and the business community, though semantic interoperability is also seen as an important issue to solve within this domain. Some of the works aimed at covering the semantic eGov domain are the following: the DIP project, the Reimdoc project, the Ontogov project, the Egov project, HOPS project, and the WEBOCRAT project.

Domain Considerations

The eGov scenario is a promising application field for the ontologies underlying the legal engineered knowledge. Quite a few ontologies have been built in the legal domain but not all of them are available or modelled just for a specific domain. The research efforts made in the legal domain by the AI community have contributed to the making of ontologies such as: LLD (McCarty, 1989), NORMA (Stamper,

1980; Stamper, 1991), FOL (Valente, 1995), FBO (Kralingen, 1995; Visser, 1995) and LRI-Core Legal Ontology (Breuker, 2004).

The emergence of legal ontologies as part of the Semantic Web initiative has provided a new opportunity for the research community and has brought about a solution to retrieve documents within the eGov domain.

CONCLUSION

In this paper we have presented the eGov Documentation Ontology (EGODO), even though this ontology is at initial state, it has well defined goals as supporting semantic applications to retrieve government documentation and, on the other, at delivering services from the public administration (within the government) to citizens. Also, we have presented two applications that use EGODO, in one hand, the EgoIR, which aims to retrieve legal documents from government and in the other, Egoster, which helps to interchange and share documents across government. The eGov domain does still have many needs: knowledge, for instance, has not been modelled at all. These needs represent real challenges for researchers. We must add here, that the legal domain is very complex and evolving and its complexity provides a different situation than that provided by domains such as physics or mathematics, and this fact will bring about the deployment of future eGov ontologies. We will be focus on further enhancement and evaluation of the EGODO; we will be centered on the reasoning capabilities of EGODO to provide advanced features to semantic applications.

REFERENCES

Breuker, J. (2004). *Constructing a Legal Core Ontology: LRI-Core*. University of Amsterdam, Leibniz Center for Law.

Curtain, G., Sommer, M., & Vis-Sommer, V. (Eds.). (2003). *The World of E-Government*. The Haworth Press. doi:10.1300/J199v02n03_01

Gómez-Pérez, A., Fernández-López, M., & Corcho, O. (2003). *Ontological Engineering*. Springer Verlag.

Gómez-Pérez, A., Ortiz-Rodríguez, F., & Villazón-Terrazas, B. (2005). Legal Ontologies for the Spanish e-Government. In *CAEPIA'05 Post-Proceedings Volume*. Springer Verlag.

Gómez-Pérez, A., Ortiz-Rodríguez, F., & Villazón-Terrazas, B. (2006). Ontology-Based Legal Information Retrieval to Improve the Information Access in e-Government. WWW2006. doi:10.1145/1135777.1135988

Hendler, J., Berners-Lee, T., & Miller, E. (2002). *Integrating Applications on the Semantic Web*. http://www.w3.org/2002/07/swint.html

Kralingen, R. W. (1995). *Frame-based Conceptual Models of Statute Law. Computer/ Law Series. No.16*. Kluwer Law International.

McCarty, L. T. (1989). A Language for Legal Discourse, I. Basic Features. *Proceedings of the Second International Conference on Artificial Intelligence and Law*, 180-189. 10.1145/74014.74037

Ortiz-Rodríguez, F., Chaparro Pelaez, J., & Pascual, F. J. (2009). Semantic model approach for eGovernment to improve sharing, retrieving and exchanging documentation across back-office. *Proceedings of the 10th Annual International Conference on Digital Government, DG.O*.

Ortiz-Rodríguez, F., Palma, R., & Villazón-Terrazas, B. (2006). Semantic based P2P System for local e-Government. In Proceedings of Informatik 2006. GI-Edition-Lecture Notes in Informatics (LNI).

Ortiz-Rodríguez, F., & Villazón-Terrazas, B. (2006). EGO Ontology Model: law and regulation approach for E-Government. *WS Proceedings of the 3rd European Semantic Web Conference*.

Panchal, R., Swaminarayan, P., Tiwari, S., & Ortiz-Rodriguez, F. (2021). AISHE-Onto: A Semantic Model for Public Higher Education Universities. *DG.O2021: The 22nd Annual International Conference on Digital Government Research.*

Stamper, R. K. (1980). *LEGOL: Modelling Legal Rules by Computer. In Computer Science and Law.* Cambridge University Press.

Stamper, R. K. (1991). The Role of Semantics in Legal Expert Systems and Legal Reasoning. *Ratio Juris*, *4*(2), 219–244. doi:10.1111/j.1467-9337.1991.tb00094.x

Valente, A. (1995). Legal Knowledge Engineering; A Modelling Approach. University of Amsterdam.

Valente, A. (2005). *Types and Roles of Legal Ontologies*. Law and the Semantic Web.

Villazón-Terrazas, B., Ortiz-Rodríguez, F., Tiwari, S., & Shandilya, S. (2020). Knowledge Graphs and Semantic Web. *Communications in Computer and Information Science*, *1232*, 1–225. doi:10.1007/978-3-030-65384-2

Visser, P. R. S. (1995). *Knowledge Specification for Multiple Legal Tasks. A Case Study of the Interaction Problem in the Legal Domain. Computer / Law Series. No. 17*. Kluwer Law International.

ENDNOTE

1 https://www.w3.org/2001/sw/

Compilation of References

Consolvo, S., Smith, I. E., Matthews, T., LaMarca, A., Tabert, J., & Powledge, P. (2005). Location disclosure to social relations: why, when, & what people want to share. *Proceedings of the ACM CHI Conference on Human Factors in Computing Systems*. 10.1145/1054972.1054985

Statista. (n.d.). https://www.statista.com/statistics/271644/worldwide-free-and-paid-mobile-app-store-downloads/

Felt, Ha, Egelman, Haney, Chin, & Wagner. (n.d.). *Android Permissions: User Attention, Comprehension, and Behavior*. Academic Press.

Enck, W., Ongtang, M., & McDaniel, P. (2009a). On lightweight mobile phone application certification. *Proceedings of the ACM Conference on Computer and Communication Security (CCS)*.

Fuchs, A., Chaudhuri, A., & Foster, J. (2009). *SCanDroid: Automated Security Certification of Android Applications. Technical report*. University of Maryland.

Howell, J., & Schechter, S. (2010). What you see is what they get. *Proceedings of the IEEE Workshop on Web 2.0 Security and Privacy (W2SP)*.

Roesner, Kohno, Moshchuk, Parno, Wang, & Cowan. (n.d.). *User-Driven Access Control: Rethinking Permission Granting in Modern Operating Systems*. Academic Press.

Enck, W., Ongtang, M., & McDaniel, P. (2009b). Understanding Android Security. *IEEE Security & Privacy Magazine*, *7*(1), 50–57. doi:10.1109/MSP.2009.26

European Commission. (n.d.). *JRC Publications Repository*. Retrieved from https://ec.europa.eu/jrc/en/publication/dealing-user-privacy-mobile-apps-issues-and-mitigation

Alenezi, M., & Almomani, I. (2017). Abusing Android permissions: A security perspective. *2017 IEEE Jordan Conference on Applied Electrical Engineering and Computing Technologies (AEECT)*, 16. 10.1109/AEECT.2017.8257772

Su, M., Chen, S., Wu, T., Chang, H., & Liu, Y. (2019). Permission Abusing by Ad Libraries of Smartphone Apps. *2019 Eleventh International Conference on Ubiquitous and Future Networks (ICUFN)*, 475-477. 10.1109/ICUFN.2019.8806157

Micro, T. (n.d.). *Threat Encyclopedia*. Retrieved from http://about-threats.trendmicro.com/us/library/image-gallery/12-most-abused-android-app-permissions

Bendre, R. (2018). Ignoring app permissions could hurt your privacy more than you think. *BGR.in*. Retrieved from https://www.bgr.in/features/ignoring-app-permissions-could-hurt-your- privacy-more-than-you-think-550748/

Khatoon, A., & Corcoran, P. (2017). Android permission system and user privacy — A review of concept and approaches. *2017 IEEE 7th International Conference on Consumer Electronics - Berlin (ICCEBerlin)*, 153-158. 10.1109/ICCE-Berlin.2017.8210616

Wu, S., & Liu, J. (2019). Overprivileged Permission Detection for Android Applications. *ICC 2019 - 2019 IEEE International Conference on Communications (ICC)*, 1-6. 10.1109/ICC.2019.8761572

Ngobeni, A., & Mhlongo, S. (2019). Towards Enhancing Security in Android Operating Systems – Android Permissions & User Unawareness. *2019 2nd International Conference on Computer Applications & Information Security (ICCAIS)*, 1-6. 10.1109/CAIS.2019.8769531

Jia, P., He, X., Liu, L., Gu, B., & Fang, Y. (2015). A framework for privacy information protection on Android. *2015 International Conference on Computing, Networking and Communications (ICNC)*, 1127-1131. 10.1109/ICCNC.2015.7069508

Bao, L., Lo, D., Xia, X., & Li, S. (2016). What Permissions Should This Android App Request? *2016 International Conference on Software Analysis, Testing and Evolution (SATE)*, 36-41. 10.1109/SATE.2016.13

Bagheri, H., Sadeghi, A., Garcia, J., & Malek, S. (2015, September 1). COVERT: Compositional Analysis of Android Inter-App Permission Leakage. *IEEE Transactions on Software Engineering*, *41*(9), 866–886. doi:10.1109/TSE.2015.2419611

Iqbal, M. (2022). *App Download Data (2022)*. Business of Apps. Retrieved from https://www.businessofapps.com/data/app-statistics/#:~:text=As%20of%20the%20first%20quarter,from%20178%20billion%20in%202017

Burguera, I., Zurutuza, U., & Nadjm-Tehrani, S. (2011). Crowdroid: Behavior-Based Malware Detection System for Android. *SPSM*, *11*, 15–26. doi:10.1145/2046614.2046619

Faruki, P., Ganmoor, V., Laxmi, V., Gaur, M., & Bharmal, A. (2013). AndroSimilar: robust statistical feature signature for Android malware detection. *SIN 2013 - Proceedings of the 6th International Conference on Security of Information and Networks*. 10.1145/2523514.2523539

Rashidi, B., Fung, C., & Vu, T. (2015). *RecDroid- An Android Application Security Framework through Crowdsourcing*. Academic Press.

Russello, G., Jimenez, A., Naderi, H., & van der Mark, W. (2013). FireDroid: hardening security in almost-stock Android. *ACM International Conference Proceeding Series*, 319-328. 10.1145/2523649.2523678

Yan, L., & Yin, H. (2012). DroidScope: Seamlessly Reconstructing the OS and Dalvik Semantic Views for Dynamic Android Malware Analysis. *USENIX Security Symposium.*

Jing, Y., Ahn, G.-J., Zhao, Z., & Hu, H. (2014). RiskMon: Continuous and automated risk assessment of mobile applications. *CODASPY 2014 - Proceedings of the 4th ACM Conference on Data and Application Security and Privacy*, 99-110. 10.1145/2557547.2557549

Zhou, Y., Wang, Z., Zhou, W., & Jiang, X. (2012). Hey, You, Get Off of My Market: Detecting Malicious Apps in Official and Alternative Android Markets. *Proceedings of the 19th Network and Distributed System Security Symposium NDSS 2012.*

Felt, A. P., Chin, E., Hanna, S., Song, D., & Wagner, D. (2011). Android permissions demystified. *Proceedings of the 18th ACM conference on Computer and Communications Security, CCS 11*, 627-638.

android-market-api. (n.d.). *Android Market for all Developers!* http://code.google.com/p/android-market-api/

IDC. (2021). *Smartphone Market Share*. Retrieved from https://www.idc.com/promo/smartphone-market-share/os

Neisse, R., Steri, G., Geneiatakis, D., & Fovino, I. N. (2016). *A privacy enforcing framework for Android applications* (Vol. 62). Computers and Security. https://www.sciencedirect.com/science/article/pii/S0167404816300840

Tang, Jin, He, & Jiang. (n.d.). *Extending Android Security Enforcement with A Security Distance Model*. Academic Press.

Developers, A. (n.d.). *App Manifest Overview*. Retrieved from https://developer.android.com/guide/topics/manifest/manifest-intro

MindSea Team. (n.d.). *25 Mobile App Statistics To Know In 2022*. MindSea. Retrieved from https://mindsea.com/app-stats/

Statista. (2022). *Number of mobile app downloads worldwide from 2016 to 2021*. Retrieved from https://www.statista.com/statistics/271644/worldwide-free-and-paid- mobile-app-store-downloads/

Barkhuus, L., & Dey, A. (2003). Location-based services for mobile telephony: a study of users' privacy concerns. *Proceedings of the International Conference on Human-Computer Interaction.*

AbdulQawy, A., Elkhouly, R., & Sallam, E. (2018). Approaching rutted road-segment alert using smartphone. *13th International Conference on Computer Engineering and Systems (ICCES)*, 341–346.

Adner, R., & Levinthal, D. (2001). Demand heterogeneity and technology evolution: Implications for product and process innovation. *Management Science*, *2001*(47), 611–628. doi:10.1287/mnsc.47.5.611.10482

Afsar, B., Badir, Y., & Kiani, U. S. (2016). Linking spiritual leadership and employee pro-environmental behavior: The influence of workplace spirituality, intrinsic motivation, and environmental passion. *Journal of Environmental Psychology, 2016*(45), 79–88. doi:10.1016/j.jenvp.2015.11.011

Agiwal, M., Saxena, N., & Roy, A. (2019). Towards connected living: 5G enabled internet of things (IoT). *IETE Technical Review, 36*(2), 190–202. doi:10.1080/02564602.2018.1444516

Ahamed, H. A. (2018). HOG-CNN Based Real Time Face Recognition. In *International Conference on Advancement in Electrical and Electronic Engineering*. IEEE.

Ahn, J., Lee, S., & Yun, S. (2018). Leaders' core self-evaluation, ethical leadership, and employees' job performance: The moderating role of employees' exchange ideology. *Journal of Business Ethics, 148*(2), 457–470. doi:10.100710551-016-3030-0

Akgün, A. E., Keskin, H., & Byrne, J. (2009). Organizational emotional capability, product and process innovation, and firm performance: An empirical analysis. *Journal of Engineering and Technology Management, 2009*(26), 103–130. doi:10.1016/j.jengtecman.2009.06.008

Akkiraju, R., Farrell, J., Miller, J. A., Nagarajan, M., Sheth, A. P., & Verma, K. (2005). *Web service semantics-wsdl-s*. Academic Press.

Al Omar, A., Rahman, M. S., Basu, A., & Kiyomoto, S. (2017). Medibchain: A blockchain based privacy preserving platform for healthcare data. *International conference on security, privacy and anonymity in computation, communication and storage*, 534–543.

Albino, V., Balice, A., & Dangelico, R. M. (2009). Environmental strategies and green product development: An overview on sustainability driven companies. *Business Strategy and the Environment, 2009*(18), 83–96. doi:10.1002/bse.638

Albort-Morant, G., Henseler, J., Cepeda-Carrión, G., & Leal-Rodríguez, A. L. (2018). Potential and realized absorptive capacity as complementary drivers of green product and process innovation performance. *Sustainability, 2018*(10), 381. doi:10.3390u10020381

Al-Falahy, N., & Alani, O. Y. (2017). Technologies for 5G networks: Challenges and opportunities. *IT Professional, 19*(1), 12–20. doi:10.1109/MITP.2017.9

Ali, G. N., Mikkilineni, A. K., Allebach, J. P., Delp, E. J., Chiang, P. J., & Chiu, G. T. (2003, January). Intrinsic and extrinsic signatures for information hiding and secure printing with electrophotographic devices. In NIP & Digital Fabrication Conference (Vol. 2003, No. 2, pp. 511-515). Society for Imaging Science and Technology.

Alt, E., & Spitzeck, H. (2016). Improving environmental performance through unit-level organizational citizenship behaviors for the environment: A capability perspective. *Journal of Environmental Management, 2016*(182), 48–58. doi:10.1016/j.jenvman.2016.07.034 PMID:27454096

Ambec, S., & Lanoie, P. (2008). Does it pay to be green? A systematic overview. *The Academy of Management Perspectives, 22*(4), 45–62. doi:10.5465/amp.2008.35590353

Amores-Salvadó, J., Martín-de Castro, G., & Navas-López, J. E. (2014). Green corporate image: Moderating the connection between environmental product innovation and firm performance. *Journal of Cleaner Production, 83*, 356–365. doi:10.1016/j.jclepro.2014.07.059

Amrutha, V. N., & Geetha, S. N. (2020). A systematic review on green human resource management: Implications for social sustainability. *Journal of Cleaner Production, 247*, 119131. doi:10.1016/j.jclepro.2019.119131

Anderson, N., Potočnik, K., & Zhou, J. (2014). Innovation and creativity in organizations: A state-of-the-science review, prospective commentary, and guiding framework. *Journal of Management, 40*(5), 1297–1333. doi:10.1177/0149206314527128

Anderson, R. (2009). *Confessions of a Radical Revolutionist.* St Martin's Press.

Ángel del Brío, J., Junquera, B., & Ordiz, M. (2008). Human resources in advanced environmental approaches–a case analysis. *International Journal of Production Research, 2008*(46), 6029–6053. doi:10.1080/00207540701352094

Anicham, Diviya, Kaviya, & Gajalakshmi. (n.d.). An Emergency Notification Android Application Using Number Plate Recognition. *International Journal of Advanced Networking & Applications*, 246-250.

Arago'n-Correa, J. A., & Rubio-Lo'pez, E. A. (2007). Proactive corporate environmental strategies: Myths and misunderstandings. *Long Range Planning, 40*, 357–381. doi:10.1016/j.lrp.2007.02.008

Aragon-Correa, J. A., Martin-Tapia, I., & Hurtado-Torres, N. E. (2013). Proactive environmental strategies and employee inclusion: The positive effects on information sharing and promoting collaboration and the influence of uncertainty. *Organization & Environment, 40*(2), 1–23. doi:10.1177/1086026613489034

Ardanuy, J. (2012). *Breve Introducción a la Bibliometría.* University of Barcelona.

Assiri & Bieman. (n.d.). Fault localization for automated program repair: effectiveness, performance, repair correctness. *Software Qual J.* DOI 10.1007/ s11219-016-9312-z

Assiri, F. Y., & Bieman, J. M. (n.d.). Fault localization for automated program repair: Effectiveness, performance, repair correctness. *Software Quality Journal.* Advance online publication. doi:10.100711219-016-9312-z

Astillo, P. V., Choudhary, G., Duguma, D. G., Kim, J., & You, I. (2021). Trmaps: Trust management in specification-based misbehavior detection system for imd-enabled artificial pancreas system. *IEEE Journal of Biomedical and Health Informatics.*

Averbakh, A., Krause, D., & Skoutas, D. (2009, October). Exploiting user feedback to improve semantic web service discovery. In *International Semantic Web Conference* (pp. 33-48). Springer. 10.1007/978-3-642-04930-9_3

Azarafzaa, M., Azaraf, M., & Akgun, H. (2021). Clustering method for spread pattern analysis of corona-virus (COVID-19) infection in Iran. *Journal of Applied Science, Engineering, Technology, and Education*, *3*(1), 1–6. doi:10.35877/454RI.asci31109

Baker, S. B., Xiang, W., & Atkinson, I. (2017). Internet of things for smart healthcare: Technologies, challenges, and opportunities. *IEEE Access: Practical Innovations, Open Solutions*, *5*, 26521–26544. doi:10.1109/ACCESS.2017.2775180

Balabdaoui, F., & Mohr, D. (2020, May 13). Age-stratified model of the COVID-19 epidemic to analyze the impact of relaxing lockdown measures: nowcasting and forecasting for Switzerland. medRxiv. doi:10.1101/2020.05.08.20095059

Baliga, A., Subhod, I., Kamat, P., & Chatterjee, S. (2018). *Performance evaluation of the quorum blockchain platform*. arXiv preprint arXiv:1809.03421.

Bandura, A. (1986). The explanatory and predictive scope of self-eficacy theory. *Journal of Social and Clinical Psychology*, *1986*(4), 359–373. doi:10.1521/jscp.1986.4.3.359

Bandura, A. (2000). Exercise of human agency through collective efficacy. *Curr. Dir. Psychol.*, *2000*(9), 75–78. doi:10.1111/1467-8721.00064

Banerjee, S. (2002). Corporate environmentalism: The construct and its measurement. *Journal of Business Research*, *55*(3), 177–191. doi:10.1016/S0148-2963(00)00135-1

Bansal, P. (2005). Evolving sustainably: A longitudinal study of corporate sustainable development. *Strategic Management Journal*, *26*(3), 197–218. doi:10.1002mj.441

Baocai, Y., Huirong, Y., Pengbin, F., Liheng, G., & Mingli, L. (2010, July). A framework and QoS based web services discovery. In *2010 IEEE International Conference on Software Engineering and Service Sciences* (pp. 755-758). IEEE. 10.1109/ICSESS.2010.5552261

Baqa, H., Truong, N. B., Crespi, N., Lee, G. M., & Le Gall, F. (2019, September). Semantic smart contracts for blockchain-based services in the Internet of Things. In *2019 IEEE 18th International Symposium on Network Computing and Applications (NCA)* (pp. 1-5). IEEE. 10.1109/NCA.2019.8935016

Barin Cruz, L., & Avila Pedrozo, E. (2009). Corporate social responsibility and green management: Relation between headquarters and subsidiary in multinational corporations. *Management Decision*, *47*(7), 1174–1199. doi:10.1108/00251740910978368

Barling, J., Christie, A., & Hoption, C. (2010). Leadership. In S. Zedeck (Ed.), *APA handbook of industrial and organizational psychology* (Vol. 1). American Psychological Association.

Barling, J., Weber, T., & Kelloway, K. E. (1996). Effects of transformational leadership training on attitudinal and financial outcomes: A field experiment. *The Journal of Applied Psychology*, *81*(6), 827–832. doi:10.1037/0021-9010.81.6.827

Barney, J. (1991). Firm resources and sustained competitive advantage. *Journal of Management*, *17*(1), 99–120. doi:10.1177/014920639101700108

Barsade, S. G., & Gibson, D. E. (2007). Why does affect matter in organizations? *The Academy of Management Perspectives*, *21*(1), 36–59. doi:10.5465/amp.2007.24286163

Bass, B. M., & Stogdill, R. M. (1990). *Bass & Stogdill's handbook of leadership: Theory, research, and managerial applications*. Simon and Schuster, Free Press.

Bastani Zadeh, R., Ghatee, M., & Eftekhari, H. R. (2018). Three-phases smartphone-based warning sys- tem to protect vulnerable road users under fuzzy conditions. *IEEE Transactions on Intelligent Transportation Systems*, *19*(7), 2086–2098. doi:10.1109/TITS.2017.2743709

Batarseh, F. A., & Gonzalez, A. J. (n.d.). Predicting failures in agile software development through data analytics. *Software Quality Journal*. Advance online publication. doi:10.100711219-015-9285-3

Bavik, Y. L., Tang, P. M., Shao, R., & Lam, L. W. (2018). Ethical leadership and employee knowledge sharing: Exploring dual-mediation paths. *The Leadership Quarterly*, *29*(2), 322–332. doi:10.1016/j.leaqua.2017.05.006

Beamon, B. M. (1999). Designing the Green Supply Chain. *Logistics Information Management*, *12*(4), 332–342. doi:10.1108/09576059910284159

Bebbington, J. (2001). Sustainable development: A review of the international development, business, and accounting literature. *Accounting Forum*, *25*(2), 128–157. doi:10.1111/1467-6303.00059

Bergmiller, G. G., & McCright, P. R. (2009). *Parallel models for Lean and green operations*. Paper presented at the Industrial Engineering Research Conference, Miami, FL. Available at: https://pdfs.semanticscholar.org/0a67/c7f0f6d60ba0506e037aaf17cfe8b3a89e7d.pdf

Berrone, P., & Gomez-Mejia, L. R. (2009). Environmental performance and executive compensation: An integrated agency-institutional perspective. *Academy of Management Journal*, *52*(1), 103–126. doi:10.5465/amj.2009.36461950

Beveridge, S., & Nelson, C. R. (1981). A new approach to decomposition of economic time series into permanent and transitory components with particular attention to measurement of the 'business cycle'. *Journal of Monetary Economics*, *7*(2), 151–174. doi:10.1016/0304-3932(81)90040-4

Bhandari, R., Raman, B., & Padmanabhan, V. (2019). Fullstop: A camera-assisted system for characterizing unsafe bus stopping. *IEEE Trans. Mob. Comput.*

Bhatt, Y., Ghuman, K., & Dhir, A. (2020). Sustainable manufacturing. Bibliometrics and content analysis. *Journal of Cleaner Production*, *260*, 120988. doi:10.1016/j.jclepro.2020.120988

BhushanS. (2021). https://medium.com/@bkshashi9/ambulance-stuck-in-traffic-scary-right-7b539ce518b9

Bibi, M., Hamid, A., Moetesum, M., & Siddiqi, I. (2019, September). Document Forgery Detection using Printer Source Identification—A Text-Independent Approach. In *2019 International Conference on Document Analysis and Recognition Workshops (ICDARW)* (Vol. 8, pp. 7-12). IEEE.

Bissing-Olson, M. J., Iyer, A., Fielding, K. S., & Zacher, H. (2013). Relationships between daily affect and pro-environmental behavior at work: The moderating role of pro-environmental attitude. *Journal of Organizational Behavior, 34*(2), 156–175. doi:10.1002/job.1788

Blondeau, V., Etien, A., Anquetil, N., Cresson, S., Croisy, P., & Ste'phane, D. (n.d.). Test case selection in industry: An analysis of issues related to static approaches. *Software Quality Journal.* Advance online publication. doi:10.100711219-016-9328-4

Bohdanowicz, P., Zientara, P., & Novotna, E. (2011). International hotel chains and environmental protection: An analysis of Hilton's we care! programme (Europe, 2006–2008). *Journal of Sustainable Tourism, 19*(7), 797–816. doi:10.1080/09669582.2010.549566

Boiral, O. (2002). Tacit knowledge and environmental management. *Long Range Planning, 35*(3), 291–317. doi:10.1016/S0024-6301(02)00047-X

Boiral, O. (2009). Greening the corporation through organizational citizenship behaviors. *Journal of Business Ethics, 87*(2), 221–236. doi:10.100710551-008-9881-2

Boiral, O., Baron, C., & Gunnlaugson, O. (2014). Environmental leadership and consciousness development: A case study among Canadian SMEs. *Journal of Business Ethics, 123*(3), 363–383. doi:10.100710551-013-1845-5

Boiral, O., & Paillé, P. (2012). Organizational citizenship behaviour for the environment: Measurement and validation. *Journal of Business Ethics, 2012*(109), 431–445. doi:10.100710551-011-1138-9

Boiral, O., Talbot, D., & Paillé, P. (2015). Leading by example: A model of organizational citizenship behavior for the environment. *Business Strategy and the Environment, 2015*(24), 532–550. doi:10.1002/bse.1835

Bossink, B. A. (2007). Leadership for sustainable innovation. *International Journal of Technology Management & Sustainable Development, 2007*(6), 135–149. doi:10.1386/ijtm.6.2.135_1

Braz, A., López-López, M., & García-Ruiz, C. (2013). Raman spectroscopy for forensic analysis of inks in questioned documents. *Forensic Science International, 232*(1-3), 206–212. doi:10.1016/j.forsciint.2013.07.017 PMID:24053882

Breuker, J. (2004). *Constructing a Legal Core Ontology: LRI-Core.* University of Amsterdam, Leibniz Center for Law.

Brito, J. M. (2018). Technological Trends for 5G Networks Influence of E-Health and IoT Applications. *Int. J. Health Med Commun., 9*(1), 1–22. doi:10.4018/IJEHMC.2018010101

Brown, M. E., Treviño, L. K., & Harrison, D. A. (2005). Ethical leadership: A social learning perspective for construct development and testing. *Organizational Behavior and Human Decision Processes, 97*(2), 117–134. doi:10.1016/j.obhdp.2005.03.002

Buckley, C., Wall, D. P., Moran, B., O'Neil, S., & Murphy, P. N. (2016). Farm gate level nitrogen balance and use efficiency changes post implementation of the EU Nitrates Directive. *Nutrient Cycling in Agroecosystems, 2016*(104), 1–13. doi:10.100710705-015-9753-y

Burliai, A., Burliai, O., Revutska, A., Smolii, L., & Klymenko, L. (2021). Organizational and economic risks of ecologization of agriculture. *Agricultural and Resource Economics: International Scientific E-Journal, 7*(1), 96–114. doi:10.51599/are.2021.07.01.06

Byun, G., Karau, S. J., Dai, Y., & Lee, S. (2018). A three-level examination of the cascading effects of ethical leadership on employee outcomes: A moderated mediation analysis. *Journal of Business Research, 88*, 44–53. doi:10.1016/j.jbusres.2018.03.004

Cabral, I., Grilo, A., & Cruz-Machado, V. (2012). A decision-making model for Lean, Agile, Resilient and Green supply chain management. *International Journal of Production Research, 50*(17), 4830-4845.

Cai, W., Lysova, E. I., Khapova, S. N., & Bossink, B. A. (2018). Servant leadership and innovative work behavior in Chinese high-tech firms: A moderated mediation model of meaningful work and job autonomy. *Frontiers in Psychology, 2018*(9), 1767. doi:10.3389/fpsyg.2018.01767 PMID:30327623

Casler, A., Gundlach, M. J., Persons, B., & Zivnuska, S. (2010). Sierra Nevada Brewing Company's thirty-year journey toward sustainability. *People & Strategy, 33*, 44–51.

Chakraborti, T., McCane, B., Mills, S., & Pal, U. (2018). Loop descriptor: Local optimal-oriented pattern. *IEEE Signal Processing Letters, 25*(5), 635–639.

Chandu, V. (n.d.). *Identification of spatial variations in COVID-19 epidemiological data using KMeans clustering algorithm: A global perspective*. doi:10.13140/RG.2.2.18339.48163

Chang, C. H., & Chen, Y. S. (2013). Green organizational identity and green innovation. *Management Decision, 2013*(51), 1056–1070. doi:10.1108/MD-09-2011-0314

Chang, T. W., Chen, Y. S., Yeh, Y. L., & Li, H. X. (2020). Sustainable consumption models for customers: Investigating the significant antecedents of green purchase behavior from the perspective of information asymmetry. *Journal of Environmental Planning and Management, 2020*, 1–21.

Chang, T. W., Yeh, Y. L., & Li, H. X. (2020). How to shape an organization's sustainable green management performance: The mediation effect of environmental corporate social responsibility. *Sustainability, 12*(21), 9198. doi:10.3390u12219198

Chatterjee, J. M., Sujath, R., & Hassanien, A. E. (2020, May 30). A machine learning forecasting model for COVID-19 pandemic in India. *Stochastic Environmental Research and Risk Assessment, 34*, 959–972. doi:(0123456789() doi:10.1007/s00477-020-01827-8(0123456789().,-volV)

Chaudhary, R. (2020). Green human resource management and employee green behavior: An empirical analysis. *Corporate Social Responsibility and Environmental Management, 2020*(27), 630–641. doi:10.1002/csr.1827

Chen, F., Lu, C., Wu, H., & Li, M. (2017). A semantic similarity measure integrating multiple conceptual relationships for web service discovery. *Expert Systems with Applications*, *67*, 19–31. doi:10.1016/j.eswa.2016.09.028

Chen, G., Farh, J. L., Campbell-Bush, E. M., Wu, Z., & Wu, X. (2013). Teams as innovative systems: Multilevel motivational antecedents of innovation in R&D teams. *The Journal of Applied Psychology*, *2013*(98), 1018–1027. doi:10.1037/a0032663 PMID:23565898

Chen, H. S., Jarrell, J. T., Carpenter, K. A., Cohen, D. S., & Huang, X. (2019). Blockchain in Healthcare: A Patient-Centered Model. *Biomed J Sci Tech Res.*, *20*(3), 15017–15022.

Chen, J., Hong, H., & Stein, J. C. (2001). Forecasting crashes: Trading volume, past returns, and conditional skewness in stock prices. *Journal of Financial Economics*, *61*(3), 345–381. doi:10.1016/S0304-405X(01)00066-6

Chen, Y. S. (2011). Green organizational identity: Sources and consequence. *Management Decision*, *2011*(49), 384–404. doi:10.1108/00251741111120761

Chen, Y. S., & Chang, C. H. (2013). The determinants of green product development performance: Green dynamic capabilities, green transformational leadership, and green creativity. *Journal of Business Ethics*, *116*(1), 107–119. doi:10.100710551-012-1452-x

Chen, Y. S., Chang, C. H., & Lin, Y. H. (2014). Green transformational leadership and green performance: The mediation effects of green mindfulness and green self-efficacy. *Sustainability*, *6*(10), 6604–6621. doi:10.3390u6106604

Chen, Y. S., Chang, C. H., & Lin, Y. H. (2014). The determinants of green radical and incremental innovation performance: Green shared vision, green absorptive capacity, and green organizational ambidexterity. *Sustainability*, *2014*(6), 7787–7806. doi:10.3390u6117787

Chen, Y. S., Chang, C. H., Yeh, S. L., & Cheng, H. I. (2015). Green shared vision and green creativity: The mediation roles of green mindfulness and green self-efficacy. *Quality & Quantity*, *2015*(49), 1169–1184. doi:10.100711135-014-0041-8

Chen, Y. S., Chang, T. W., Lin, C. Y., Lai, P. Y., & Wang, K. H. (2016). The influence of proactive green innovation and reactive green innovation on green product development performance: The mediation role of green creativity. *Sustainability*, *2016*(8), 966. doi:10.3390u8100966

Chen, Y. S., Lai, S. B., & Wen, C. T. (2006). The influence of green innovation performance on corporate advantage in Taiwan. *Journal of Business Ethics*, *67*(4), 331–339. doi:10.100710551-006-9025-5

Chhun, S., Moalla, N., & Ouzrout, Y. (2016). QoS ontology for service selection and reuse. *Journal of Intelligent Manufacturing*, *27*(1), 187–199. doi:10.100710845-013-0855-6

Chou, C.-J. (2014). Hotels' environmental policies and employee personal environmental beliefs: Interactions and outcomes. *Tourism Management*, *40*, 436–446. doi:10.1016/j.tourman.2013.08.001

Choudhary, G., Astillo, P. V., You, I., Yim, K., Chen, R., & Cho, J.-H. (2020). Lightweight misbehavior detection management of embedded iot devices in medical cyber physical systems. *IEEE eTransactions on Network and Service Management*, *17*(4), 2496–2510.

Chuang, S. P., & Huang, S. J. (2015). Effects of business greening and green IT capital on business competitiveness. *Journal of Business Ethics*, *128*(1), 221–231. doi:10.100710551-014-2094-y

Chuang, S. P., & Huang, S. J. (2018). The effect of environmental corporate social responsibility on environmental performance and business competitiveness: The mediation of green information technology capital. *Journal of Business Ethics*, *150*(4), 991–1009. doi:10.100710551-016-3167-x

Connelly, C. E., Ford, D. P., Turel, O., Gallupe, B., & Zweig, D. (2014). 'I'm busy (and competitive)!' Antecedents of knowledge sharing under pressure. *Knowledge Management Research and Practice*, *2014*(12), 74–85. doi:10.1057/kmrp.2012.61

Çop, S., Olorunsola, V. O., & Alola, U. V. (2021). Achieving environmental sustainability through green transformational leadership policy: Can green team resilience help? *Bus Strat Env.*, *30*, 671–682.**Error! Hyperlink reference not valid.** https://onlinelibrary.wiley.com/doi/epdf/10.1002/bse.2646 doi:10.1002/bse.2646

Cunningham, P., & Delany, S. (2007). k-Nearest neighbour classifiers. *Mult Classif Syst.*

Curtain, G., Sommer, M., & Vis-Sommer, V. (Eds.). (2003). *The World of E-Government*. The Haworth Press. doi:10.1300/J199v02n03_01

D'Mello, S., Ones, D., Klein, R., Wiernik, B., & Dilchert, S. (2011). Green company rankings and reporting of pre-environmental efforts in organizations. *Proceedings of the Poster session presented at the 26th Annual Conference of the Society for Industrial and Organizational Psychology.*

Daily, B. F., Bishop, J. W., & Govindarajulu, N. (2009). A conceptual model for organizational citizenship behavior directed toward the environment. *Business Strategy and the Environment*, *13*(4), 209–222.

Daily, B. F., Bishop, J. W., & Govindarajulu, N. (2009). Conceptual model for organizational citizenship behavior directed toward the environment. *Business & Society*, *48*(2), 243–256. doi:10.1177/0007650308315439

Daily, B. F., Bishop, J. W., & Steiner, S. (2007). The mediating role of EMS teamwork as it pertains to HR factors and perceived environmental performance. *Journal of Applied Business Research*, *23*, 95–109.

Dangelico, R. M. (2016). Green product innovation: Where we are and where we are going. *Business Strategy and the Environment*, *2016*(25), 560–576. doi:10.1002/bse.1886

Dangelico, R. M., & Pujari, D. (2010). Mainstreaming green product innovation: Why and how companies integrate environmental sustainability. *Journal of Business Ethics*, *2010*(95), 471–486. doi:10.100710551-010-0434-0

Davies, N. G., Klepac, P., Liu, Y., Prem, K., Jit, M., Pearson, C. A. B., Quilty, B. J., Kucharski, A. J., Gibbs, H., Clifford, S., Gimma, A., van Zandvoort, K., Munday, J. D., Diamond, C., Edmunds, W. J., Houben, R. M. G. J., Hellewell, J., Russell, T. W., Abbott, S., ... Eggo, R. M. (2020). Age-dependent effects in the transmission and control of COVID-19 epidemics. *Nature Medicine*, *26*(8), 1205–1211. doi:10.103841591-020-0962-9 PMID:32546824

De Aguiar. (2020). A survey of blockchain-based strategies for healthcare. *ACM Computing Surveys*, *53*(2), 1–27.

De Bruijn, J., Bussler, C., Domingue, J., Fensel, D., Hepp, M., Kifer, M., & Stollberg, M. (2005). Web service modeling ontology (wsmo). *Interface: a Journal for and About Social Movements*, *5*(1), 50.

de Jong, M. (2021). Inclusive capitalism. The emergence of a new purpose paradigm in economics and business administration and its implications for public policy. *Global Public Policy and Governance*, *1*, 159–174. https://link.springer.com/content/pdf/10.1007/s43508-021-00020-z.pdf doi:10.1007/s43508-021-00020-z

Deloitte. (2019). *Global Health Care Outlook Shaping the Future*. Deloitte.

Derviş, H. (2020). Bibliometric Analysis using Bibliometrix an R Package. *Journal of Scientometric Research*, *8*(3), 156–160. doi:10.5530/jscires.8.3.32

Devi, M. U., Rao, C. R., & Jayaram, M. (2014). Statistical measures for differentiation of photocopy from print technology forensic perspective. *International Journal of Computers and Applications*, *105*(15).

Di Stefano, G., Peteraf, M., & Veronay, G. (2010). Dynamic capabilities deconstructed: A bibliographic investigation into the origins, development, and future directions of the research domain. *Industrial and Corporate Change*, *19*(4), 1187–1204. doi:10.1093/icc/dtq027

Dias-Neto, A. C., Matalonga, S., Martı'n, S., Robiolo, G., & Travassos, G. H. (n.d.). Toward the characterization of software testing practices in South America: Looking at Brazil and Uruguay. *Software Quality Journal*. Advance online publication. doi:10.100711219-016-9329-3

Ding, D., Zhang, M., Pan, X., Yang, M., & He, X. (2019, July). Modeling extreme events in time series prediction. In *Proceedings of the 25th ACM SIGKDD International Conference on Knowledge Discovery & Data Mining* (pp. 1114-1122). ACM.

Dubose, J. (2000). Sustainability and performance at Interface Inc. *Interfaces*, *30*(3), 190–201.

Dubovitskaya, A., Xu, Z., Ryu, S., Schumacher, M., & Wang, F. (2018). Secure and Trustable Electronic Medical Records Sharing using Blockchain. *AMIA Annu Symp Proc.,* 650-659.

Dües, C. M., Tan, K. H., & Lim, M. (2013). Green as the new Lean: How to use Lean practices as a catalyst to be greening your supply chain. *Journal of Cleaner Production*, *40*, 93–100. doi:10.1016/j.jclepro.2011.12.023

Dumont, J., Shen, J., & Deng, X. (2017). Effects of green HRM practices on employee workplace green behavior: The role of psychological green climate and employee green values. *Human Resource Management, 56*(4), 613–627. doi:10.1002/hrm.21792

Dutta, K., Umashankar, V., Choi, G., & Parsa, H. G. (2008). A comparative study of consumers' green practice orientation in India and the United States: A study from the restaurant industry. *Journal of Foodservice Business Research, 11*(3), 269–285. doi:10.1080/15378020802316570

Dvir, T., Eden, D., Avolio, B. J., & Shamir, B. (2002). Impact of transformational leadership on follower development and performance: A field experiment. *Academy of Management Journal, 45*, 735–744.

Eagly, A. H. (1987). *Sex differences in social behavior: A social role interpretation.* Erlbaum.

Eagly, A. H., & Wood, W. (2012). Social role theory. In P. A. M. Van Lange, A. W. Kruglanski, & E. T. Higgins (Eds.), *Handbook of theories of social psychology* (pp. 458–476). Sage. doi:10.4135/9781446249222.n49

ECA (European Court of Auditors). (2014). *Special Report. Integration of EU Water Policy Objectives with the CAP: A Partial Success*. European Court of Auditors. Available online: https://www.eca.europa.eu/Lists/ECADocuments/SR14_04/SR14_04_EN.pdf

ECA (European Court of Auditors). (2017). *Greening: A More Complex Income Support Scheme, Not Yet Environmentally.* Effective; European Court of Auditors. Available online: https://www.eca.europa.eu/Lists/ECADocuments/SR17_21/SR_GREENING_EN.pdf

Eco Measures Report. (2010). Interface Inc.

Edmondson, A. C., & Nembhard, I. M. (2009). Product development and learning in project teams: The challenges are the benefits. *Journal of Product Innovation Management, 2009*(26), 123–138. doi:10.1111/j.1540-5885.2009.00341.x

Egri, C. P., & Herman, S. (2000). Leadership in the North American environmental sector: Values, leadership styles, and contexts of environmental leaders and their organizations. *Academy of Management Journal, 2000*(43), 571–604.

Eisenhardt, K. M., & Martin, J. A. (2000). Dynamic capabilities: What are they? *Strategic Management Journal, 32*(10-11), 1105–1121. doi:10.1002/1097-0266(200010/11)21:10/11<1105::AID-SMJ133>3.0.CO;2-E

Eisler, I., Le Grange, D., & Asen, K. (2003). Family interventions. In J. Treasure, U. Schmidt, & E. van Furth (Eds.), *Handbook of eating disorders: Theory, treatment and research* (pp. 291–310). Wiley.

Ekblaw, Azaria, Halamka, & Lippman. (2016). *A case study for blockchain in healthcare: " medrec " prototype for electronic health records and medical research data.* Academic Press.

Elenkov, D. S., & Manev, I. M. (2005). Top management leadership and influence on innovation: The role of sociocultural context. *Journal of Management, 2005*(31), 381–402. doi:10.1177/0149206304272151

El-Gayar, O. F., Ambati, L. S., & Nawar, N. (2020). Wearables, Artificial intelligence, and the Future of Healthcare. In *AI and Big Data's Potential for Disruptive Innovation* (pp. 104–129). IGI Global. doi:10.4018/978-1-5225-9687-5.ch005

Elkasrawi, S., & Shafait, F. (2014, April). Printer identification using supervised learning for document forgery detection. In *2014 11th IAPR International Workshop on Document Analysis Systems* (pp. 146-150). IEEE.

Engstro, E., & Petersen, K. (n.d.). SERP-test: A taxonomy for supporting industry–academia communication. *Software Quality Journal*. Advance online publication. doi:10.100711219-016-9322-x

Epitropaki, O., Kark, R., Mainemelis, C., & Lord, R. G. (2017). Leadership and followership identity processes: A multilevel review. *The Leadership Quarterly, 2017*(28), 104–129. doi:10.1016/j.leaqua.2016.10.003

Esty, D. C., & Winston, A. S. (2006). *Green to Gold: How Smart Companies Use Environmental Strategy to Innovate Create Value and Build Competitive Advantage*. Yale University Press.

Fan, K., Wang, S., Ren, Y., Li, H., & Yang, Y. (2018). MedBlock: Efficient, and Secure Medical Data Sharing Via Blockchain. *J Med Syst., 42*(8), 136.

Feddaoui, I., Felhi, F., Algarni, F., & Akaichi, J. (2021). QoS-Based Collaborative Filtering for Web Service Mining. *International Journal of Web Portals, 13*(1), 40–61. doi:10.4018/IJWP.2021010103

Fernández, E., Junquera, B., & Ordiz, M. (2006). Managers' profile in environmental strategy: A review of the literature. *Corporate Social Responsibility and Environmental Management, 2006*(13), 261–274. doi:10.1002/csr.109

Ferreira, A., Bondi, L., Baroffio, L., Bestagini, P., Huang, J., Dos Santos, J. A., ... Rocha, A. (2017). Data-driven feature characterization techniques for laser printer attribution. *IEEE Transactions on Information Forensics and Security, 12*(8), 1860–1873.

Flannery, B. L., & May, D. R. (1994). Prominent factors influencing environmental activities: Application of the environmental leadership model (ELM). *The Leadership Quarterly, 1994*(5), 201–221. doi:10.1016/1048-9843(94)90012-4

Fleitas, P. E., Paz, J. A., Simoy, M. I., Vargas, C., Cimino, R. O., Krolewiecki, A. J., & Aparicio, J. P. (2020, October). Understanding the value of clinical symptoms of COVID-19. A logistic regression model. *MedRvix*. doi:10.1101/2020.10.07.20207019

Florida, R., & Davison, D. (2001). Gaining from Green Management: Environmental management systems inside and outside the factory. *California Management Review, 43*(3), 64–84. doi:10.2307/41166089

Forman, M., & Jorgensen, S. (2001). The social shaping of participation of employees in environmental work within enterprises—Experiences from a Danish context. *Technology Analysis and Strategic Management, 13*, 71–90. doi:10.1080/09537320120040455

Francoeur, V., Paillé, P., Yuriev, A., & Boiral, O. (2021). The Measurement of Green Workplace Behaviors: A Systematic Review. *Organization & Environment, 34*(1), 18–42. doi:10.1177/1086026619837125

Franses, P. H. (1991). Seasonality, nonstationarity and the forecasting of monthly time series. *International Journal of Forecasting, 7*(2), 199–208. doi:10.1016/0169-2070(91)90054-Y

Frooman, J. (1999). Stakeholder influence strategies. *Academy of Management Review, 24*(2), 191–205. doi:10.5465/amr.1999.1893928

Fuad, A., Lee, Y-C. G., & Hsu, C-Y. (2020). Bibliometric Analysis of Bioscience Trends Journal (2007-2017): Knowledge dynamics and visualization. *Library Philosophy and Practice (ejournal).* 4076.

Fukushima, T., Kamei, Y., McIntosh, S., Yamashita, K., & Ubayashi, N. (2014). An empirical study of just-in-time defect prediction using cross-project models. *Proceedings of the 11th Working Conference on Mining Software Repositories.*

Galeazzo, A., Furlan, A., & Vinelli, A. (2014). Lean and green in action: Interdependencies and performance of pollution prevention projects. *Journal of Cleaner Production, 85*, 191–200. doi:10.1016/j.jclepro.2013.10.015

Gario, A., Andrews, A., & Hagerman, S. (n.d.). Fail-safe testing of safety-critical systems: A case study and efficiency analysis. *Software Quality Journal.* Advance online publication. doi:10.100711219-015-9283-5

Garousi, V., Eskandar, M. M., & Herkilog˘lu, K. (n.d.). Industry–academia collaborations in software testing: Experience and success stories from Canada and Turkey. *Software Quality Journal.* Advance online publication. doi:10.100711219-016-9319-5

Gawriljuk, V. O., Zin, P. K., Foil, D. H., Bernatchez, J., Beck, S., Beutler, N., . . . Ekins, S. (2020). Machine Learning Models Identify Inhibitors of SARS-CoV-2. bioRxiv. doi:10.1101/2020.06.16.154765

Gioia, D.A., Patvardhan, S.D., Hamilton, A.L., & Corley, K.G. (2013). Organizational identity formation and change. *Acad. Manag. Ann., 7*(1), 123-193.

Goel, R., Jain, A., Verma, K., Bhushan, M., Kumar, A., & Negi, A. (2020). Mushrooming Trends and Technologies to Aid Visually Impaired People. In *International Conference on Emerging Trends in Information Technology and Engineering (ic-ETITE'20).* IEEE.

Goldstein, N. J., Cialdini, R. B., & Griskevicius, V. (2008). A room with a viewpoint: Using social norms to motivate environmental conservation in hotels. *The Journal of Consumer Research, 2008*(35), 472–482. doi:10.1086/586910

Gómez-Pérez, A., Ortiz-Rodríguez, F., & Villazón-Terrazas, B. (2006). Ontology-Based Legal Information Retrieval to Improve the Information Access in e-Government. WWW2006. doi:10.1145/1135777.1135988

Gómez-Pérez, A., Fernández-López, M., & Corcho, O. (2003). *Ontological Engineering*. Springer Verlag.

Gómez-Pérez, A., Ortiz-Rodríguez, F., & Villazón-Terrazas, B. (2005). Legal Ontologies for the Spanish e-Government. In *CAEPIA'05 Post-Proceedings Volume*. Springer Verlag.

González García, E., Colomo Magaña, E., & Cívico Ariza, A. (2020). Quality Education as a Sustainable Development Goal in the Context of 2030 Agenda: Bibliometric Approach. *Sustainability*, *12*(15), 5884. doi:10.3390u12155884

González-Benito, J., & González-Benito, O. (2006). A review of determinant factors of environmental proactivity. *Business Strategy and the Environment*, *15*(2), 87–102. doi:10.1002/bse.450

Graci, S., & Dodds, R. (2008). Why goes green? The business case for environmental commitment in the Canadian hotel industry. *Anatolia*, *19*(2), 251–270. doi:10.1080/13032917.2008.9687072

Granger, C. W., & Joyeux, R. (1980). An introduction to longmemory time series models and fractional differencing. *Journal ofTime Series Analysis*, *1*(1), 15–29. doi:10.1111/j.1467-9892.1980.tb00297.x

Greenwood, L., Rosenbeck, J., & Scott, J. (2012). The role of the environmental manager in advancing environmental sustainability and social responsibility in the organization. *Journal of Environmental Sustainability*, *2*(2), 59–75.

Griggs, K.N., Ossipova, O., Kohlios, C.P., Baccarini, A.N., Howson, E.A., & Hayajneh, T. (2018). Healthcare blockchain system using smart contracts for secure automated remote patient monitoring. *J Med Syst.*, *42*(7), 130.

Gropper, A. (2016). Powering the physician-patient relationship with hie of one blockchain health it. In *ONC/NIST use of Blockchain for healthcare and research workshop*. ONC/NIST.

Gupta, N. P. S. (2020). Automated Attendance System Using OpenCV. In *2020 8th International Conference on Reliability, Infocom Technologies and Optimization (Trends and Future Directions) (ICRITO)*, (pp. 1226-1230). IEEE.

Gupta, A., & Jha, R. K. (2015). A survey of 5G network: Architecture and emerging technologies. *IEEE Access: Practical Innovations, Open Solutions*, *3*, 1206–1232. doi:10.1109/ACCESS.2015.2461602

Hallam, C., & Contreras, C. (2016). Integrating lean and green management. *Management Decision*, *54*(9), 2157–2187. doi:10.1108/MD-04-2016-0259

Halme, M., Anttonen, M., Kuisma, M., Kontoniemi, N., & Heino, E. (2007). Business models for material efficiency services: conceptualization and application. *Ecol. Econ.*, *63*(1), 126-137.

Hamdouch, A., & Depret, M. H. (2010). Policy integration strategy and the development of the 'green economy': Foundations and implementation patterns. *Journal of Environmental Planning and Management, 2010*(53), 473–490. doi:10.1080/09640561003703889

Hanna, M. D., Rocky Newman, W. R., & Johnson, P. (2000). Linking operational and environmental improvement through employee involvement. *International Journal of Operations & Production Management, 20*(2), 148–165. doi:10.1108/01443570010304233

Harris, L. C., & Crane, A. (2002). The greening of organizational culture: Management views on the depth, degree, and diffusion of change. *Journal of Organizational Change Management, 15*(3), 214–234. doi:10.1108/09534810210429273

Hart, S.L., & Milstein, M.B. (2003). Creating sustainable value. *Acad. Manag. Exec., 17*(2), 56-69.

Hart, S. (1997). Beyond greening: Strategies for a sustainable world. *Harvard Business Review, 75*, 66–76.

Hart, S. L. (1995). A natural-resource-based view of the firm. *Academy of Management Review, 20*(4), 986–1014. doi:10.5465/amr.1995.9512280033

Hart, S. L. (2005). *Capitalism at the Crossroads: The Unlimited Business Opportunities in Solving the World's Most Difficult Problems*. Wharton School Publishing.

Harvey, A. C., & Todd, P. H. J. (1983). Forecasting economic time series with structural and BoxJenkins models: A case study. *Journal of Business & Economic Statistics, 1*(4), 299–307.

Hathaliya, J., Sharma, P., Tanwar, S., & Gupta, R. (2019). Blockchain-based remote patient monitoring in healthcare 4.0. *2019 IEEE 9th International Conference on Advanced Computing (IACC)*, 87–91.

Hendler, J., Berners-Lee, T., & Miller, E. (2002). *Integrating Applications on the Semantic Web*. http://www.w3.org/2002/07/swint.html

Hersey, K. (1998). A close look at ISO 14000. *Professional Safety, 43*, 26–29.

Hii, P. C., & Chung, W. Y. (2011). A comprehensive ubiquitous healthcare solution on an AndroidTM mobile device. *Sensors (Basel), 11*(7), 6799–6815. doi:10.3390110706799 PMID:22163986

Hilton, O. (1992). *Scientific examination of questioned documents*. CRC Press.

Hofer, C., Cantor, D. E., & Dai, J. (2012). The competitive determinants of a firm's environmental management activities: Evidence from US manufacturing industries. *Journal of Operations Management, 30*(1-2), 69–84. doi:10.1016/j.jom.2011.06.002

Holliday, C., (2001). Sustainable growth the DuPont way. *Harv. Bus. Rev., 79*(8), 129-132.

Holt, D., (2011). Where are they now? Tracking the longitudinal evolution of environmental businesses from the 1990s. *Bus. Strat. Env., 20*(4), 238-250.

Honcharuk, I.V., & Tomashuk, I.V. (2017). Vplyv ekoloho-ekonomichnoho faktoru na osoblyvosti orhanizatsiino-ekonomichnoho mekhanizmu vykorystannia resursnoho potentsialu silskykh terytorii. *Ekonomika, finansy, menedzhment: aktualni pytannia nauky i praktyky, 4*(20), 55-62.

Honcharuk, I.V., & Tomashuk, I.V. (2018). Derzhavne rehuliuvannia rozvytku resursnoho potentsialu silskykh terytorii: zahalni aspekty. *Ekonomika, finansy, menedzhment: aktualni pytannia nauky i praktyky, 4*(32), 19-30.

Hosaain, S., & Batcha, D. (2020). Authorship Pattern on Dentistry Research During 1999-2018: A Scientometric Profile. *Library Philosophy and Practice (e-journal)*. 4036. https://digitalcommons.unl.edu/libphilprac/4036

Hossain, E., & Hasan, M. (2015). *5G cellular: Key enabling technologies and research challenges*. arXiv:1503.00674.

Hossain, M. S., & Muhammad, G. (2017). Emotion-aware connected healthcare big data towards 5G. *IEEE Internet Things J.*, *5*(4), 2399–2406. doi:10.1109/JIOT.2017.2772959

Imasiku, K. (2021). *Organizational Insights, Challenges and Impact of Sustainable Development in Developing and Developed Nations*. IntechOpen. https://www.intechopen.com/chapters/73366 doi:10.5772/intechopen.93748

Ismail, L., Materwala, H., & Zeadally, S. (2019). Lightweight blockchain for healthcare. *IEEE Access: Practical Innovations, Open Solutions*, *7*, 149935–149951.

Istepanian, R. S., Hu, S., Philip, N. Y., & Sungoor, A. (2011). The potential of Internet of m-health Things "m-IoT" for non-invasive glucose level sensing. *Proceedings of the 2011 Annual International Conference of the IEEE Engineering in Medicine and Biology Society*, 5264–5266. 10.1109/IEMBS.2011.6091302

Iwendi, C., Bashir, A. K., Peshkar, A., Sujatha, R., Chatterjee, J. M., Pasupuleti, S., Mishra, R., Pillai, S., & Jo, O. (2020, July). COVID-19 Patient Health Prediction Using Boosted Random Forestlgorithm. *Frontiers in Public Health*, *8*, 357. Advance online publication. doi:10.3389/fpubh.2020.00357 PMID:32719767

Jabbour, C. J. C. (2011). How green are HRM practices, organizational culture, learning and teamwork? A Brazilian study. *Industrial and Commercial Training*, *43*(2), 98–105. doi:10.1108/00197851111108926

Jabbour, C. J. C. (2013a). Environmental training in organisations: From a literature review to a framework for future research. *Resources, Conservation and Recycling*, *74*, 144–155. doi:10.1016/j.resconrec.2012.12.017

Jabbour, C. J. C. (2013b). Environmental training and environmental maturity of Brazilian companies with ISO14001: Empirical evidence. *Journal of Cleaner Production*, *43*(2), 1–8.

Jabbour, C. J. C., Almada Santos, F. C., Azevedo Fonseca, S., & Seido Nagano, M. (2013). Green teams: Understanding their roles in the environmental management of companies located in Brazil. *Journal of Cleaner Production*, *46*, 58–66. doi:10.1016/j.jclepro.2012.09.018

Jabbour, C. J. C., Santos, F. C. A., & Nagano, M. S. (2010). Contribution of HRM throughout the stages of environmental management: Methodological triangulation applied to companies in Brazil. *International Journal of Human Resource Management*, *21*(7), 1049–1089. doi:10.1080/09585191003783512

Jabbour, Ch., & Chiappetta, J. (2015). Environmental training and environmental management maturity of Brazilian companies with ISO14001: Empirical evidence. *Journal of Cleaner Production*, *96*, 331–338. doi:10.1016/j.jclepro.2013.10.039

Jabid, T., Kabir, M. H., & Chae, O. (2010, August). Gender classification using local directional pattern (LDP). In *2010 20th International conference on pattern recognition* (pp. 2162-2165). IEEE.

Jackson, S. E., Renwick, D. W. S., Jabbour, C. J. C., & Muller-Camen, M. (2011). State-of-the-art and future directions for green human resource management: Introduction to the special issue. *German Journal of Research in Human Resource Management*, *25*(2), 99–116. doi:10.1177/239700221102500203

Jackson, S. E., & Seo, J. (2010). The greening of strategic HRM scholarship. *Organizational Management Journal*, *7*(4), 278–290. doi:10.1057/omj.2010.37

Jain, H., Joshi, S., Gupta, G., & Khanna, N. (2020). Passive classification of source printer using text-line-level geometric distortion signatures from scanned images of printed documents. *Multimedia Tools and Applications*, *79*(11), 7377–7400.

Jalal, S. (2019). Co-authorship and co-occurrences analysis using BibliometrixR package: A case study of India and Bangladesh. *Annals Of Library And Information Studies*, *66*, 57–64.

Jiang, S., Cao, J., Wu, H., Yang, Y., Ma, M., & He, J. (2018). Blochie: a blockchain-based platform for healthcare information exchange. 2018 IEEE international conference on smart computing (smartcomp), 49–56.

Jiao, H., Zhang, J., Li, J. H., & Shi, J. (2017). Research on cloud manufacturing service discovery based on latent semantic preference about OWL-S. *International Journal of Computer Integrated Manufacturing*, *30*(4-5), 433–441.

Jin, X., & Han, J. (2011). *K-Means Clustering*. . doi:10.1007/978-0-387-30164-8_425

Jing, X.-Y., Ying, S., Zhang, Z.-W., Wu, S.-S., & Liu, J. (2014). Dictionary learning based software defect prediction. *Proceedings of the 36th International Conference on Software Engineering.*

Johansen, D. (1998). Interface Inc: Taking the lead toward sustainability. *Corporate Environmental Strategy*, *5*(3), 54. doi:10.1016/S1066-7938(00)80100-1

Jurado, M., Jose, P., & Fuentes, J. M. (2014). Lean management, supply chain management and sustainability: A literature review. *Journal of Cleaner Production*, *85*, 134–150. doi:10.1016/j.jclepro.2013.09.042

Kao, M. R., Liu, C. Y., Huang, Y. C., & Chang, N. J. (2010). Research of the relationship among business green management, environmental performance, and competitive advantage. *Journal of Management & Systems, 17*(2), 255–278.

Kaouan, M., Bouchiha, D., Benslimane, S. M., & Boukli-Hacene, S. (2020, December). Towards Service Ontology for Web Services Storage and Discovery. In *2020 4th International Symposium on Informatics and its Applications (ISIA)* (pp. 1-6). IEEE. 10.1109/ISIA51297.2020.9416537

Kar, N. A. (2012). Study of implementing automated attendance system using face recognition technique. *International Journal of Computer and Communication Engineering*, 100.

Kartun-Giles, A., Jayaprakasam, S., & Kim, S. (2018). Euclidean matchings in ultra-dense networks. *IEEE Communications Letters, 22*(6), 1216–1219. doi:10.1109/LCOMM.2018.2799207

Kee-hung, L., Cheng, T., & Tang, A. (2010). Green retailing: Factors for successes. *California Management Review, 52*(2), 6–31. doi:10.1525/cmr.2010.52.2.6

Kennedy, E. H., & Kmec, J. (2018). Reinterpreting the gender gap in household pro-environmental behaviour. *Environmental Sociology, 4*(3), 299–310. doi:10.1080/23251042.2018.1436891

Khanday, A. M., Rabani, S. T., Khan, K. R., Rouf, N., & Mohi Ud Din, M. (2020, June). Machine learning based approaches for detecting COVID-19 using clinical text data. *International Journal of Information Technology*. doi:10.1007/s41870-020-00495-9

Khan, M. A. S., Jianguo, D., Ali, M., Saleem, S., & Usman, M. (2019). Interrelations Between Ethical Leadership, Green Psychological Climate, and Organizational Environmental Citizenship Behavior: A Moderated Mediation Model. *Frontiers in Psychology, 10*, 1977. doi:10.3389/fpsyg.2019.01977 PMID:31555173

Khanna, N., Mikkilineni, A. K., Chiu, G. T. C., Allebach, J. P., & Delp, E. J. (2008, August). Survey of scanner and printer forensics at purdue university. In *International Workshop on Computational Forensics* (pp. 22-34). Springer. 10.1007/978-3-540-85303-9_3

Khusro, S., Ali, Z., & Ullah, I. (2016). *Recommender Systems: Issues*. Challenges, and Research Opportunities. doi:10.1007/978-981-10-0557-2_112

Kim, S., Whitehead, E. J., & Zhang, Y. (2008). Classifying software changes: Clean or buggy? *Software Engineering, IEEE Transactions on, 34*(2), 181-196.

Kim, S., Zhang, H., Wu, R., & Gong, L. (2011). *Dealing with noise in defect prediction*. Paper presented at the Software Engineering (ICSE), 2011 33rd International Conference on.

Kim, A., Kim, Y., Han, K., Jackson, S. E., & Ployhart, R. E. (2017). Multilevel influences on voluntary workplace green behavior: Individual diferences, leader behavior, and coworker advocacy. *Journal of Management, 2017*(43), 1335–1358. doi:10.1177/0149206314547386

Kim, Y. J., Woo, G. K., Hyung-Min, Ch., & Kullada, P. (2019). The effect of green human resource management on hotel employees' ecofriendly behavior and environmental performance. *International Journal of Hospitality Management, 76*, 83–93. doi:10.1016/j.ijhm.2018.04.007

Kitazawa, S., & Sarkis, J. (2000). The relationship between ISO 14001 and continuous source reduction programs. *International Journal of Operations & Production Management*, *20*(2), 225–248. doi:10.1108/01443570010304279

Klusch, M., Fries, B., & Sycara, K. (2006, May). Automated semantic web service discovery with OWLS-MX. In *Proceedings of the fifth international joint conference on Autonomous agents and multiagent systems* (pp. 915-922). 10.1145/1160633.1160796

Kodinariya, T., & Makwana, P. (2013). Review on Determining of Cluster in K-means Clustering. *International Journal of Advance Research in Computer Science and Management Studies.*, *1*, 90–95.

Kopecký, J., Vitvar, T., Bournez, C., & Farrell, J. (2007). Sawsdl: Semantic annotations for wsdl and xml schema. *IEEE Internet Computing*, *11*(6), 60–67. doi:10.1109/MIC.2007.134

Kralingen, R. W. (1995). *Frame-based Conceptual Models of Statute Law. Computer/Law Series. No.16*. Kluwer Law International.

Krawiec, R., Housman, D., White, M., Filipova, M., Quarre, F., & Barr, D. (2018). *Blockchain: Opportunities for health care*. Proc.

Kulkarni & Ade. (2017). Intelligent Traffic Control System Implementation for Traffic Violation Control, Congestion Control and Stolen Vehicle Detection. IJES, 5(2).

Kumar, J., Harshavardhan, A., Bhukya, H., & Prasad, A. K. (n.d.). *Advanced machine learning-based analytics on COVID-19 data using generative adversarial networks. In Materials Today: Proceedings*. Elsevier. doi:10.1016/j.matpr.2020.10.053

Kumar, V., & Christodoulopoulou, A. (2014). Sustainability and branding: An integrated perspective. *Industrial Marketing Management*, *43*(1), 6–15. doi:10.1016/j.indmarman.2013.06.008

Kuo, Kim, & Ohno-Machado. (2017). Blockchain distributed ledger technologies for biomedical and health care applications. *Journal of the American Medical Informatics Association*, *24*(6), 1211–1220.

Laksham, S., Surulinathi, M., Balasubramani, R., & Srinivasaragavan, S. (2020). Mapping the Research output on Coronavirus: A Scientometric Study. *Gedrag & Organisatie Review*, *33*(2), 163–186.

Lampert, C. H., Mei, L., & Breuel, T. M. (2006, November). Printing technique classification for document counterfeit detection. In *2006 International Conference on Computational Intelligence and Security* (Vol. 1, pp. 639-644). IEEE.

Lampikoski, T. (2012). Green, innovative, and profitable: a case study of managerial capabilities at Interface Inc. *Tech. Inn. Manag. Rev.*, *2*(11), 4-12.

Lampikoski, T., (2012). Green, innovative, and profitable: a case study of managerial capabilities at Interface Inc. *Tech. Inn. Manag. Rev., 2*(11), 4-12.

Li, Z., Jing, X., & Zhu, X. (2018). Progress on approaches to software defect prediction. IET Software, 12(3), 161-175. doi:10.1049/iet-sen.2017.0148

Li, Zhu, Shen, Gao, Tao, & Liu. (2018). Blockchain-based data preservation system for medical data. *Journal of Medical Systems*, *42*, 6.

Li, C. P., Jiang, J., Chen, W., Ji, T., & Smee, J. (2017). 5G ultra-reliable and low-latency systems design. *Proceedings of the 2017 European Conference on Networks and Communications (EuCNC)*, 1–5.

Likas, A., Vlassis, N., & Verbeek, J. J. (2003). The global k-means clustering algorithm. *Pattern Recognition, 36*(2), 451-461.

Lim, B. C., & Ployhart, R. E. (2004). Transformational leadership: Relations to the five-factor model and team performance in typical and maximum contexts. *The Journal of Applied Psychology, 2004*(89), 610–621. doi:10.1037/0021-9010.89.4.610 PMID:15327348

Lin, B., Jones, C., & Hsieh, C. (2001). Environmental practices and assessment: A process perspective. *Industrial Management & Data Systems, 101*(2), 71–80. doi:10.1108/02635570110384348

Li, S., Da Xu, L., & Zhao, S. (2018). 5G Internet of Things: A survey. *J. Ind. Inf. Integr.*, *10*, 1–9.

Li, S., Ngniatedema, T., & Chen, F. (2017). Understanding the impact of green initiatives and green performance on financial performance in the US. *Business Strategy and the Environment, 26*(6), 776–790. doi:10.1002/bse.1948

Liu, H., & Song, B. (2017). Stock trends forecasting by multi-layer stochastic ANN bagging. *2017 IEEE 29th International Conference on Tools with Artificial Intelligence (ICTAI)*, 322–329.

Liu, C., Hoi, S. C., Zhao, P., & Sun, J. (2016, February). Online arima algorithms for time series prediction. *Thirtieth AAAI conference on artificial intelligence.*

Liu, J., Chen, J., & Tao, Y. (2015). Innovation Performance in New Product Development Teams in China's Technology Ventures: The Role of Behavioral Integration Dimensions and Collective Efficacy. *Journal of Product Innovation Management, 2015*(32), 29–44. doi:10.1111/jpim.12177

Liu, J., Yuan, Y., & Zhang, J. (2018). Environmental leadership in organizations: A literature review and prospects. *Economic Management Journal, 40*(10), 193–208.

Liu, S. H., Lin, C. B., Chen, Y., Chen, W., Huang, T. S., & Hsu, C. Y. (2019). An EMG Patch for the Real-Time Monitoring of Muscle-Fatigue Conditions During Exercise. *Sensors (Basel), 19*(14), 3108. doi:10.339019143108 PMID:31337107

Liu, X., Jia, M., Zhang, X., & Lu, W. (2018). A novel multichannel Internet of things based on dynamic spectrum sharing in 5G communication. *IEEE Internet Things J.*, *6*(4), 5962–5970. doi:10.1109/JIOT.2018.2847731

Liu, X., & Zhang, X. (2019). NOMA-based Resource Allocation for Cluster-based Cognitive Industrial Internet of Things. *IEEE Transactions on Industrial Informatics, 16*(8), 5379–5388. doi:10.1109/TII.2019.2947435

Li, X., Zheng, Z., & Dai, H. N. (2021). When services computing meets blockchain: Challenges and opportunities. *Journal of Parallel and Distributed Computing, 150*, 1–14. doi:10.1016/j.jpdc.2020.12.003

Loknath, Y. & Azeem, B. (2017). *Green Management-Concept and Strategies. National Conference on Marketing and Sustainable Development.*

LOTKA's Law of Productivity - Venus Iconography. (2020). Retrieved 1 September 2020, from https://sites.google.com/site/venusiconography/home/research-papers/lotka-s-law-of-productivity

Lu, C. (2021, March). A Novel Web Service Discovery Method Combining Semantic Interface Similarity and Context Similarity. *IOP Conference Series. Earth and Environmental Science, 693*(1), 012006. doi:10.1088/1755-1315/693/1/012006

Luu, T. T. (2019). Green human resource practices and organizational citizenship behavior for the environment: The roles of collective green crafting and environmentally specific servant leadership. *Journal of Sustainable Tourism, 2019*(27), 1167–1196. doi:10.1080/09669582.2019.1601731

Lyon, T. P., & Maxwell, J. W. (2008). Corporate social responsibility and the environment: A theoretical perspective. *Review of Environmental Economics and Policy, 2*(2), 240–260. doi:10.1093/reep/ren004

Mangla, Kumar, Mehta, Bhushan, & Mohanty. (2022). *Real-Life Applications of the Internet of Things: Challenges, Applications, and Advances*. CRC Press.

Marcus, A. A., & Fremeth, A. R. (2009). Green management matters regardless. *The Academy of Management Perspectives, 8*, 17–26.

Marquis, C., Jackson, S. E., & Li, Y. (2015). Building sustainable organizations in China. *Management and Organization Review, 2015*(11), 427–440. doi:10.1017/mor.2015.37

Marshall, M. E., & Mayer, D. W. (1992). Environmental training: It's good business. *Business Horizons, 35*(2), 54–58. doi:10.1016/S0007-6813(05)80197-1

Martin, D., Paolucci, M., McIlraith, S., Burstein, M., McDermott, D., McGuinness, D., & Sycara, K. (2004, July). Bringing semantics to web services: The OWL-S approach. In *International Workshop on Semantic Web Services and Web Process Composition* (pp. 26-42). Springer.

Mason, K., & Spring, M. (2011). The sites and practices of business models. *Industrial Marketing Management, 40*(6), 1032–1041. doi:10.1016/j.indmarman.2011.06.032

Matukova, H., Khrapkina, V., Bahashova, N., Matukova-Yaryha, D., & Levitskaia, A. (2021). Eco-Controlling: Prospects for the Environmental Policy in Enterprise Management. *SHS Web of Conferences, 100*. doi:10.1051/shsconf/202110005011

Ma, Y., Zhang, Z., Chen, S., Yu, Y., & Tang, K. (2019). A comparative study of aggressive driving behav- ior recognition algorithms based on vehicle motion data. *IEEE Access: Practical Innovations, Open Solutions*, *7*, 8028–8038. doi:10.1109/ACCESS.2018.2889751

McCarty, L. T. (1989). A Language for Legal Discourse, I. Basic Features. *Proceedings of the Second International Conference on Artificial Intelligence and Law*, 180-189. 10.1145/74014.74037

McDonagh, P., & Prothero, A. (1997). *Green management: A reader*. Dryden Press.

Meditskos, G., & Bassiliades, N. (2009). Structural and role-oriented web service discovery with taxonomies in OWL-S. *IEEE Transactions on Knowledge and Data Engineering*, *22*(2), 278–290. doi:10.1109/TKDE.2009.89

Mehta, P. A. (2016). An Efficient Attendance Management Sytem based on Face Recognition using Matlab and Raspberry Pi 2. *International Journal of Engineering Technology Science and Research*, 71-78.

Menzies, T., Dekhtyar, A., Distefano, J., & Greenwald, J. (2007). Problems with precision: A response to "comments on 'data mining static code attributes to learn defect predictors'". *IEEE Transactions on Software Engineering, 33*(9), 637.

Meschede, C. (2020). The Sustainable Development Goals in Scientific Literature: A Bibliometric Overview at the Meta-Level. *Sustainability*, *12*(11), 4461. doi:10.3390u12114461

Mikkilineni, A. K., Arslan, O., Chiang, P. J., Kumontoy, R. M., Allebach, J. P., Chiu, G. T. C., & Delp, E. J. (2005, January). Printer forensics using svm techniques. In NIP & Digital Fabrication Conference (Vol. 2005, No. 1, pp. 223-226). Society for Imaging Science and Technology.

Mittal, S., & Dhar, R. L. (2016). Effect of green transformational leadership on green creativity: A study of tourist hotels. *Tourism Management*, *57*, 118–127. doi:10.1016/j.tourman.2016.05.007

Mockus, A., & Votta, L. G. (2000). Identifying reasons for software changes using historic databases. *Software Maintenance, 2000. Proceedings. International Conference on.*

Mohamed, M., Khafagy, M., & Ibrahim, M. (2019). *Recommender Systems Challenges and Solutions Survey*. . doi:10.1109/ITCE.2019.8646645

Mohamed, C. V. (2012). Fingerprint attendance system for classroom needs. In *2012 Annual IEEE India Conference (INDICON)* (pp. 433-438). Kochi: IEEE. 10.1109/INDCON.2012.6420657

Mohammad, A., Virdee, B. S., See, C. H., Abd-Alhameed, R. A., Falcone, F., & Ernesto, L. (2019). High-Gain Metasurface in polyimide on-chip Antenna Based on cRLH-tL for Sub-terahertz integrated circuits. *IETE Technical Review*, *36*, 190–202.

Muhammad, L. J., Islam, M., Sharif, U. S., & Ayon, I. S. (2020). *Predictive Data Mining Models for Novel Coronavirus (COVID-19)*. Infected Patients Recovery. Research Square., doi:10.21203/rs.3.rs-33247/v1

Muhammad, L. J., Islam, M., Usman, S. S., & Ayon, S. I. (2020). Predictive Data Mining Models for Novel Coronavirus (COVID-19) Infected Patients' Recovery. *SN Computer Science*, *1*(4), 206. doi:10.100742979-020-00216-w PMID:33063049

Mullen, J. E., & Kelloway, E. K. (2009). Safety leadership: A longitudinal study of the effects of transformational leadership on safety outcomes. *Journal of Occupational and Organizational Psychology*, *82*(2), 253–272. doi:10.1348/096317908X325313

Munim, Z. H., Dushenko, M., Jimenez, V. J., Shakil, M. H., & Imset, M. (2020). Big data and artificial intelligence in the maritime industry: A bibliometric review and future research directions. *Maritime Policy & Management*, *47*(5), 577–597. doi:10.1080/03088839.2020.1788731

MuthuKalyani, K V. (2013). Smart application for AMS using face recognition. *Computer Science & Engineering: An International Journal.*

N, J., S, V., P, K., S, S., & S, N. (2021, June). IOT and Computer Vision based Digital Signaling and Security for Indian Traffic Signal. *International Research Journal of Engineering and Technology*, *8*(6), 2336–2339.

Nam, J., Pan, S. J., & Kim, S. (2013). Transfer defect learning. *Proceedings of the 2013 International Conference on Software Engineering.*

Nasir, A., Shaukat, K., Hameed, I. A., Luo, S., Alam, T. M., & Iqbal, F. (2020). A Bibliometric Analysis of Corona Pandemic in Social Sciences: A Review of Influential Aspects and Conceptual Structure. *IEEE Access: Practical Innovations, Open Solutions*, *8*, 133377–133402. doi:10.1109/ACCESS.2020.3008733 PMID:34812340

Nelson, C. R., & Plosser, C. R. (1982). Trends and random walks in macroeconomic time series: Some evidence and implications. *Journal of Monetary Economics*, *10*(2), 139–162. doi:10.1016/0304-3932(82)90012-5

Neuendorf, K. A. (2002). *The Content Analysis Guidebook*. Sage.

Ng, P. F., Butt, M. M., Khong, K. W., & Ong, F. S. (2014). Antecedents of Green Brand Equity. *Journal of Business Ethics, Springer*, *121*(2), 203–215. doi:10.100710551-013-1689-z

Nguyen, K. T., Laurent, M., & Oualha, N. (2015). Survey on secure communication protocols for the Internet of Things. *Ad Hoc Networks*, *32*, 17–31. doi:10.1016/j.adhoc.2015.01.006

Nichol & Brandt. (2016). *Co-Creation of Trust for Healthcare: The Cryptocitizen Framework for Interoperability with Blockchain*. Academic Press.

Nidumolu, R., Prahalad, C.K., & Rangaswami, M.R., (2009). Why sustainability is now thekey driver of innovation. *Harv. Bus. Rev.*, *87*(9), 2-9.

Nitalaksheswara Rao, K., & Satyananda Reddy, C. (2019). *A novel under sampling strategy for efficient software defect analysis of skewed distributed data*. doi:10.1007/s12530-018-

Norton, T. A., Parker, S. L., Zacher, H., & Ashkanasy, N. M. (2015). Employee green behavior: A theoretical framework, multilevel review, and future research agenda. *Organization & Environment, 2015*(28), 103–125. doi:10.1177/1086026615575773

Norton, T. A., Zacher, H., & Ashkanasy, N. M. (2014). Organisational sustainability policies and employee green behavior: The mediating role of work climate perceptions. *Journal of Environmental Psychology, 38*, 49–54. doi:10.1016/j.jenvp.2013.12.008

Norton, T. A., Zacher, H., Parker, S. L., & Ashkanasy, N. M. (2017). Bridging the gap between green behavioral intentions and employee green behavior: The role of green psychological climate. *Journal of Organizational Behavior, 38*(7), 996–1015. doi:10.1002/job.2178

Noura, M., Atiquzzaman, M., & Gaedke, M. (2019). Interoperability in internet of things: Taxonomies and open challenges. *Mobile Networks and Applications, 24*(3), 796–809. doi:10.100711036-018-1089-9

Ofualagba Godswill, O. O. (2018). Automated Student Attendance Management Using Face Recognition. *International Journal of Educational Research and Information Science*, 31-37.

Ojala, T., Pietikainen, M., & Harwood, D. (1994, October). Performance evaluation of texture measures with classification based on Kullback discrimination of distributions. In *Proceedings of 12th international conference on pattern recognition* (Vol. 1, pp. 582-585). IEEE.

Ojaroudi, P. (2019). Recent developments of reconfigurable antennas for current and future wireless communication systems. *Electronics (Basel), 8*(2), 128. doi:10.3390/electronics8020128

Ojaroudiparchin, N., Shen, M., & Pedersen, G. F. (2016). Wide-scan phased array antenna fed by coax-to-microstrip lines for 5G cell phones. *Proceedings of the 2016 21st International Conference on Microwave, Radar and Wireless Communications (MIKON),* 1–4.

Ones, D. S., & Dilchert, S. (2012a). Employee green behaviors. In D. S. S. E. Jackson (Ed.), *Managing human resource for environmental sustainability* (pp. 85–116). Jossey-Bass.

Ortiz-Rodríguez, F., & Villazón-Terrazas, B. (2006). EGO Ontology Model: law and regulation approach for E-Government. *WS Proceedings of the 3rd European Semantic Web Conference.*

Ortiz-Rodríguez, F., Palma, R., & Villazón-Terrazas, B. (2006). Semantic based P2P System for local e-Government. In Proceedings of Informatik 2006. GI-Edition-Lecture Notes in Informatics (LNI).

Ortiz-Rodríguez, F., Chaparro Pelaez, J., & Pascual, F. J. (2009). Semantic model approach for eGovernment to improve sharing, retrieving and exchanging documentation across back-office. *Proceedings of the 10th Annual International Conference on Digital Government, DG.O.*

Ottman, J. A., & ... . (2006). Avoiding Green Marketing Myopia. *Environment, 2006*, 48.

Owen, D. (1992). *Green reporting: Accountancy and the challenge of the nineties*. Chapman Hall.

Pagiola, S., Honey-Roses, J., & Freire-Gonzalez, J. (2020). Assessing the permanence of land-use change induced by payments for environmental services: Evidence from Nicaragua. *Tropical ConservationScience, 13*, 194008292092267–194008292092215. doi:10.1177/1940082920922676

Paille´, P., Chen, Y., Boiral, O., & Jin, J. (2014). The impact of human resource management on environmental performance: An employee-level study. *Journal of Business Ethics, 121*, 451–466.

Paillé, P., Chen, Y., Boiral, O., & Jin, J. (2014). The impact of human resource management on environmental performance: An employee-level study. *Journal of Business Ethics, 2014*(121), 451–466. doi:10.100710551-013-1732-0

Palattella, M. R., Dohler, M., Grieco, A., Rizzo, G., Torsner, J., Engel, T., & Ladid, L. (2016). Internet of things in the 5G era: Enablers, architecture, and business models. *IEEE Journal on Selected Areas in Communications, 34*(3), 510–527. doi:10.1109/JSAC.2016.2525418

Pan, C., & Tian, H. (2017). Environmental leadership, green organizational identity, and corporate green innovation performance. *Chinese Journal of Management, 14*(6), 832–841.

Panchal, R., Swaminarayan, P., Tiwari, S., & Ortiz-Rodriguez, F. (2021). AISHE-Onto: A Semantic Model for Public Higher Education Universities. *DG.O2021: The 22nd Annual International Conference on Digital Government Research.*

Pane Haden, S. S., Oyler, J. D., & Humphreys, J. H. (2009). Historical, practical, and theoretical perspectives on green management: An exploratory analysis. *Management Decision, 47*(7), 1041–1055. doi:10.1108/00251740910978287

Pang, Z., Tian, J., & Chen, Q. (2014). Intelligent packaging and intelligent medicine box for medication management towards the Internet-of-Things. *Proceedings of the 16th International Conference on Advanced Communication Technology*, 352–360. 10.1109/ICACT.2014.6779193

Paolucci, M., Kawamura, T., Payne, T. R., & Sycara, K. (2002, June). Semantic matching of web services capabilities. In *International semantic web conference* (pp. 333-347). Springer.

Parchin, N. O., & Abd-Alhameed, R. A. (2018). A compact Vivaldi antenna array for 5G channel sounding applications. *Proceedings of the 12th European Conference on Antennas and Propagation (EuCAP 2018)*, 9–13. 10.1049/cp.2018.1205

Pathan, SBai, ABhushan, M. (2020). A Study on Health Care using Data Mining Techniques. *Journal of Critical Reviews, 7*(19), 7877–7890. doi:10.31838/jcr.07.19.896

Patra & Van Hamme. (2020). *Detecting Vehicles' Relative Position on Two-Lane Highways Through a Smartphone-Based Video Overtaking Aid Application.* Academic Press.

Pearce, C. L., & Ensley, M. D. A. (2004). reciprocal and longitudinal investigation of the innovation process: The central role of shared vision in product and process innovation teams (PPITs). *Journal of Organizational Behavior, 2004*(25), 259–278. doi:10.1002/job.235

Peattie, K. (1992). *Green Marketing*. Pitman.

Peng, Y., & Lin, S. (2008). Local responsiveness pressure, subsidiary resources, green management adoption, and subsidiaries performance: Evidence from Taiwanese manufacturers. *Journal of Business Ethics*, *79*(1-2), 199–212. doi:10.100710551-007-9382-8

Pham, N. T., Tučková, Z., & Chiappetta Jabbour, C. J. (2019). Greening the hospitality industry: How do green human resource management practices influence organizational citizenship behavior in hotels? A mixed-methods study. *Tourism Management*, *72*, 386–399. doi:10.1016/j.tourman.2018.12.008

Pham, R., Kiesling, S., & Singer, L. (n.d.). On boarding inexperienced developers: Struggles and perceptions regarding automated testing. *Software Quality Journal*. Advance online publication. doi:10.100711219-016-9333-7

Phillips, L. (2007). Go green to gain the edge over rivals. *People Management*, *13*, 9.

Podsakoff, P. M., MacKenzie, S. B., & Bommer, W. H. (1996). Transformational leader behaviors and substitutes for leadership as determinants of employee satisfaction, commitment, trust, and organizational citizenship behaviors. *Journal of Management*, *22*(2), 259–298. doi:10.1177/014920639602200204

Poon, J. (2000). Enhancing corporate image in organisations. *Management Research News*, *23*(5/6), 28–54. doi:10.1108/01409170010782037

Porter, M., & Van der Linde, C. (1995). Green and competitive: Ending the stalemate. *Harvard Business Review*, *73*(5), 120–134.

Post, C., Rahman, N., & Rubow, E. (2011). Green governance: Boards of directors' composition and environmental corporate social responsibility. *Business & Society*, *50*(1), 189–223. doi:10.1177/0007650310394642

Pourhomayoun, M., & Shakibu, M. (2020). Predicting Mortality Risk in Patients with COVID-19 Using Artificial Intelligence to Help Medical Decision-Making. medRxiv. doi:10.1101/2020.03.30.20047308

Pouzols, F. M., & Lendasse, A. (2010). Effect of different detrending approaches on computational intelligence models of time series. *The 2010 International Joint Conference on Neural Networks (IJCNN)*, 1–8.

Prakash, K. B., Imambi, S. S., Mohammed, I., Kumar, T. P., & Pavan, Y. N. (2020). Analysis, Prediction and Evaluation of COVID-19 Datasets using Machine Learning Algorithms. *International Journal of Emerging Trends in Engineering Research*.

Prasad, P., & Elmes, M. (2005). In the name of the practical: Unearthing the hegemony of pragmatics in the discourse of environmental management. *Journal of Management Studies*, *42*, 845–867. doi:.00521.x doi:10.1111/j.1467-6486.2005

Pujari, D., Wright, G., & Peattie, K. (2003). Green and competitive: Influences on environmental new product development performance. *Journal of Business Research*, *2003*(56), 657–671. doi:10.1016/S0148-2963(01)00310-1

Purohit, L., & Kumar, S. (2019, July). Clustering based approach for web service selection using skyline computations. In *2019 IEEE international conference on web services (ICWS)* (pp. 260-264). IEEE.

Purohit, L., & Kumar, S. (2018). A classification based web service selection approach. *IEEE Transactions on Services Computing*, *14*(2), 315–328. doi:10.1109/TSC.2018.2805352

Rahman, N., & Post, C. (2012). Measurement issues in environmental corporate social responsibility (ECSR): Toward a transparent, reliable, and construct valid instrument. *Journal of Business Ethics*, *105*(3), 307–319. doi:10.100710551-011-0967-x

Ramus, C. A. (2001). Organizational support for employees: Encouraging creative ideas for environmental sustainability. *California Management Review*, *2001*(43), 85–105. doi:10.2307/41166090

Ramus, C. A., & Killmer, A. B. (2007). Corporate greening through prosocial extra role behaviors—A conceptual framework for employee motivation. *Business Strategy and the Environment*, *2007*(16), 554–570. doi:10.1002/bse.504

Ramus, C. A., & Steger, U. (2000). The roles of supervisory support behaviors and environmental policy in employee 'eco initiatives' at leading-edge European companies. *Academy of Management Journal*, *43*(4), 605–626. doi:10.2307/1556357

Rao & Reddy. (2018). An Efficient Software Defect Analysis Using Correlation-Based Oversampling. *Arabian Journal for Science and Engineering*.

Rathod, H. Y. W. (2017). Automated attendance system using machine learning approach. In *2017 International Conference on Nascent Technologies in Engineering (ICNTE)* (pp. 1-5). IEEE. 10.1109/ICNTE.2017.7947889

Reilly, A. H., & Hynan, K. A. (2014). Corporate communication, sustainability, and social media: It's not easy (really) being green. *Business Horizons*, *57*(6), 747–758. doi:10.1016/j.bushor.2014.07.008

Renwick, D. W. S., Redman, T., & Maguire, S. (2013). Green Human Resource Management: A review and research agenda. *International Journal of Management Reviews*, *15*(1), 1–14. doi:10.1111/j.1468-2370.2011.00328.x

Robertson, J. L., & Barling, J. (2013). Greening organizations through leaders' influence on employees' pro-environmental behaviors. *Journal of Organizational Behavior*, *34*(2), 176–194. doi:10.1002/job.1820

Robertson, J. L., & Barling, J. (2017). Toward a new measure of organizational environmental citizenship behavior. *Journal of Business Research*, *75*, 57–66. doi:10.1016/j.jbusres.2017.02.007

Rodríguez-Soler, R., Uribe-Toril, J., & De Pablo Valenciano, J. (2020). Worldwide trends in the scientific production on rural depopulation, a bibliometric analysis using bibliometrix R-tool. *Land Use Policy*, *97*, 104787. doi:10.1016/j.landusepol.2020.104787

Rubik, F., Frankl, P., Pietroni, L., & Scheer, D. (2007). Eco-labelling and consumers: Towards a re-focus and integrated approaches. *International Journal of Technology Management & Sustainable Development*, *2007*(2), 175–191. doi:10.1504/IJISD.2007.016932

Russo, M. V., & Fouts, P. A. (1997). A resource-based perspective on corporate environmental performance and profitability. *Academy of Management Journal*, *40*(3), 534–559. doi:10.2307/257052

Ruta, M., Scioscia, F., Ieva, S., Capurso, G., Pinto, A., & Di Sciascio, E. (2018, August). A Blockchain Infrastructure for the Semantic Web of Things. SEBD.

Ryu, Jang, & Baik. (n.d.). A transfer cost-sensitive boosting approach for cross-project defect prediction. Software Quality Journal. Advance online publication. 10.1007/s11219-015-9287-1

Sachs, J. D. (2012). From millennium development goals to sustainable development goals. *Lancet*, *379*(9832), 2206–2211. doi:10.1016/S0140-6736(12)60685-0 PMID:22682467

Saeed, B. B., Afsar, B., Hafeez, S., Khan, I., Tahir, M., & Afridi, M. A. (2018). Promoting employee's proenvironmental behavior through green human resource management practices. *Corporate Social Responsibility and Environmental Management*, *26*(2), 424–438. doi:10.1002/csr.1694

Salehi, A. W., Baglat, P., & Gupta, G. (2020, June 12). Review on machine and deep learning models for the detection and prediction of Coronavirus. *Materials Today: Proceedings*, *33*, 3896–3901. Advance online publication. doi:10.1016/j.matpr.2020.06.245 PMID:32837918

Sangers, J., Frasincar, F., Hogenboom, F., & Chepegin, V. (2013). Semantic web service discovery using natural language processing techniques. *Expert Systems with Applications*, *40*(11), 4660–4671. doi:10.1016/j.eswa.2013.02.011

Sarmir, S., Islam, M. T., Salwa, M., Hossain, Z., Shah, M., Hasan, N. M., . . . Haque, A. M. (2021, May 31). Duration and Risk Factors of Post-COVID Symptoms Following Recovery Among the Medical Doctors in Bangladesh. *Cureus*. Advance online publication. doi:10.7759/cureus.15351

Sathya, R., & Abraham, A. (2013). Comparison of Supervised and Unsupervised Learning Algorithms for Pattern Classification. *International Journal of Advanced Research in Artificial Intelligence.*, *2*. Advance online publication. doi:10.14569/IJARAI.2013.020206

Sawhney, S. K. K. (2019). Real-Time Smart Attendance System using Face Recognition Techniques. In *2019 9th International Conference on Cloud Computing, Data Science & Engineering (Confluence)* (pp. 522-525). IEEE.

Schahn, J., & Holzer, E. (1990). Studies of individual environmental concern: The role of knowledge, gender, and background variables. *Environment and Behavior*, *22*(6), 767–786. doi:10.1177/0013916590226003

Schreyer, M., Schulze, C., Stahl, A., & Effelsberg, W. (2009, March). Intelligent Printing Technique Recognition and Photocopy Detection for Forensic Document Examination. In Informatiktage (Vol. 8, pp. 39-42). Academic Press.

Shang, S., Memon, N., & Kong, X. (2014). Detecting documents forged by printing and copying. *EURASIP Journal on Advances in Signal Processing*, *2014*(1), 1–13.

Shanthini & Sreeja. (2016). Improved Traffic Control Systems For Emergency Vehicle Clearance And Stolen Vehicle Detection. *International Research Journal of Engineering and Technology*, *3*(3), 630–635.

Sharma, S. (2000). Managerial interpretations and organizational context as predictors of corporate choice of environmental strategy. *Academy of Management Journal*, *43*(4), 681–697.

Sharma, S. (2009). The mediating effect of information availability between organizational design variables and environmental practices in the Canadian hotel industry. *Business Strategy and the Environment*, *18*(4), 266–276. doi:10.1002/bse.603

Shatnawi, R. (2014). Empirical study of fault prediction for open-source systems using the Chidamberand Kemerer metrics. *IET Software, 8*(3), 113-119. Doi:10.1049/iet-sen.2013.0008

Shatouri, R. M., Omar, R., Igusa, K., & Flávio de São, P. F. (2013). Embracing green technology innovation through strategic human resource management: A case of an automotive company. *American Journal of Economics and Business Administration*, *5*(2), 65–73. doi:10.3844/ajebasp.2013.65.73

Shobayo, O. (2020). Development of Smart Plate Number Recognition. *Applied Computational Intelligence and Soft Computing*.

Shrivastava, P. (1995). The role of corporations in achieving ecological sustainability. *Academy of Management Review*, *20*(4), 936–960. doi:10.2307/258961

Siami-Namini, S., Tavakoli, N., & Namin, A. S. (2018, December). A comparison of ARIMA and LSTM in forecasting time series. In *2018 17th IEEE international conference on machine learning and applications (ICMLA)* (pp. 1394-1401). IEEE.

Siegmund, B., Michael, P. M. T., & Hirschfeld, R. (2014). Studying the Advancement in Debugging Practice of Professional Software Developers. *IEEE International Symposium on Software Reliability Engineering Workshops*. DOI 10.1109/ISSREW.2014.36

Singh, N. P., Sonbhadra, S. K., & Agarwal, S. (2020, June 1). COVID-19 Epidemic Analysis using Machine Learning and Deep Learning Algorithms. medRxiv. doi:10.1101/2020.04.08.20057679

Singh, V. J., Bhushan, M., Kumar, V., & Bansal, K. L. (2015). Optimization of Segment Size Assuring Application Perceived QoS in Healthcare. *Lecture Notes in Engineering and Computer Science, 2217*(1), 274-278.

Singh, P., Pramanik, D., Dey, A. K., & Choudhury, P. (2021). Recommender Systems: An Overview, Research Trends, and Future Directions. *International Journal of Business and Systems Research.*, *15*(1), 14–52. doi:10.1504/IJBSR.2021.111753

Skoutas, D., Simitsis, A., & Sellis, T. (2007, July). A ranking mechanism for semanticweb service discovery. In *2007 IEEE Congress on Services (Services 2007)* (pp. 41-48). IEEE. 10.1109/SERVICES.2007.8

Smola, A., & Vishwanathan, S. V. N. (2008). *Introduction to machine learning*. Cambridge University.

So¨zer, Ceren¸ & Gebizli. (n.d.). Model-based testing of digital TVs: An industry-as laboratory Approach. *Software Quality Journal*. Advance online publication. doi:10.100711219-016-9321-y

Somasekar, J., Kumar, P. P., Sharma, A., & Ramesh, G. (2020, September 14). *Machine learning and image analysis applications in the fight against COVID-19 pandemic: Datasets, research directions, challenges and opportunities*. doi:10.1016/j.matpr.2020.09.352

Sridevi, S., Karpagam, G. R., & Kumar, B. V. (2021). Incorporating blockchain for semantic web service selection (SWSS) method. *Sadhana*, *46*(2), 1–14. doi:10.100712046-021-01619-y

Srivastava, A., Bartol, K. M., & Locke, E. A. (2006). Empowering leadership in management teams: Effects on knowledge sharing, efficacy, and performance. *Academy of Management Journal*, *2006*(49), 1239–1251. doi:10.5465/amj.2006.23478718

Sroufe, R., & Gopalakrishna-Remani, V. (2018). Management, social sustainability, reputation, and financial performance relationships: An empirical examination of US firms. *Organization & Environment*, *32*(3), 331–362. doi:10.1177/1086026618756611

Staats, H., Harland, P., & Wilke, H. A. M. (2004). Effecting durable change: A team approach to improve environmental behavior in the household. *Environment and Behavior*, *36*(3), 341–367. doi:10.1177/0013916503260163

Stafford-Smith, M., Griggs, D., Gaffney, O., Ullah, F., Reyers, B., Kanie, N., Stigson, B., Shrivastava, P., Leach, M., & O'Connell, D. (2017). Integration: The key to implementing the sustainable development goals. *Sustainability Science*, *12*(6), 911–919. doi:10.100711625-016-0383-3 PMID:30147763

Stamper, R. K. (1980). *LEGOL: Modelling Legal Rules by Computer. In Computer Science and Law*. Cambridge University Press.

Stamper, R. K. (1991). The Role of Semantics in Legal Expert Systems and Legal Reasoning. *Ratio Juris*, *4*(2), 219–244. doi:10.1111/j.1467-9337.1991.tb00094.x

Starik, M., Throop, G. M., Doody, J. R., & Joyce, M. E. (1996). Growing an environmental strategy. *Business Strategy and the Environment*, *1996*(5), 12–21. doi:10.1002/(SICI)1099-0836(199603)5:1<12::AID-BSE37>3.0.CO;2-X

https://www.statista.com/statistics/746954/india-number-of-road-accidents/

Stone, L. J. (2000). When case studies are not enough: The influence of corporate culture and employee attitudes on the success of cleaner production initiatives. *Journal of Cleaner Production*, *8*(5), 353–359. doi:10.1016/S0959-6526(00)00037-8

Strauss, A. L., & Corbin, J. M. (1998). *Basics of Qualitative Research. Techniques and Procedures for Developing Grounded Theory*. Sage.

Stringer, L. (2009). *The Green workplace. Sustainable strategies that benefit employees, the environment, and the bottom line*. Macmillan.

Sunil Aryala, R. S. (2019). Automatic Attendance System Using Deep Learning. *International Conference on Sustainable Computing in Science, Technology and Management (SUSCOM).*

Sun, K., Chen, J., & Viboud, C. (2020, April). Early epidemiological analysis of the coronavirus disease 2019 outbreak based on crowdsourced data: A population-level observational study. *The Lancet. Digital Health*, *2*(4), e201–e208. doi:10.1016/S2589-7500(20)30026-1 PMID:32309796

Su, X., Xu, A., Lin, W., Chen, Y., Liu, S., & Xu, W. (2020, April-June). Environmental Leadership, Green Innovation Practices, Environmental Knowledge Learning, and Firm Performance. *SAGE Open*, *10*(2), 1–14. doi:10.1177/2158244020922909

Sweileh, W. M. (2020). Bibliometric analysis of scientific publications on "sustainable development goals" with emphasis on "good health and well-being" goal (2015–2019). *Globalization and Health*, *16*(1), 68. doi:10.118612992-020-00602-2 PMID:32723366

Szabo, N. (1996). Smart contracts: building blocks for digital markets. *EXTROPY: The Journal of Transhumanist Thought*, *18*(2).

Tabish, R., Ghaleb, A. M., Hussein, R., Touati, F., Mnaouer, A. B., Khriji, L., & Rasid, M. F. A. (2014). 3G/WiFi-enabled 6LoWPAN-based U-healthcare system for ubiquitous real-time monitoring and data logging. *Proceedings of the 2014 Middle East Conference on Biomedical Engineering (MECBME)*, 277–280. 10.1109/MECBME.2014.6783258

Tatoglu, E., Bayraktar, E., Sahadev, S., Demirbag, M., & Glaister, K. W. (2014). Determinants of voluntary environmental management practices by MNE subsidiaries. *Journal of World Business*, *49*(4), 536–548. doi:10.1016/j.jwb.2013.12.007

Teixeira, A. A., Jabbour, C. J. C, & Jabbour, A. B. L. (2012). *Relationship between green management and environmental training in companies located in Brazil*. Academic Press.

Teixeira, A. A., Jabbour, C. J. C., & Jabbour, A. B. L. (2012). Relationship between green management and environmental training in companies located in Brazil: A theoretical framework and case studies. *International Journal of Production Economics*, *140*(1), 318–329. doi:10.1016/j.ijpe.2012.01.009

Tejas, K., Reddy, K. A., Reddy, D. P., Bharath, K., Karthik, R., & Kumar, M. R. (2019). *Efficient license plate recognition system with smarter Interpretation through IoT. In Soft Computing for Problem Solving*. Springer.

Tektas, M. (2010). Weather forecasting using ANFIS and ARIMA mod els. *Environmental Research, Engineering and Management*, *51*(1), 5–10.

Temminck, E., Mearns, K., & Fruhen, L. (2015). Motivating employees towards sustainable behaviour. *Business Strategy and the Environment, 24*(6), 402–412. doi:10.1002/bse.1827

Testa, F., Corsini, F., Gusmerotti, N. M., & Iraldo, F. (2018). Predictors of organizational citizenship behavior in relation to environmental and health & safety issues. *International Journal of Human Resource Management*. Advance online publication. doi:10.1080/09585192.2017.1423099

Theodouli, A., Arakliotis, S., Moschou, K., Votis, K., & Tzovaras, D. (2018). On the design of a blockchain-based system to facilitate healthcare data sharing. *2018 17th IEEE International Conference On Trust, Security And Privacy In Computing And Communications/12th IEEE International Conference On Big Data Science And Engineering (TrustCom/BigDataSE)*, 1374–1379.

Toktay, L. B., Selhat, L., & Anderson, R. (2006). Doing well by doing good: interface's vision of being the first industrial company in the world to attain sustainability. In Enterprise Transformation: Understanding and Enabling Fundamental Change. Wiley.

Too, L., & Bajracharya, B. (2015). Sustainable Campus: Engaging the community in sustainability Int. *Journal on Sustainability in Higher Education, 16*(1), 57–71. doi:10.1108/IJSHE-07-2013-0080

Tosun, A., Bener, A. B., & Akbarinasaji, S. (n.d.). A systematic literature review on the applications of Bayesian networks to predict software quality. *Software Quality Journal*. Advance online publication. doi:10.100711219-015-9297-z

Trivedi, G. (2020). *Visualization and Scientometric Mapping of Global Agriculture Big Data Research*. DigitalCommons@University of Nebraska - Lincoln. Retrieved 9 September 2020, from https://digitalcommons.unl.edu/libphilprac/2478/

Tsai, M. J., & Liu, J. (2013, May). Digital forensics for printed source identification. In *2013 IEEE International Symposium on Circuits and Systems (ISCAS)* (pp. 2347-2350). IEEE.

Tsai, M. J., & Yuadi, I. (2016, September). Printed source identification by microscopic images. In *2016 IEEE International Conference on Image Processing (ICIP)* (pp. 3927-3931). IEEE.

United Nations. (1987). Informe Brundtland, "Nuestro Futuro en Común." Available online: www.un.org/es/comun/docs/?sym-bol=A/42/427

United Nations. (n.d.). *Sustainable development goals*. https://www.un.org/sustainabledevelopment/sustainable-development-goals/

Unnikrishnan, S., & Hegde, D. S. (2007). Environmental training and cleaner production in Indian industry—A micro-level study. *Resources, Conservation and Recycling, 50*(4), 427–441. doi:10.1016/j.resconrec.2006.07.003

US Colleges and Universities. (n.d.). *Opendatasoft*. https://data.opendatasoft.com/explore/dataset/us-colleges-and-universities%40public/information /

Valencia, D. N. (2020, March 24). Brief Review on COVID-19: The 2020. *Cureus*. Advance online publication. doi:10.7759/cureus.7386 PMID:32337113

Valente, A. (1995). Legal Knowledge Engineering; A Modelling Approach. University of Amsterdam.

Valente, A. (2005). *Types and Roles of Legal Ontologies*. Law and the Semantic Web.

Verma, K., Bhardwaj, S., Arya, R., Mir, S. U. I., Bhushan, M., Kumar, A., & Samant, P. (2019). Latest Tools for Data Mining and Machine Learning. *International Journal of Innovative Technology and Exploring Engineering, 8*(9S), 18-23. doi:10.35940/ijitee.I1003.0789S19

Verma, K., Bhardwaj, S., Arya, R., Ul Islam, M. S., Bhushan, M., Kumar, A., & Samant, P. (2019). Latest tools for data mining and machine learning. *International Journal of Innovative Technology and Exploring Engineering, 8*(9S), 18-23.

Verrier, B., Rose, B., Caillaud, E., & Remita, H. (2014). Combining organizational performance with sustainable development issues: The lean and green project benchmarking Repository. *Journal of Cleaner Production, 85*, 83–93. doi:10.1016/j.jclepro.2013.12.023

Vicente-Molina, M. A., Fernández-Sainz, A., & Izagirre-Olaizola, J. (2018). Does gender make a difference in pro-environmental behavior? The case of the Basque Country university students. *Journal of Cleaner Production, 176*, 89–98. doi:10.1016/j.jclepro.2017.12.079

Villazón-Terrazas, B., Ortiz-Rodríguez, F., Tiwari, S., & Shandilya, S. (2020). Knowledge Graphs and Semantic Web. *Communications in Computer and Information Science, 1232*, 1–225. doi:10.1007/978-3-030-65384-2

Viriyasitavat, W., Da Xu, L., & Bi, Z. (2020). Specification Patterns of Service-Based Applications Using Blockchain Technology. *IEEE Transactions on Computational Social Systems, 7*(4), 886–896. doi:10.1109/TCSS.2020.2999574

Viriyasitavat, W., Da Xu, L., Bi, Z., Hoonsopon, D., & Charoenruk, N. (2019). Managing qos of internet-of-things services using blockchain. *IEEE Transactions on Computational Social Systems, 6*(6), 1357–1368. doi:10.1109/TCSS.2019.2919667

Visser, P. R. S. (1995). *Knowledge Specification for Multiple Legal Tasks. A Case Study of the Interaction Problem in the Legal Domain. Computer/Law Series. No. 17*. Kluwer Law International.

Vojislav, Miˇsi'c, Miˇsi'c, & Chang. (2019). Towards a blockchain-based healthcare information system. *2019 IEEE/CIC International Conference on Communications in China (ICCC)*, 13–18.

Wagner, M. (2011). Environmental management activities and sustainable HRM in German Manufacturing firms. Incidence, determinants, and outcomes. *German Journal of Research in Human Resource Management, 25*(2), 157–177. doi:10.1177/239700221102500206

Wang, X., Vitvar, T., Kerrigan, M., & Toma, I. (2006, December). A qos-aware selection model for semantic web services. In *International Conference on Service-Oriented Computing* (pp. 390-401). Springer. 10.1007/11948148_32

Warin. (n.d.). https://warin.ca/shiny/bibliometrix/#section-bibliometrix

Wehrmeyer, W. (1996). *Greening people: Human resources and environmental management.* Greenleaf.

Whitmarsh, L., & O'Neill, S. (2010). Green identity, green living? The role of pro-environmental self-identity in determining consistency across diverse pro-environmental behaviours. *Journal of Environmental Psychology, 2010*(30), 305–314. doi:10.1016/j.jenvp.2010.01.003

Wininger, F., Ezzati-Jivan, N., & Dagenais, M. R. (n.d.). A declarative framework for stateful analysis of execution traces. *Software Quality Journal.* Advance online publication. doi:10.100711219-016-9311-0

Witt, A., Kurths, J., & Pikovsky, A. (1998). Testing stationarity in time series. *Physical Review E: Statistical Physics, Plasmas, Fluids, and Related Interdisciplinary Topics, 58*(2), 1800–1810. doi:10.1103/PhysRevE.58.1800

Woosnam, K. M., Strzelecka, M., Nisbett, G. S., & Keith, S. J. (2019). Examining millennials' global citizenship attitudes and behavioral intentions to engage in environmental volunteering. *Sustainability, 11*(8), 2324. doi:10.3390u11082324

Wu, Y., Kong, X., & Guo, Y. (2009, November). Printer forensics based on page document's geometric distortion. In *2009 16th IEEE International Conference on Image Processing (ICIP)* (pp. 2909-2912). IEEE.

Wu, Y., Pan, Z., Luo, X., Gao, J., & Zhang, Y. (2016). A hybrid forecasting method of electricity consumption based on trend extrapolation theory and LSSVM. *2016 IEEE PES AsiaPacific Power and Energy Engineering Conference (APPEEC),* 2333–2337.

Wu, P. J., Wu, T. J., & Yuan, K. S. (2019). "Green" Information Promotes Employees' Voluntary Green Behavior via Work Values and Perceived Accountability. *Sustainability, 2019*(11), 6335. doi:10.3390u11226335

Wurmser. (2019). US Time Spent with Mobile 2019: Smartphones Gain Minutes, but New Challengers Emerge. *Yoram Emarketer, 30.* https://www.emarketer.com/content/us-time-spent-with-mobile-2019

Xiao, C., & McCright, A. M. (2015). Gender differences in environmental concern: Revisiting the institutional trust hypothesis in the USA. *Environment and Behavior, 47*(1), 17–37. doi:10.1177/0013916513491571

Xue, Y., Zhang, C., & Ji, Y. (2015, September). RESTful web service matching based on WADL. In *2015 International Conference on Cyber-Enabled Distributed Computing and Knowledge Discovery* (pp. 364-371). IEEE. 10.1109/CyberC.2015.89

Xu, J., Guan, J., & Lin, Y. (2017). Institutional pressures, top managers' environmental awareness and environmental innovation practices: An institutional theory and upper echelons theory perspective. *Management Review, 29*(9), 72–83.

Xu, X., Yu, J., Chen, Y., Zhu, Y., Qian, S., & Li, M. (2018). Leveraging audio signals for early recognition of inattentive driving with smartphones. *IEEE Transactions on Mobile Computing, 17*(7), 1553–1567. doi:10.1109/TMC.2017.2772253

Yang, J., Li, L., Shi, Y., & Xie, X. (2018). *An ARIMA model with adaptive orders for predicting blood glucose concentrations and hypoglycemia. IEEE J. Biomed. Health Inf.*

You, Yim, Sharma, Choudhary, Chen, & Cho. (2018). On iot misbehavior detection in cyber physical systems. *2018 IEEE 23rd Pacific Rim International Symposium on Dependable Computing (PRDC)*, 189–190.

Young, W., Davis, M., McNeill, I. M., Malhotra, B., Russell, S., Unsworth, K., & Clegg, C. (2015). Changing behavior: Successful environmental programmes in the workplace. *Business Strategy and the Environment, 24*(8), 689–703. doi:10.1002/bse.1836

Yuan, C., & Yang, H. (2019). *Research on K-Value Selection Method of K-Means Clustering Algorithm.* doi:10.3390/j2020016

Yusliza, M. Y., Ramayah, T., & Othman, N.-Z. (2015). While examining adoption factors, HR role and attitude towards using e-HRM is the start-off in determining the successfulness of green HRM? *Journal of Advanced Management Science, 3*, 337–343. Retrieved November 5, 2014, from www.joams.com

Zhang, B., Wang, Z., & Lai, K. H. (2015). Mediating effect of managers' environmental concern: Bridge between external pressures and firms' practices of energy conservation in China. *Journal of Environmental Psychology, 43*, 203–215. doi:10.1016/j.jenvp.2015.07.002

Zhang, P. (2018). Fhirchain: Applying blockchain to securely and scalably share clinical data. *Computational and Structural Biotechnology Journal, 16*, 267–278.

Zheng, J. (2010). Cost-sensitive boosting neural networks for software defect prediction. *Expert Systems with Applications, 37*(6), 4537-4543.

Zhou, S., Zhang, D., Lyu, C., & Zhang, H. (2018). Does seeing "mind acts upon mind" affect green psychological climate and green product development performance? The role of matching between green transformational leadership and individual green values. *Sustainability, 10*(9), 3206. doi:10.3390u10093206

Zoogah, D. (2011). The dynamics of Green HRM behaviors: A cognitive social information processing approach. *Zeitschrift fur Personalforschung, 25*(2), 117–139. doi:10.1177/239700221102500204

About the Contributors

Fernando Ortiz-Rodriguez is a Full Professor and is the Social Science Research Center Director at Tamaulipas Autonomous University. Before this, he was the Executive Director at the International Institute of Studies (IIES); he created the First Business School in Tamaulipas, Mexico, and before this, he was the Information Technology Manager at Emerson Electric, where he developed more than 40 software's, some of them used globally in Emerson and achieved technology convergence implementing the first efforts on IoT and Industry 4.0 He is a member of National Systems Researchers (SNI) of the National Council of Science and Technology (CONACYT). He published several articles, book chapters, and books in prestigious editorials. He has guided more than fifteen bachelors and five postgraduate theses. Additionally, he has more than ten software patents. He received a Latin-American award (U-Gob) 2021 and 2022 for implementing and developing software to help with Covid-19 diagnosis in Mexico.

Sanju Mishra Tiwari has worked as a Research Associate for a sponsored research project "Intelligent Real time Situation Awareness and Decision Support System for Indian Defence" funded by DRDO in Department of Computer Applications, National Institute of Technology, Kurukshetra. Her current research interests include Knowledge Based Systems, Ontologism, Intelligent Decision Support, Semantic/ Intelligent Query Search Engines, Intrusion Detection, Swarm Intelligence. She has to-date published 19 research papers one 1 book chapter with international and national publishers. She has 10 years experience in teaching field and 2 years for research field for developing ontologies. She is the member of IEEE. She is working as an organizing committee member of MIR Labs, USA.

Sailesh Iyer has a Ph.D. (In computer Science) and currently serving as a Professor and Dean with Rai University, Ahmedabad. He has more than 22 years of experience in Academics, Industry, and Corporate Training out of which 18 years are in core Academics. He has Patents to his credit and is involved as an Editor for various book projects with IGI Global (USA), Taylor and Francis (UK) and

Bentham Science (UAE). A hardcore Academician and Administrator, he has excelled in Corporate Training, Delivered Expert Talk in various AICTE sponsored STTPs, ATAL FDPs, Reputed Universities, Government organized Workshops, Orientation, and Refresher Courses organized by HRDC, Gujarat University. Research Contributions include reputed Publications, Track Chair at ICDLAIR 2020 (Springer Italy), icSoftComp 2020, IEMIS 2020 (Springer), ICRITO 2020 (IEEE), ARISE-2021, FTSE-2021, and TPC members of various reputed International and National Conferences, Reviewer of International Journals like Multimedia Tools and Applications (Springer), Journal of Computer Science (Scopus Indexed), International Journal of Big Data Analytics in Healthcare (IGI Global), Journal of Renewable Energy and Environment, and Editor in various Journals. Expert Talk on Research-based topics in various Universities and Conferences in addition to guiding Research Scholars as supervisors. He has also been invited as a Judge for various events, Examiner for Reputed Universities, is a Computer Society of India Lifetime Member and also serving as Managing Committee (MC) Member, CSI Ahmedabad Chapter from 2018-2020.

Jose Melchor Medina-Quintero is a Doctor in Business Information Systems from the Polytechnic University of Madrid (Spain) and expert in Information Technologies and Systems. Leader of the research group Information Technologies and Strategy (UAT-CA-132), for which he has published 22 papers in high-impact journals (Scopus and JCR) both in Mexico and internationally: Journal of Computer Information Systems, Journal of Information Systems and Technology Management, Innovar Journal, Cuadernos de Administración, Management Studies, Nova Scientia, Accounting and Administration, among others. He is the co-author of 20 book chapters on information technologies and systems. He wrote the book Information Systems Project Planning and co-authored a leading-edge statistical book such as Multivariate Analysis Techniques. Applied to the social sciences, co-author in 6 books with a business administration approach, and coordinator of 5 books. He has supervised 73 theses between graduate, master, and doctorate. Some of the research projects in which Dr. Medina has participated: i) Study of the administrative profile of the SMEs in the state of Tamaulipas, ii) Consolidation of the scientific and technological infrastructure for the exploration and sustainable exploitation of unconventional hydrocarbons, oil/gas shale in Mexico, iii) Institutional Repository of the Autonomous University of Tamaulipas, iv) Information technologies and systems in public and private organizations in the state of Tamaulipas, v) e-Government: Tax collector, vi) e-Banking for customer loyalty and satisfaction, among others. Since 2010, he is a member of the National System of Researchers in Mexico.

* * *

Anshu Agarwal is currently working as Assistant Professor, SoIT, AURO University. Her areas of interest are Data Mining, Data Science, Data Analysism Machine Learning, Software Engineering, etc.

Riten Bhagra is currently pursuing under-graduation in Computer Science and Engineering (Specialisation in Cyber Security and Digital Forensics) from Vellore Institute of Technology. He describes himself as a passionate programmer, Machine Learning Novice, Mathematics enthusiast, and inquisitive learner. Currently, he plans to pursue his M.S from University of Southern California, USA in Applied Data Science. His areas of interest include machine learning, network security, artificial intelligence and blockchain.

Harshita Chadha is an academically strong engineering student from New Delhi pursuing a degree in computer science. She has experience working in both research and academia on a wide range of problems ranging from network security to machine learning. She is also the founder of her sole proprietary innovation lab, Meraki. An avid researcher with many inventions in the patent pipeline, she aims to build a lifetime career in the field of artificial intelligence for healthcare.

Rupak Chakravarty is currently serving as a Faculty member at the Department of Library and Information Science, Panjab University, Chandigarh since 2006. His areas of research include Open Access, Library Technologies and Technology-enabled Library Services, Plagiarism, Reference Management Software, Cloud Technologies, Informetrics, etc. He has successfully guided 4 Research Scholars and currently 7 scholars are pursuing Ph.D. under his supervision. He has authored more than 60 research publications published in national and international journals and has delivered more than 100 invited talks at Human Resource Development Centres, Skill development Training programmes, capacity building workshops, conferences/seminars and Webinars. He has a passionate interest in free and open source software (FOSS), cloud applications and Educational Technology (ET) and keeps on experimenting with them. He has also served as Associate Editor of the Directory of Open Access Journals (DOAJ) for three consecutive years. He is currently the Chief Editor of the Journal International Journal of Information Studies and Libraries (IJISL).

Gaurav Choudhary received a Ph.D. in Information Security Engineering from Soon- chunhyang University, South Korea. He has done a Master of Technology in Cyber Security from the Sardar Patel University of Police and received a Chancellor Gold Medal for Academic Excellence. He is presently working as a Security Researcher at DTU Compute, Department of Applied Mathematics and Computer

Science, Tech- nical University of Denmark (DTU). Prior to joining DTU, he has also worked as an Assistant Professor in the School of Computer Science, University of Petroleum and Energy Studies (UPES), and School of Computer Science and Engineering (SCSE) at VIT Bhopal University. His current research interests include Threat Intelligence, IoT and CPS Security, Cyber Se- curity, Vulnerability Assessment, 5G Security, Drone Security, and Cryptography. He has authored or co-authored many reputed SCI journal/conference papers and book chapters.

Shruti Gupta has a keen interest in data science and AI. This interest fuels her enthusiasm and desire to learn, study as well as create and contribute. As a curious and hardworking person, she wants to make a significant contribution in her field. She aspires to work in a firm with a professional work driven environment where she can utilize and apply her knowledge and skills which would enable her to grow while fulfilling organizational goals. She is a very friendly and enthusiastic person who believes that communication and community is essential part of any successful business.

Abdul Mueed Hafiz completed his B.Tech. in Electronics & Communication Engineering, from the Dept. of ECE, National Institute of Technology, Srinagar, in 2005; his M.Tech. in Communication & Information Technology, from the Dept. of ECE, NIT Srinagar, in 2008; his Ph.D. in Artificial Intelligence, from the Dept. of E&IT, University of Kashmir, Main Campus, Srinagar, in 2018. Currently he is employed as Assistant Professor, Dept. of ECE, Institute of Technology, University of Kashmir, Zakura Campus, Hazratbal, Srinagar, J&K, India. His research interests include Computer Vision, Pattern Recognition, and Artificial Intelligence. He has 20+ publications in international journals/book chapters/international conferences. He is reviewer for SCI journals from IEEE, IET, ACM, Springer, etc. He is also a member of A.C.M. and I.E.E.E. societies.

Sunita Jalal has done MTech from Banasthali Vidyapith University and pursuing her Ph.D. from MNNIT, Allahabad. Currently she is working as Assistant Professor in Department of Computer Engineering, College of Technology, Govind Ballabh Pant University of Agriculture & Technology, Pantnagar. She has over 15 years of experience in the academics.

Chetan Negi has done MTech from IIT Kharagpur and pursuing his Ph.D. from College of Technology, Govind Ballabh Pant University Of Agriculture & Technology, Pantnagar. Currently he is working as Assistant Professor in Department of Computer Engineering, College of Technology, Govind Ballabh Pant University

Of Agriculture & Technology, Pantnagar. He has over 16 years of experience in the academics.

José G. Vargas-Hernández, M.B.A., Ph.D., Member of the National System of Researchers of Mexico and a research professor at Tecnológico Mario Molina Unidad Zapopan formerly at University Center for Economic and Managerial Sciences, University of Guadalajara. Professor Vargas-Hernández has a Ph. D. in Public Administration and a Ph.D. in Organizational Economics. He has undertaken studies in Organisational Behaviour and has a Master of Business Administration, published four books and more than 200 papers in international journals and reviews (some translated to English, French, German, Portuguese, Farsi, Chinese, etc.) and more than 300 essays in national journals and reviews. He has obtained several international Awards and recognition.

Reeta Verma has done Ph.D. from College of Technology, Govind Ballabh Pant University Of Agriculture & Technology, Pantnagar. Currently she is working as Assistant Professor in Department of Electronics & Communication Engineering, College of Technology, Govind Ballabh Pant University Of Agriculture & Technology, Pantnagar. She has over 18 years of experience in the academics.

Rakesh Kumar Yadav has done his Ph.D. from IFTM University, Moradabad, Uttar Pradesh INDIA and is currently working as Assistant Professor in the Department of Computer Science and Engineering, IFTM University. He is currently working as an Assistant Professor in Department of Computer Science & Engineering, IFTM University, UP. He has Published more than 25+ research papers in reputed international journals and conferences. He has been reviewed many research papers of journals and conferences. He has published many Patents. He has guided various M.Tech and PhD students. He has written many books and examination series also. He has also taken many guest lectures. His main research work focuses on Machine Learning, Biometric, Image Processing, Computer Vision, Soft Computing and Artificial Intelligence. He has 15+ years of teaching experience in higher education.

Index

www.ingramcontent.com/pod-product-compliance
Lightning Source LLC
LaVergne TN
LVHW061549290926
849162LV00004B/952